Extrapolation of Dosimetric Relationships for Inhaled Particles and Gases

Editorial Review Board

The Editors are most grateful to the members of the Review Board for their contributions toward the peer review of the chapters in this book.

Extrapolation of Dosimetric Relationships for Inhaled Particles and Gases

Edited by

James D. Crapo
Center for Extrapolation Modeling
Duke University Medical Center
Durham, North Carolina

Frederick J. Miller
United States Environmental Protection Agency
Research Triangle Park, North Carolina

Elaine D. Smolko
Center for Extrapolation Modeling
Duke University Medical Center
Durham, North Carolina

Judith A. Graham
United States Environmental Protection Agency
Research Triangle Park, North Carolina

A. Wallace Hayes
RJR Nabisco, Inc.
Winston-Salem, North Carolina

Academic Press, Inc.
Harcourt Brace Jovanovich, Publishers
San Diego New York Berkeley Boston London Sydney Tokyo Toronto

Academic Press, Inc.
San Diego, California 92101

United Kingdom Edition published by
Academic Press Limited
24–28 Oval Road, London NW1 7DX

Library of Congress Cataloging in Publication Data

Symposium on Extrapolation Modeling of Inhaled Particles and Gases:
Lung Dosimetry (1987 : Duke University Medical Center)
Extrapolation of dosimetric relationships for inhaled particles and gases / [edited by] James D. Crapo ... [et al.].
p. cm.
"An outgrowth of the Symposium on Extrapolation Modeling of Inhaled Particles and Gases: Lung Dosimetry, October 8-10, 1987, Duke University Medical Center; sponsored by the Duke University Center for Extrapolation Modeling."
Includes bibliographies and index.
ISBN 0-12-196780-8 (alk. paper)
1. Aerosols--Toxicology--Congresses. 2. Toxicity testing--Congresses. 3. Toxicology, Experimental--Congresses. 4. Chemical dosimetry--Congresses. 5. Toxicology--Animal models--Congresses. 6. Air--Pollution--Research--Methodology--Congresses. I. Crapo, James D. II. Duke University. Center for Extrapolation Modeling. III. Title.
RA1270.A34S96 1987
615.9'1--dc19 88-34696
CIP

Printed in the United States of America
89 90 91 92 9 8 7 6 5 4 3 2 1

Contents

Part Three. Experimental Dosimetry of Inhaled Particles and Gases

Part Four. New Methods for Determining Dosimetry of Inhaled Particles and Gases

Part Five. Modeling Approaches to Predicting Dosimetry of Inhaled Particles and Gases

This book is based on presentations at the

Symposium on
Extrapolation Modeling of Inhaled Particles and Gases: Lung Dosimetry

October 8–10, 1987
Duke University Medical Center

Sponsored by the Duke University Center for Extrapolation Modeling

Support provided by:
RJR Nabisco, Inc.
Southern California Edison Company
U.S. Environmental Protection Agency

Preface

The environmental pollutants to which humans are exposed are increasing rapidly in terms of number and complexity. One of the great challenges in environmental medicine is to define more accurately the adverse health effects likely to be encountered by exposures to these pollutants, since reducing exposure to zero is not politically or economically achievable. For many common pollutants, accurate extrapolations from animal toxicologic studies to humans and from short-term to long-term chronic effects are essential. The problems that need to be addressed are extremely important, not only to the academic and medical communities that discuss and treat the toxicological problems, but also to governmental agencies that must set regulatory standards, to industry whose products and manufacturing processes contribute to many of the problems, and to the general population that enjoys the benefits of a highly urbanized society but must directly or indirectly pay the health and economic costs associated with either tolerating or removing the environmental pollutants.

The Center for Extrapolation Modeling at Duke University was established in 1986 to coordinate research efforts directed at refining models that more accurately allow data with respect to dose distribution, injury distribution, magnitude of injury, and patterns of repair to be extrapolated from one animal species to another. Extrapolation models which can be verified and have characteristics that can be generalized across classes of pollutants offer the only feasible approach for obtaining the data needed to assess effects of the large number of pollutants in our environment. Physical testing of the acute and chronic effects in multiple animal species for large numbers of different pollutants is not an economically realistic approach.

This conference addressed various topics related to inhaled particles and gases, with emphasis on oxidant air pollutants and modeling of the respiratory system. Some of the most extensive data bases which can be applied to the relatively young field of extrapolation modeling exist in these areas. The successful development of effective new models requires a combined input from creative investigators in a broad number of fields. One of the primary purposes of this conference has been to assemble these types of investigators in a forum that could focus future research efforts in areas of common interest and facilitate the development and refinement of extrapolation models.

We greatly appreciate contributions from those who participated in the conference and who prepared the resulting manuscripts, all of which have undergone peer review and appear in this volume. We also appreciate the broad support for this conference provided by private industry, academic institutions, and the United States Environmental Protection Agency.

James D. Crapo
Elaine D. Smolko
Frederick J. Miller
Judith A. Graham
A. Wallace Hayes

Part One

Introduction

Chapter 1

Extrapolation Modeling: Advancements and Research Issues in Lung Dosimetry

F. J. Miller
Inhalation Toxicology Division
Health Effects Research Laboratory
U.S. Environmental Protection Agency
Research Triangle Park, North Carolina 27711

E. D. Smolko
J. D. Crapo
Center for Extrapolation Modeling
Duke University Medical Center
Durham, North Carolina 27710

I. GOALS OF EXTRAPOLATION MODELING

One of the dilemmas currently encountered when evaluating health effects due to environmental toxicant exposure is a lack of sufficient information on which to base quantitative assessments. Most data on exposures and subsequent effects are available from animal toxicology studies, but more limited data are available from human clinical and epidemiological studies. Though animal toxicology studies provide an extensive base of information, the data are generally used only qualitatively to assess potential health effects in humans. Controlled human exposure studies provide the best information for potential effects, however, ethical considerations prohibit exposures to toxicants under conditions that might elicit a strong response or require highly invasive measurement procedures for assessing effects. Extrapolation modeling provides a mechanism for combining information from human and animal exposures into an overall assessment of

Disclaimer: The research described in this article has been reviewed by the Health Effects Research Laboratory, U.S. Environmental Protection Agency, and approved for publication. Approval does not signify that the contents necessarily reflect the views and policies of the Agency nor does mention of trade names or commercial products constitute endorsement or recommendation for use.

toxicity, thereby taking advantage of the strengths of both approaches while diminishing limitations. Relationships between effects that are discovered through scientific studies must be used to predict effects that are unknown and for which data cannot be accumulated. Specifically, this involves developing experimental correlations between the known acute effects in man and animals as well as between the acute and chronic effects in animals. These relationships can then be used to extrapolate from acute to chronic effects in man and from chronic effects in animals to chronic effects in man.

To accomplish quantitative extrapolations, three major categories of information are required: dosimetry, species sensitivity, and health effects. Dosimetry refers to the amount of a pollutant which reaches specific target sites after exposure to a given concentration. Species sensitivity refers to potential variations in response of different species receiving the same dose of a pollutant. The linking of dosimetry and sensitivity data allows the quantitative extrapolation of effective pollutant concentrations between animals and man. To complete the extrapolation, however, the health effects being considered must be adequately understood.

Accumulating all data required for a quantitative extrapolation is a massive undertaking that can be accomplished only with a carefully planned multidisciplinary approach. The intent of this paper is not to address all issues related to extrapolation modeling, but to focus on lung dosimetry and provide an overview of accomplishments and goals. Some specific aspects of lung dosimetry to be discussed are factors affecting dosimetric relationships, methods for determining lung dose, and approaches for modeling or estimating dose.

II. RECENT ACCOMPLISHMENTS

For quantitative extrapolation modeling to become a reality, goals must be clearly defined to ensure the coordination of research projects. Before identifying the issues that will be addressed, the base of information currently available must be assessed. Much progress has been made in obtaining information required for examining interspecies dosimetry of inhaled compounds, particularly in the areas of physiological variables needed as model input parameters (EPA, 1988) and anatomical structure of the lungs of animals and humans (Mercer *et al.,* 1987; Mortensen *et al.,* see Chap. 6, this volume). New data are available on the influence of airway size as a function of age, from a dosimetry viewpoint (Phalen *et al.,* 1985; Phalen and Oldham, 1987) and relative to a basic understanding of lung growth (Mortensen *et al.,* see Chap. 6, this volume). Research on quantitating mucous layer thickness for animals and humans within specific regions of the respiratory tract is being conducted by Dr. Robert Mercer and colleagues at the Duke University Center for Extrapolation Modeling. Dr. Pi-Wan Cheng and colleagues at the University of North Carolina at Chapel Hill are developing microtechniques to collect mucus for quantitation in animals and humans within specific regions of the respiratory tract. A new methodology, the aerosol bolus technique (Stahlhofen *et al.,* 1987), offers promise in understanding lung function in individuals with chronic lung disease.

While progress is being made on validating models, further refinements are needed in technological approaches that may be useful in model validation, such as isotope ratio mass spectroscopy (Hatch and Aissa, 1987) and positron emission tomography (Coleman *et al.,* see Chap. 18, this volume). Examples of recent advancements towards model validation are the measurement of total respiratory tract uptake of ozone in various animal species (Wiester *et al.,* 1988) and the determination of regional respiratory tract uptake of ozone in humans (Gerrity *et al.,* 1988).

Though dosimetry models for mixtures of gases and particles are not yet available, specific aspects of developing these models are being addressed. Progress to date includes research on individual variations in dose and species (e.g., age dependency) (Phalen *et al.,* 1985; Martonen *et al.,* 1987, and Chap. 25, this volume). Models are available to investigate cellular dose distribution within a given airway (Overton, 1988). Development of theoretical constructs is essential for advancement in this area, since most situations of actual exposures involve mixtures.

Current research continues to focus on identifying the portion of the respiratory tract that evidences the greatest effect of an inhaled toxicant. The respiratory tract is the portal of entry for many environmental pollutants. Since the pollutant is continuously removed from the inhaled airstream, the concentration in the lumen of lung airways declines distally from the trachea. Simultaneously, the liquid layer lining the tracheobronchial airways that protects underlying tissue decreases in thickness and may undergo changes in composition. Thus, the pattern of cellular injury is often seen more distally in the airways, as with the case of ozone (Fig. 1-1). For many inhaled toxicants, such as the oxidant gases, the site of greatest lung injury occurs near the bronchiole–alveolar duct (BAD) junction (Barry *et al.,* 1985; Chang *et al.,* 1986, 1988). The point of maximal effect, however, can vary with different compounds. Specifically, the area of maximal effect is likely to depend upon the nature of the toxicant and its effect, as well as the organ in which toxicity is observed. The accomplishments described in later chapters reflect procedures which apply to multiple endpoints and organs, and can help identify regions of maximal effect. An integration of multifaceted data will advance the understanding of environmental contaminant toxicity and further the goals of quantitative extrapolation modeling.

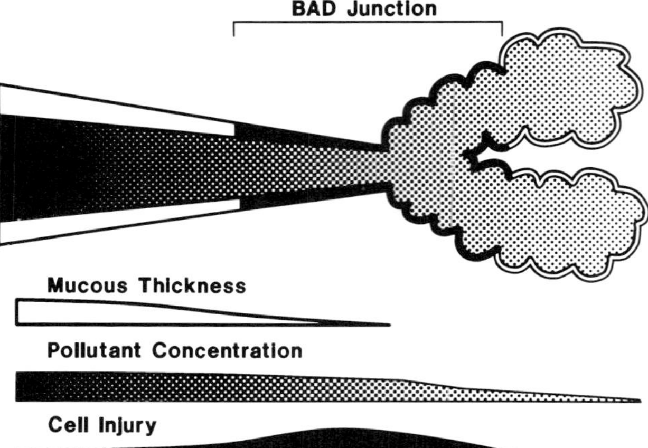

Fig. 1-1 Lung injury by inhaled oxidants. The BAD junction is the bronchiole-alveolar duct junction, a site at which a number of inhaled pollutants have been shown to exhibit their maximal toxic effects.

III. INTERDISCIPLINARY SYNTHESIS OF CONCEPTS AND KNOWLEDGE

To estimate human toxicity from exposure to chemicals and to characterize risk, an interdisciplinary synthesis of concepts and knowledge is essential. The process for achieving these goals involves several components (Fig. 1-2). One component is the development of a data base of basic information regarding exposures and effects reported in the literature, e.g., the Critical Toxicity Reference System (Smolko *et al.,* 1986). Another component involves using the information system to compile data regarding anatomical, physiological, biochemical, and physical parameters for species of interest. This information can then be used to develop physiologically-based dosimetry models for predicting dose to specific target sites. Dose–response relationships are established by correlating dose (from physiologically-based dosimetry models and reported exposure regimens) with effects resulting from the exposure regimens. Such relationships from animal studies may require translation to expected dose–response patterns for humans. The sensitivity of a particular species to effects of interest must also be factored into the analyses. This type of algorithm allows the estimation of human toxicity when sufficient data are available.

Alternatively, when sufficient data are not available, the process shown in Fig. 2 can be used to identify research gaps and to design appropriate

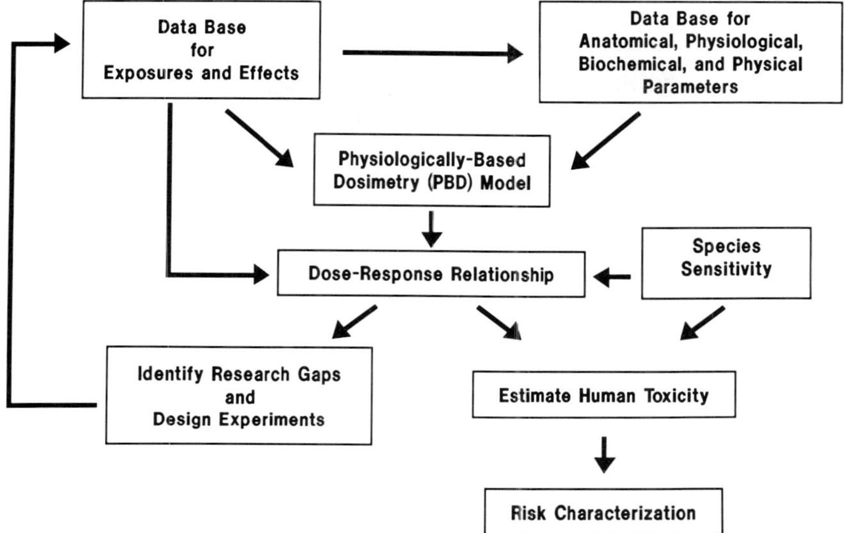

Fig. 1-2 Process for synthesizing interdisciplinary concepts and knowledge to characterize human risk from exposure to toxicants (adapted from Miller *et al.*, 1987).

experiments addressing needs. As exemplified by the research projects that have been conducted on nasopharyngeal removal of ozone in man and laboratory animals, interdisciplinary research is critical for filling data gaps and for furthering quantitative extrapolation modeling. Physiologically-based dosimetry models have been developed for predicting the lower respiratory tract uptake of ozone in man (Miller *et al.*, 1985) and in rats (Overton *et al.*, 1987), however, these dosimetry models require that the ozone concentration just above the trachea be specified. Several years ago, such data were not available.

Interactions among scientists regarding animal and human physiology, animal and human morphometry, aerobiology, physics, biochemistry, analytical chemistry, electrical engineering, inhalation engineering, and biostatistics were necessary to obtain the nasopharyngeal removal data. While this undertaking was extremely complex, some major accomplishments evolved from these scientific interactions. Research activities included:

1. Modifying an ozone analyzer to achieve a sufficiently rapid response time for breath-by-breath measurements in humans;
2. Performing concentration–response studies in human subjects and developing a method to deconvolute the signal from instrumentation used to measure uptake;

3. Developing an exposure system which minimizes dead space and allows mass balance expressions to include adjustments for lag-time responsiveness of the instrumentation;
4. Developing an improved restraining system for exposure of rats in a sensitive mass balance exposure apparatus; and
5. Applying isotope–ratio mass spectroscopy in biochemical studies to determine the relative proportionality of $^{18}O_3$ removed in the heads of rats and using these findings to adjust total respiratory tract uptake data obtained in the mass balance studies for estimates of nasopharyngeal uptake.

Experimental factors (e.g., protocols, techniques) had to be coordinated among scientists to ensure data comparability and utility for input into ozone dosimetry models. As a result of these efforts, the original data base has been expanded, and the dosimetry models can be used to improve characterization of human risk from ozone exposure. While this procedural application has focused on ozone, the principles involved relate space, in general, to extrapolation modeling of inhaled chemicals.

IV. FUTURE DIRECTIONS

One of the major issues to be addressed over the next several years relative to extrapolation models is validation. A model can be most useful to predict across different species and exposure scenarios if its accuracy and limitations are known. Model validation has been extremely difficult for some reactive gases. Since the data set employed to develop a model should not be used to test the validity of that model, new data are needed. The utilization of new experimental methodologies can have a significant impact on model validation, however, a greater and continued effort is required in many areas.

Comparative biology is one area which should receive special emphasis. The incorporation of actual biological values into models rather than best estimates, would contribute significantly to the utility of these models. An enhanced understanding of the quantitative differences between various animal species and man will improve judgments related to defining the most appropriate species to be used for studies, understanding limitations of available data, and synthesizing information across experimental studies of different designs. The *American Review of Respiratory Disease* (Vol. 128, No. 2, 1983) is an excellent resource for the types of comparative biological data which influence interspecies judgments on effects of inhaled material on respiratory tract issues. This journal supplement is based on a

1982 workshop on the comparative biology of the lung sponsored by the National Heart, Lung, and Blood Institute of the National Institutes of Health. Continued research in this area is essential for developing and refining extrapolation models.

Another major research issue is to determine whether or not subgroups that are especially susceptible to the effects of exposure to particular environmental pollutants exist within the general population. The majority of information currently available relates exposure and effect to the normal population, with some emphasis on the effects in children, exercising adults, or individuals with a preexisting disease (EPA, 1982, 1986a). Some data showing deposition of particles relative to various disease states also have been generated (Stahlhofen *et al.,* 1987), however, much of the morphological data that would provide input for model construction are not available. Groups that should be studied further include, but are not limited to, individuals with chronic bronchitis or emphysema, and smokers. Though some limited data on deposition of particles in smokers are available (Lippmann *et al.,* 1970; Lippmann, 1977), specific information is needed on changes in lung structure and function that will affect pollutant gas distributions in the lungs of individuals within this large subgroup. The susceptibility of any population subgroup must be given particular consideration when assessing environmental effects.

Studies on the clearance of inhaled pollutants from the lung are also necessary. Information on initial dose deposition is available (EPA, 1982, 1986b), however, retained dose, which factors in the temporal pattern of an uncleared pollutant, is more relevant when examining long-term effects. The net bioavailability of gases or particles is inextricably associated with these resulting effects. The dosimetry of hygroscopic particles will continue to be an evolving and important field of investigation, since acid aerosols are recognized as major constituents of environmental pollution.

A great deal of current research relates to issues of species sensitivity and toxicodynamics. Species sensitivity, the dose delivered to a target site which generates an equivalent biological response, is becoming increasingly important as research progresses towards a quantification of species differences essential for extrapolation modeling. Microdosimetric information is also desirable if toxicological assessments are to be focused on very specific anatomical locations within the respiratory tract. Additional data on mechanisms and structure will allow identification of localized target areas to be examined for critical effects. This should lead to improved decisions regarding intraspecies vs interspecies effects, various risk assessments, and applicable uncertainty factors. Without knowledge of the basic biochemical and molecular alterations leading to an effect, unfocused studies could generate data on effects and/or species that are vastly different and unrelated.

Studies on mechanisms should focus on determining whether similar mechanisms of toxicity are operative at both high and low concentration exposures. For example, the overloading of clearance mechanisms may cause cumulative dose effects with high exposure levels that do not occur with low levels of exposure (Morrow, 1988). The implication is that effects associated with these high exposure levels may not be the same as those observed in more realistic human exposures, or at least that they cannot be linearly extrapolated. Since the goal is defining human risk, high to low dose extrapolation issues must be addressed.

If appropriate models for extrapolation are to be developed, their particular application must be clearly described. The problem could be defined as the need to establish a National Ambient Air Quality Standard for a ubiquitous pollutant or to control point source emissions. The specific nature of each scenario will determine the analytical approach to be used and the emphasis to be placed on model validation.

The ultimate goal of defining human risk must be a common theme in all extrapolation modeling efforts. Without coordinated research, many studies that may not optimally contribute to efforts toward reaching this common goal are likely to be performed. Individuals with various areas of expertise must work together as a team to improve the data base on animal and human differences that relate to anatomical, physiological, and biochemical effects, or parameters that modulate dose. Basically, this constitutes development of physiologically-based dosimetry models that are used to synthesize information and to integrate concentration–response and exposure relationships, leading to estimates of human toxicity. Concentration–response information from animals and humans can be translated into dose–response information associated with target sites of interest for specific effects. However, dose can have different connotations, depending upon the effect being studied. Different representations of dose (e.g., mass per unit of surface area, number of particles, number of fibers, total mass) may be appropriate for estimating different toxic effects.

Extrapolation modeling involves conceptual processes that identify research gaps and facilitate the design of experiments to broaden the base of information for risk assessments. Research described in subsequent chapters contributes toward improved human risk characterizations. Specific issues in the area of dosimetry of inhaled particles and gases which will be addressed include: lung structure/function relationships that influence pollutant distribution; experimental dosimetry; new methods for determining dosimetry; and modeling approaches for predicting dosimetry. Though the data presented contribute substantially to the extrapolation effort, additional work is needed to understand fully the relationships between exposures to environmental pollutants and the resulting effects.

REFERENCES

American Review of Respiratory Disease (1983). Vol. 128, No. 2. Journal Supplement.

Barry, B. E., Miller, F. J., and Crapo, J. D. (1985). Effects of inhalation of 0.12 and 0.25 parts per million ozone on the proximal alveolar region of juvenile and adult rats. *Lab. Invest.* **53,** 692–704.

Chang, L., Graham, J. A., Miller, F. J., Ospital, J. J., and Crapo, J. D. (1986). Effects of subchronic inhalation of low concentrations of nitrogen dioxide. I. The proximal alveolar region of juvenile and adult rats. *Toxicol. Appl. Pharmacol.* **83,** 46–61.

Chang, L. Y., Mercer, R. R., Stockstill, B., Miller F. J., Graham, J. A., Ospital, J. J., and Crapo, J. D. (1988). Effects of low levels of NO_2 on terminal bronchiolar cells and its relative toxicity compared to O_3. *Toxicol. Appl. Pharmacol.* **96,** 451–464.

(EPA) U.S. Environmental Protection Agency (1982). Air quality criteria for particulate matter and sulfur oxides. Vol. III, Chap. 14. EPA/600/8-82/029c.

(EPA) U.S. Environmental Protection Agency (1986a). Air quality criteria for ozone and other photochemical oxidants. Vol. V. EPA/600/8-84/020eF.

(EPA) U.S. Environmental Protection Agency (1986b). Second addendum to air quality criteria for particulate matter and sulfur oxides (1982): Assessment of newly available health effects information. EPA/600/8-86-020F.

(EPA) U.S. Environmental Protection Agency (1988). Reference physiological parameters in pharmacokinetic modeling. EPA/600/6-88/004.

Gerrity, T. R., Weaver, R. A., Berntsen, J., House, D. E., and O'Neil, J. J. (1988). Extrathoracic and intrathoracic removal of O_3 in tidal-breathing humans. *J. Appl. Physiol.* **65,** 393–400.

Hatch, G. E., and Aissa, M. (1987). Determination of absorbed dose of ozone (O_3) in animals and humans using stable isotope (oxygen-18) tracing. *Air Pollut. Control Assoc. Annu. Meet. 80th, New York, June 21–26* Tech. Pap. 87–99.2.

Lippmann, M. (1977). Regional deposition of particles in the human respiratory tract. *In* "Handbook of Physiology, Section 9: Reactions to Environmental Agents" (D. H. K. Lee, H. L. Falk, and S. D. Murphy, eds.), pp. 213–232. American Physiological Society, Bethesda, Maryland.

Lippmann, M., Albert, R. E., and Peterson, H. T., Jr. (1970). The regional deposition of inhaled aerosols in man. *Proc. Int. Symp. Br. Occup. Hyg. Soc., London, Sept. 14–23* **1,** 105–120.

Martonen, T. B., Graham, R. C., and Hofmann, W. (1989). Human subject age and activity level: Factors addressed in a biomathematical deposition program for extrapolation modeling. *Hanford Life Sci. Symp., 26th, Health Phys.* In press.

Mercer, R. R., Laco, J. M., and Crapo, J. D. (1987). Three-dimensional reconstruction of alveoli in the rat lung for determination of pressure-volume relationships. *J. Appl. Physiol.* **62,** 1480–1487.

Miller, F. J., Overton, J. H., Jr., Jaskot, R. H., and Menzel, D. B. (1985). A model of the regional uptake of gaseous pollutants in the lung. I. The sensitivity of the uptake of ozone in the human lung to lower respiratory tract secretions and exercise. *Toxicol. Appl. Pharmacol.* **79,** 11–27.

Miller, F. J., Overton, J. H., Jr., Smolko, E. D., Graham, R. C., and Menzel, D. B. (1987). Hazard assessment using an integrated physiologically based dosimetry modeling approach: Ozone. *In* "Pharmacokinetics in Risk Assessment: Drinking Water and Health," Vol. 8. pp. 353–368. National Academy Press, Washington, D. C.

Morrow, P. E. (1988). Possible mechanisms to explain dust overloading of the lungs. *Fundam. Appl. Toxicol.* **10,** 369–384.

Overton, J. H. (1988). Respiratory tract dosimetry modeling of air toxics. U.S.-Dutch Expert Workshop on Air Toxics. Amersfoort, The Netherlands, May 16–18.

Overton, J. H., Graham, R. C., and Miller, F. J. (1987). A model of the regional uptake of gaseous pollutants in the lung. II. The sensitivity of ozone uptake in laboratory animal lungs to anatomical and ventilatory parameters. *Toxicol. Appl. Pharmacol.* **88,** 418–432.

Phalen, R. F., and Oldham, M. J. (1987). Predicted inhaled particle deposition in growing people. *Air Pollut. Control Assoc. 80th Annu. Meet., New York, June 21–26.* Tech. Pap. 87–99.6.

Phalen, R. F., Oldham, M. J., Beauckase, C. B., Crocker, T. T., and Mortensen, J. D. (1985). Predictions for particle deposition in the tracheobronchial airways of the growing human. *Anat. Rec.* **212,** 368–380.

Smolko, E. D., McKee, D. J., and Menzel, D. B. (1986). Critical toxicity reference system. I. An approach for managing quantitative toxicity data. *J. Am. College Toxicol.* **5,** 589–598.

Stahlhofen, W., Gebhart, J., and Scheuch, G. (1989). Aerosol boluses. *Proc. Conf. Suscept. Inhaled Pollut., Williamsburg, Virginia, Sept. 29–Oct. 1,* American Society for Testing and Materials, in press.

Wiester, M. J., Tepper, J. S., King, M. E., Ménache, M. G., and Costa, D. L. (1988). Comparative study of ozone (O_3) uptake in three strains of rats and in the guinea pig. *Toxicol. Appl. Pharmacol.* **96,** 140–146.

Part Two

Lung Structure–Function Relationships That Influence Distribution of Inhaled Particles and Gases

Chapter 2

Lung Structure – Function Relationships: Introduction

D. L. Dungworth
School of Veterinary Medicine
University of California
Davis, California 95616

The size and complexity of the mammalian respiratory tract provide ample opportunity for variations in the anatomic location of the principal sites of damage caused by inhaled particles and gases. The pattern of damage caused by any one agent depends on complex interactions between the nature of the agent, its site-specific dosimetry at various anatomic locations throughout the respiratory tract, the effectiveness of local defense mechanisms, and the inherent susceptibility to damage of the cells at risk in various anatomic sites. Inhalation of mixtures, particularly of particles and gases, compound these interactions.

Local dosimetry, which is the focus of this monograph, plays a major role in determining both the principal site and the nature of damage. The first chapter in this section deals with the population of cells at risk, which is the other major determinant of the pattern of damage. C. G. Plopper and colleagues (Chapter 3, this volume) emphasize the high degree of morphologic diversity of the cells comprising the epithelium of trachea, bronchi, and bronchioles among mammalian species. They note that the subgross anatomy of the pulmonary centriacinar (junctional) region differs among species. Both of these variations have an important impact on the lung's response to inhaled materials and on the extrapolation of findings from experimental animals to humans. The implications of the morphologic differences in pulmonary epithelial cell populations with respect to functional capabilities and differing susceptibilities to toxicologic damage have yet to be properly explored, as noted by Plopper and colleagues. The implications of differences in cellular content of monooxygenases relative to the effect of xenobiotics requiring metabolic activation are reasonably obvious. The implications of variations in the number and type of secretory cells with respect to interspecies differences in response to inhaled particles and gases are not well understood. One aspect that needs to be thoroughly investigated mechanistically is the extent to which phenotypic changes are induced in chronically injured respiratory epithelial cells, since there is less divergence in epithelial cell populations in chronically injured lungs than there is in unaffected lungs. There is an obvious and pressing need to develop data that will enable precise corrections for differences in pulmonary subgross anatomy and epithelial cell populations when extrapolating toxicologic responses from animals to humans.

The remaining four chapters in this section deal with selected aspects of the impact of structure on dosimetry. In the first of two chapters on the nasal cavity, D. F. Proctor emphasizes unique structural and functional characteristics of the human nasal region and cautions against direct extrapolation from results in experimental animals to expected consequences in humans. He is primarily concerned with the influence that uptake of particles and gases in the upper respiratory tract has on pulmonary dosimetry and with the changes that occur with nasal vs oronasal breathing. He stresses the need for *in vivo* studies in humans whenever possible and for quantitative linkage between findings in experimental animals and humans by using physical models.

In the second chapter on the nasal region, K. T. Morgan and colleagues describe studies specifically designed to examine the relationship between nasal airflow patterns and the topographic distribution of lesions within the nasal cavity of rats inhaling various chemical pollutants. They provide details of the methods for obtaining an accurate plastic mold of the rat nasal cavity, through the intermediate step of producing a metal alloy cast, and for mimicking nasal airflow characteristics using a water–dye system. Their demonstration that predominant flow and region of turbulence correlate well with the main site of formaldehyde-induced lesions provides convincing evidence that nasal airflow patterns are a major determinant of the site of inhalant-induced lesions. As Morgan and colleagues note, their system provides valuable comparisons between nasal characteristics in experimental animals and humans which can begin to fill in part of the data gap identified by Proctor. Morgan and colleagues exemplify the need to correlate local dosimetry and the nature of the population of cells at risk. Their specific example is that the principal target site in the rat for an agent, such as formaldehyde, is covered by epithelium with few ciliated or secretory cells, while the main site in the rhesus monkey is different and is covered by densely ciliated epithelium with fairly plentiful secretory cells.

The final two chapters in this section concern morphometric characteristics of different levels of airways. JD Mortensen and colleagues present age-related morphometric analyses of airways using silicone rubber castings of human lungs. Their findings have an important bearing on the use of dosimetry models across the age range of human populations and are particularly relevant to modeling dosimetry in infants and children who

are an especially sensitive subpopulation because of possible harmful effects of pollutants on lung growth and development. Mortensen and colleagues conclude that total airway volume and the total length of conducting airways are directly related to bodily growth parameters, such as age, height, weight, chest circumference, and body surface area. Of these, chest circumference is the most reliable predictor. In contrast, the number of airway segments and their branching patterns and angles are not significantly correlated with the measured bodily growth parameters.

A striking feature of the pathologic findings in lungs of animals exposed to relatively low concentrations of many injurious agents is the inhomogeneity of the centriacinar response. For example, there is considerable variation in the amount of damage caused by oxidant gases even among closely located bronchiole–alveolar duct junctions in the rat lung. This variation persists from acute through chronic time frames. Reasons for this inhomogeneous response are addressed by R. R. Mercer and J. D. Crapo in their chapter on anatomical modeling of microdosimetry by the use of three-dimensional reconstructions of rat airways. Their findings are an excellent example of the usefulness of three-dimensional reconstructions for providing insights into pathogenetic mechanisms. Mercer and Crapo conclude that the dose of inhaled pollutant delivered at the bronchiolo–alveolar duct junction is dependent both on airway dead space and the volume of the terminal ventilatory unit supplied by each junction. They indicate that there is probably sufficient local variation in dead space, and especially in ventilatory unit volume, to account for local variations in dosimetry and hence the observed local differences in the amount of damage. Whether this explanation can fully account for the observed inhomogeneities in the amount of centriacinar damage deserves further exploration.

Chapter 3

Species Differences in Airway Cell Distribution and Morphology

C. G. Plopper
J. St. George
A. Mariassy
S. Nishio
J. Heidsiek
A. Weir
N. Tyler
D. Wilson
D. Cranz
D. Hyde

California Primate Research Center and Department of Anatomy
School of Veterinary Medicine
University of California, Davis
Davis, California 95616

I. INTRODUCTION

The epithelial lining of respiratory conducting airways is the primary site for pulmonary injury resulting from exposure to environmental toxicants. This appears to be the case for a wide range of compounds which have been tested for their toxicity to the respiratory system. This holds true whether these compounds are administered by inhalation or other systemic means, such as ingestion. The extrapolation to humans of studies of respiratory toxicants using animal models assumes comparability in the structure and function of the respiratory system of these model species and humans. The underlying assumption is that data obtained in model species can be extrapolated to humans. The conjecture of this summary is that the tracheobronchial epithelium of most mammalian species commonly used as models does not adequately reflect the condition in human conducting airways. This study focuses on the epithelial population lining the tracheobronchial airways (trachea, bronchi, and bronchioles) of the mammalian respiratory system. We will consider the following questions concerning the epithelial populations of conducting airways:

1. Is the composition of the epithelial populations similar in the same portion of the airway in different species?
2. Does the relative and absolute abundance of the epithelial cell types vary in the same airway level in different species?
3. Is the composition of epithelial populations the same in all airway levels of the tracheobronchial tree in one species?
4. Are the secretory products contained in these epithelial cell types similar in the same airway level of different species?
5. Does the cytochemical composition of the secretory product vary in cells in different portions of the tracheobronchial tree in the same species?
6. Is the organization of the centriacinar region the same in different species?

II. CELL TYPES IN AIRWAY EPITHELIUM

A. Ultrastructure

At least eight types of epithelial cells lining the conducting respiratory airways in mammals have been identified (Breeze and Wheeldon, 1977; Jeffery, 1983; Plopper, 1983; Widdicombe and Pack, 1982). They differ not only in ultrastructure but also in both function and sensitivity to inhaled intoxicants. These cell types include ciliated cells, basal cells, mucous goblet cells, Clara cells, serous cells, small mucous granule cells, neuroendocrine cells, brush cells, and a variety of intermediate forms.

Basal cells are characterized as very small, flattened cells closely attached to the basal lamina and not extending to the airway lumen (Plopper *et al.*, 1983d). These cells have a small cytoplasm-to-nucleus ratio, and the cyto-

plasm is filled primarily with intermediate filaments. There are large numbers of desmosomes attaching the basal cell to the surrounding epithelial cells. The ciliated cell is attached to the basal lamina and extends to the luminal surface. Its apical surface contains large numbers of cilia interspersed with long microvilli. The majority of the cytoplasm is electron-lucent compared to other epithelial cell types, and the apical region of the cytoplasm is filled with mitochondria.

Several secretory cell types have been identified in various portions of the conducting airway tree. These include the mucous goblet cell, Clara cell, and serous cell. These three cell types are considered to be secretory because their cytoplasm contains relatively large numbers of spherical, membrane-bound inclusion bodies which have been identified as secretory granules. The mucous goblet cell is described as being filled with large granules of varying electron density (Jeffery, 1977; Mariassy and Plopper, 1984). The nucleus is compressed at the basal side of the cell, and few organelles are interspersed in the cytoplasm between individual granules. The Clara cell is characterized as having smaller, electron-dense granules in its apical cytoplasm (Plopper, 1983; Plopper *et al.,* 1983a; Widdicombe and Pack, 1982). The majority of the cytoplasm is filled with large numbers of organelles. These include abundant agranular endoplasmic reticulum in the apical side of the nucleus and a significant amount of granular endoplasmic reticulum in the perinuclear cytoplasm. The serous cell is characterized primarily by large numbers of electron-dense secretory granules. The cytoplasm is filled with granular endoplasmic reticulum and Golgi apparatus (Mariassy and Plopper, 1984).

B. Secretory Cell Morphometry

One of the major questions in characterizing different types of secretory cells within the conducting airway tree is the degree of difference between these cell types. One method to determine this is to morphometrically compare the relative amount of cellular volume occupied by different organelles and other cellular components. Such a study has been done for airway epithelium in sheep (Fig. 3-1) (Mariassy and Plopper, 1984). By ultrastructural criteria, four types of mucous cells (M1-4), serous cells (SC) of submucosal glands, and Clara cells (CC) have been described in sheep. Figure 3-1 illustrates the considerable variation in the proportion of cell volume which is occupied by different cellular components between different airway epithelial cell types of the sheep. The proportion of the cell volume occupied by the nucleus can vary from below 10 to almost 30%. The secretory granule content ranges from approximately 5% in Clara cells

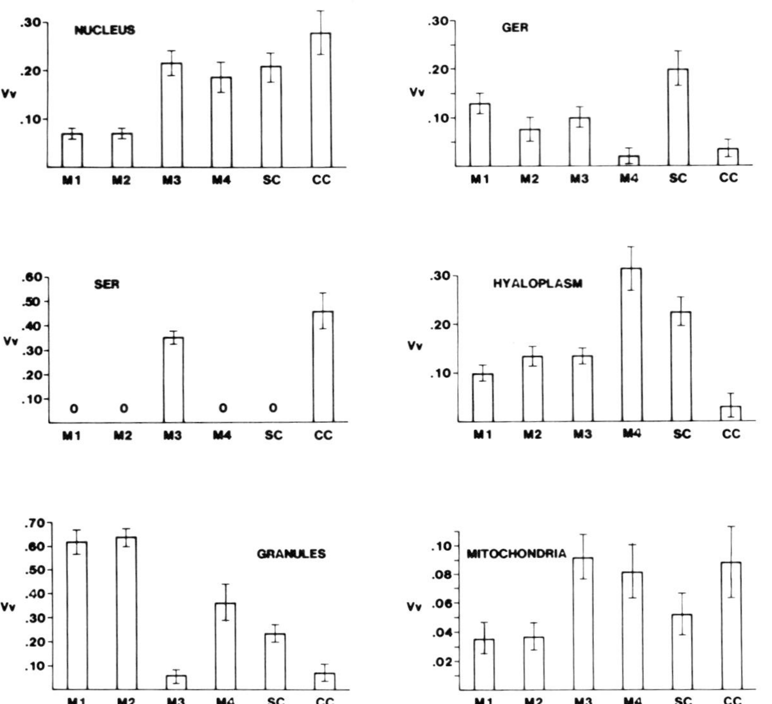

Fig. 3-1 Morphometric comparison of the proportions of cellular components in different epithelial cell types lining tracheobronchial airways of the sheep. Vv, volume fraction; SER, smooth endoplasmic reticulum; GER, granular endoplasmic reticulum. For details, see Mariassy and Plopper (1984).

and small mucous granule (M3) cells to well over 60% in some types of mucous cells. Mitochondrial abundance varies from as little as 3.5 to almost 10%. Only two of these cell types, the Clara cell and small mucous granule cell, have significant amounts of smooth endoplasmic reticulum. There are also differences in abundance of granular endoplasmic reticulum and in organelle-free cytoplasm.

III. EPITHELIAL CELL ABUNDANCE AND DENSITY IN PROXIMAL AIRWAYS

A. Trachea

The composition of the epithelial population of the trachea has been shown to differ from species to species (Breeze and Wheeldon, 1977; Jeffery, 1983; Plopper *et al.,* 1983d; Reid and Jones, 1979; Wilson *et al.,*

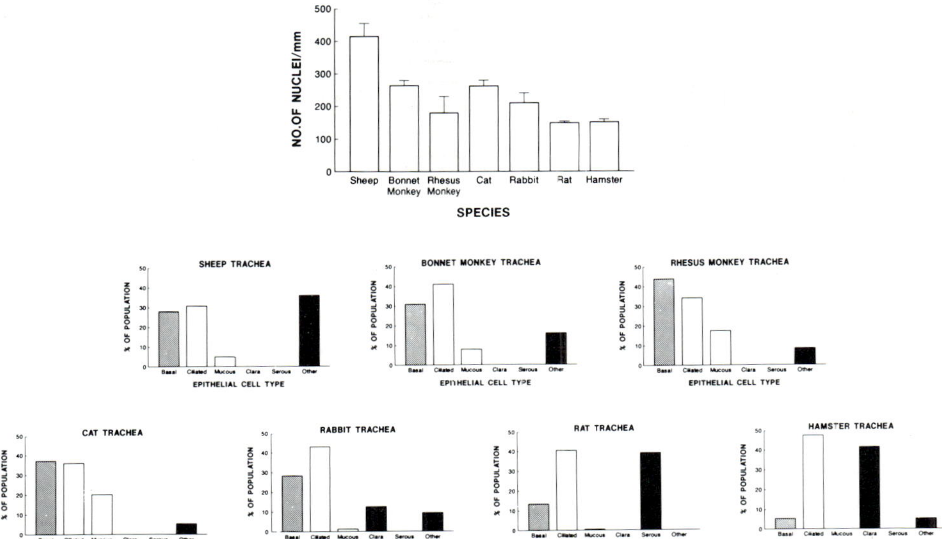

Fig. 3-2 A comparison of abundance and relative proportion of epithelial cells in the trachea of six species: sheep; bonnet monkey; rhesus monkey; cat; rabbit; rat; and hamster. For details, see Plopper *et al.* (1983d).

1984; Pack *et al.,* 1981). Figure 3-2 compares the epithelial populations of the distal third of the trachea of six different mammalian species (Plopper *et al.,* 1983d). There is a wide variability not only in the total abundance of the entire cell population, as measured by number of cells per millimeter of basal lamina, but also in the relative percentage of specific cell types within the population. The number of cells per unit length of basal lamina varies by a factor of almost four. The relative proportion of basal cells varies by a factor of almost 10. Only the proportion of ciliated cells seems to be relatively constant. The predominant secretory cell type varies in different species: mucous goblet or small mucous granule cells (sheep, bonnet monkey, rhesus monkey, cat); Clara cells (hamster, rabbit); and serous cells (rat). Even when the same cell type is present, its relative abundance is not the same in different species (compare Fig. 3-2 to Pack *et al.,* 1981).

B. Bronchi and Bronchioles

The intrapulmonary airways progress to the parenchymal gas exchange area by an extensive branching pattern. To characterize the epithelial populations in the airway tree, it is necessary to define the location of airways to be sampled in terms of branching pattern from the trachea, distance from the terminal bronchiole, and the region within a particular lung lobe. The most effective way which we have found to perform such a localization is to microdissect the airway tree from the lobar bronchus down to the centriacinar region (Plopper *et al.,* 1983c). Microdissection of intrapulmonary airways allows not only high selectivity but also demonstrates the variability in the branching pattern in different species. This approach allows comparisons of the toxic response in the same airway level in different species, in different airway levels of the same species, and in different airway levels and regions of a single lobe of one individual.

Comparing the epithelial populations of intrapulmonary airways in different species demonstrates that variation occurs both in the relative proportions of specific cell types by airway generation and within the same species. Figure 3-3 compares the relative abundance of the major cell types in the airway tree of the sheep, rabbit, and rhesus monkey (Plopper *et al.,* 1983a, 1988; Mariassy and Plopper, 1983). There are distinct distributional differences in the abundance of the basal cells, mucous cells, Clara cells, and ciliated cells. Additionally, there is considerable variation in the numbers of generations between the trachea and the most distal terminal bronchiole in these three species. The percentage of basal cells in these species generally decreases in more distal airways; however, the number of airway generations in which they are found, as well as their proportion of

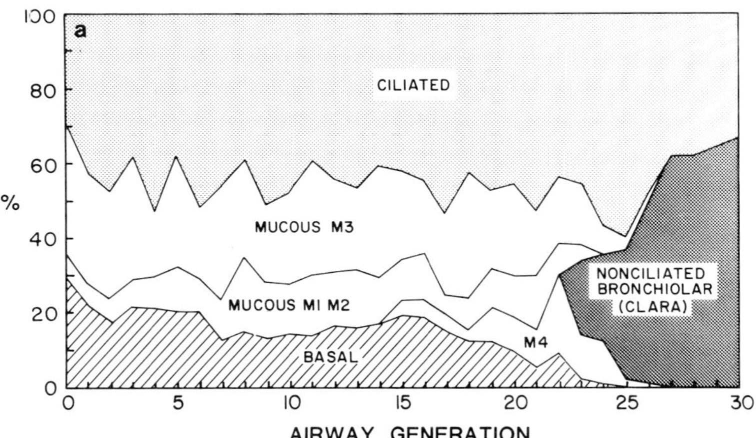

Fig. 3-3 A comparison of the relative abundance of the identifiable cell types throughout the tracheobronchial tree in three species: (a) sheep, left cranial lobe; (b) rabbit, right cranial lobe; (c) rhesus monkey, right cranial lobe. The airway generation number equals the number of generations of branching from the trachea (0) to the most distal bronchiole. For details, see Plopper *et al.* (1983a, 1988), Mariassy and Plopper (1983). *(Figure continues)*

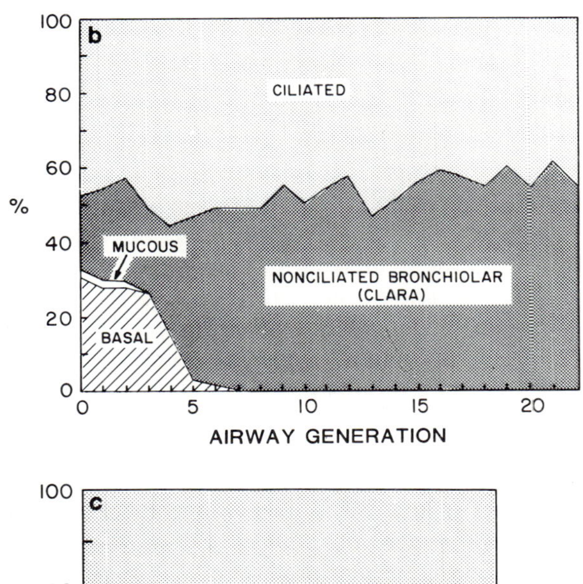

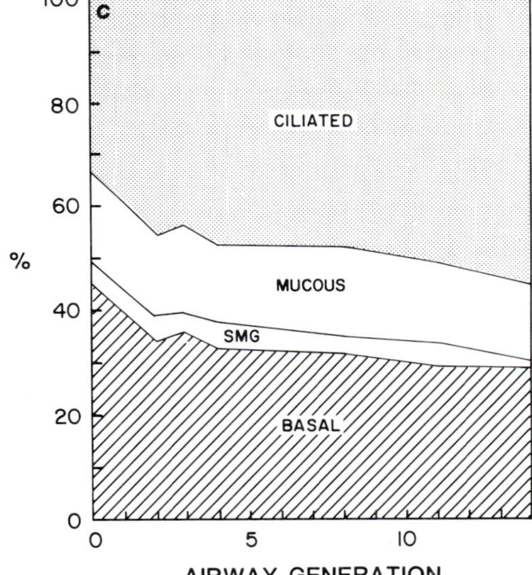

Fig. 3-3 *(continued)*

the population, varies from species to species. The most consistent feature is the presence of ciliated cells. Secretory cells differ in abundance and type of cell present. The proportion of brush cells in the rat lung also varies between the trachea (3%), lobar bronchus (0%), terminal bronchiole (1.4%), and first alveolar duct bifurcation (10.1%) (Chang *et al.,* 1986).

IV. COMPOSITION OF THE CENTRIACINAR REGION

A. Architectural Organization

The centriacinar region is the site of the junction between the most distal conducting airways and the gas exchange area or lung parenchyma. Two different arrangements for this junctional region have been defined for mammals (Tyler, 1983). In most of the smaller laboratory species, and some of the larger domesticated ones, there are several generations of nonalveolarized bronchioles and a single, very short, alveolarized bronchiole which directly joins alveolar ducts. Species which have this arrangement include the horse, ox, sheep, pig, rabbit, guinea pig, hamster, gerbil, rat, and mouse. In all of these species the epithelial population is simple cuboidal with approximately equal numbers of ciliated and nonciliated or Clara cells (Plopper, 1983; Widdecombe and Pack, 1982). A number of other species have an extensive transition zone between nonalveolarized bronchioles and the alveolar duct in the form of long, alveolarized bronchioles or respiratory bronchioles. Species which have few nonalveolarized, noncartilaginous bronchioles and several generations of alveolarized (respiratory) bronchioles include the macaque monkey (Castleman *et al.,* 1975; Tyler and Plopper, 1985), dog (Freeman *et al.,* 1973; Hyde *et al.,* 1978), and cat (Hyde *et al.,* 1985; Plopper *et al.,* 1983b). Species in which there are several generations of nonalveolarized, noncartilaginous bronchioles and several generations of alveolarized bronchioles include the human (Clara, 1937; von Hayak, 1960) and ferret (Hyde *et al.,* 1979). Respiratory bronchioles have the luminal surface lined by epithelial populations characteristic of more proximal conducting airways, i.e., simple cuboidal epithelium, interrupted by alveolar outpocketings lined by epithelial populations (type 1 and type 2 alveolar epithelial cells) characteristic of alveoli in the gas exchange area. In the three species in which this region has been studied, dog (Plopper *et al.,* 1980b), cat (Hyde *et al.,* 1985; Plopper *et al.,* 1980b, 1983b) and macaque monkey (Castleman *et al.,* 1975; Tyler and Plopper, 1985), the bronchiolar epithelial population consists almost entirely of nonciliated cells completely around the circumference of the respiratory bronchiole in the cat and the dog, and on the side opposite the pulmonary artery of the macaque monkey. On the side adjacent to the pulmonary arteriole in primates, the epithelium is pseudostratified columnar with mucous goblet cells. This population closely resembles that of more proximal bronchi and the trachea.

B. Clara Cell Ultrastructure

Interspecies variation exists in the ultrastructural characteristics of the nonciliated (Clara) cells lining in the respiratory and terminal bronchioles

(Widdicombe and Pack, 1982; Plopper *et al.*, 1980a–c). Figure 3-4 compares the ultrastructural features of the nonciliated bronchiolar cells in three species (Hyde and Plopper, 1984). In a large number of species (horse, sheep, rabbit, guinea pig, rat, hamster, and mouse), more than 40% of the cytoplasm is occupied by agranular endoplasmic reticulum (Fig. 3-4a). The nucleus is the second most prominent component followed by mitochondria. Glycogen composes approximately 10% or less of the cell volume in these species (Fig. 3-4a). In other species the Clara cell contains primarily glycogen (Fig. 3-4b). In the ox, cat, and dog, more than 60% of the cytoplasmic volume is glycogen. Other extranuclear components, including agranular endoplasmic reticulum and mitochondria, make up a relatively small proportion of the cell volume. Primate species (Fig. 3-4c) do not have an abundance of either endoplasmic reticulum or glycogen in the nonciliated cells lining respiratory bronchioles. These cells have a smaller volume than in the other species (Fig. 3-4c), and the largest component is the nucleus. While a larger percentage of the cell is organelle-free hyaloplasm (Fig. 3-4c), other organelles compose less than 10% of the cell volume.

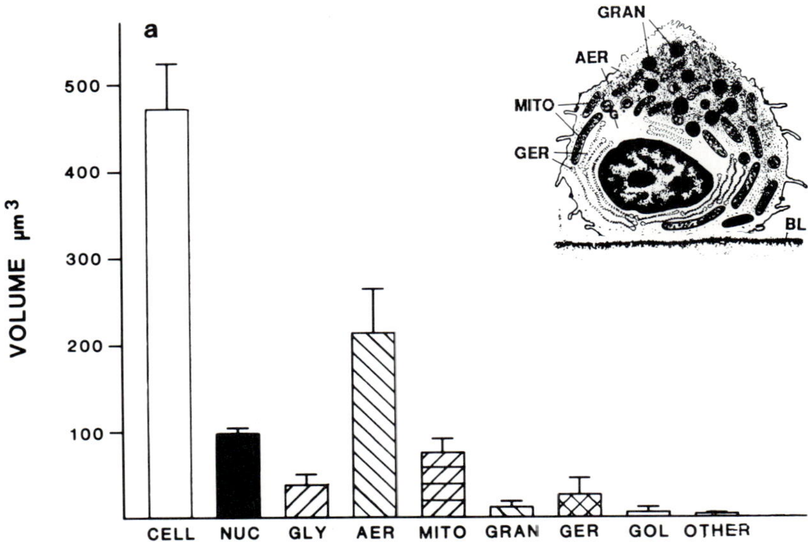

Fig. 3-4 Morphometric comparison of the volumes of cellular components in the Clara cells of three species: (a) rabbit; (b) cat; (c) bonnet monkey. Nucleus (NUC); glycogen (GLY); agranular endoplasmic reticulum (AER); mitochondria (MITO); secretory granules (GRAN); granular endoplasmic reticulum (GER); golgi apparatus (GOL); other components and hyaloplasm (OTHER); BL, basal lamina. For details, see Hyde and Plopper (1984). *(Figure continues)*

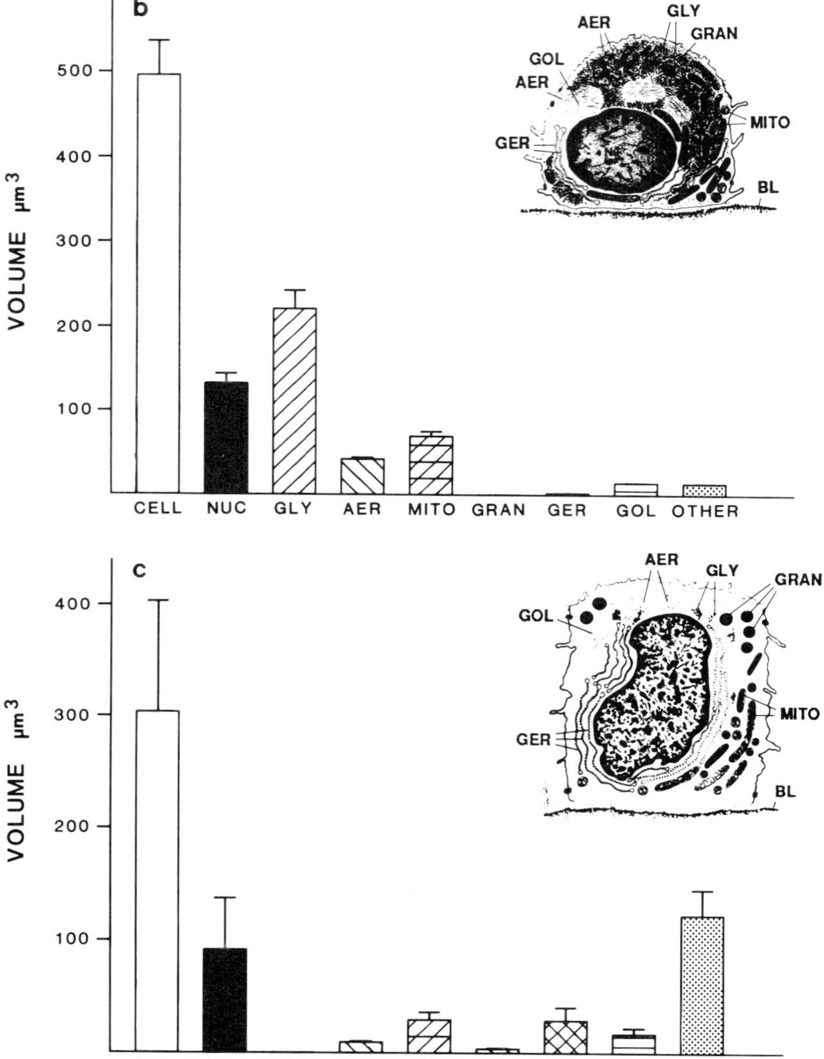

Fig. 3-4 *(continued)*

V. GLYCOCONJUGATE CONTENT OF SECRETORY GRANULES

A. Carbohydrate Cytochemistry

Conventional carbohydrate cytochemical methods have demonstrated a wide variation in the carbohydrate content of the secretory cells between species (Spicer *et al.*, 1971, 1980, 1983; Marsan *et al.*, 1978; St. George *et al.*, 1984; Plopper *et al.*, 1984). The periodic acid Schiff reaction in combination with Alcian Blue (AB/PAS) is used to distinguish between neutral (AB−/PAS+) and acidic glycoconjugates (AB+/PAS+). High-iron diamine in sequence with Alcian Blue (HID/AB) is used to distinguish between sulfated (HID+/AB+) and nonsulfated acidic (HID−/AB+) glycoconjugates. While the mucous cell is usually the predominant secretory cell, this is not the case for all species (Table 3-1). In species in which the Clara or serous cell is the predominant secretory cell in the trachea, the majority of the secretory product is a neutral glycoconjugate. In species in which mucous goblet cells are the predominant secretory cell type, the majority of the secretory product is an acidic glycoconjugate. In some species, such as the rhesus monkey, all the secretory cells contain an acidic sulfated glycoconjugate. In other species, such as the sheep, some mucous goblet cells contain sulfated glycoconjugates and others contain sialoglycoconjugates. As Table 3-2 shows, there is less variability in glycoconjugate composition in the submucosal glands. In serous cells the secretory product is neutral and in mucous cells it is either a sialo- (acidic nonsulfated) or

Table 3-1 Comparison of Carbohydrate Content of Tracheal Epithelium

			Carbohydrate Content[a]		
Species	Cell Type	Abundance	PAS	AB	HID
Hamster[b]	Clara	+++	+	−	−
Rat[c]	Serous	+++	+	−	−
	Mucous	+	+	+	−
Rabbit[d]	Mucous	+	+	+	+
	Clara	+++	+/−	−	−
Sheep	Mucous	++	+	+	+/−
Rhesus[e]	Mucous	++	+	+	+
Human[f]	Mucous	+++	+	+	+/−

[a] PAS, periodic acid Schiff; AB, Alcian Blue; HID, high-iron diamine.
[b] Emura and Mohr (1975).
[c] McCarthy and Reid (1964).
[d] Plopper *et al.* (1984).
[e] St. George *et al.* (1984).
[f] Spicer *et al.* (1971).

Table 3-2 Comparison of Carbohydrate Content of Tracheal Submucosal Glands

Species	Abundance	Secretory Cell	Carbohydrate Content[a] PAS	AB	HID
Hamster[b]	?	?	?	?	?
Rat[c]	+	Serous	+	−	−
		Mucous	+	+	+
Rabbit	+/−	Mucous	+	+	+
Sheep	++	Serous	+	−	−
		Mucous	+	+	+
Rhesus[d]	++	Serous	+	−	−
		Mucous	+	+	+
Human[e]	+++	Serous	+	−	−
		Mucous	+	+	+/−

[a] PAS, periodic acid Schiff; AB, Alcian Blue; HID, high-iron diamine.
[b] No submucosal glands described for hamsters.
[c] Plopper *et al.* (1984).
[d] St. George *et al.* (1986).
[e] Spicer *et al.* (1971).

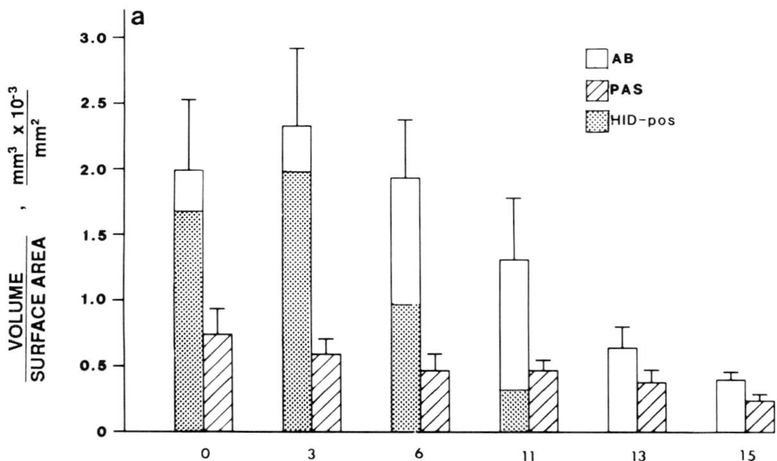

Fig. 3-5 Morphometric comparison of the volume of stored secretory product per unit basal lamina in the tracheobronchial tree of the rhesus monkey: (a) surface lining epithelium; (b) submucosal glands; (c) total secretory product; Alcian Blue (AB); periodic acid Schiff (PAS); high-iron diamine (HID); lobar bronchus (3). Sixth generation intrapulmonary bronchus (6). Eleventh generation intrapulmonary bronchus (11). Terminal bronchial (13). Respiratory bronchial (15). For details, see Plopper *et al.* (1988). *(Figure continues)*

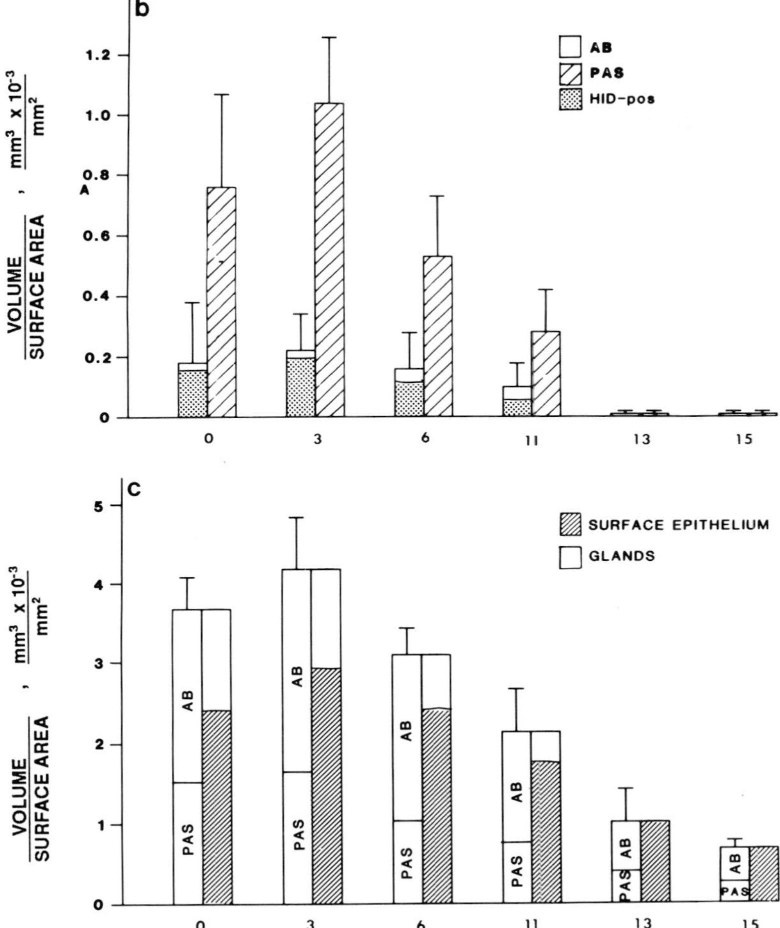

Fig. 3-5 *(continued)*

sulfomucin. The principal difference between species in airway mucosecretory apparatus is in the abundance and distribution of submucosal glands (Breeze and Wheeldon, 1977; Jeffery, 1977, 1983; Plopper *et al.*, 1983a; Mariassy and Plopper, 1983; Castleman *et al.*, 1975; Hyde *et al.*, 1979; Spicer *et al.*, 1980).

B. Quantitative Comparison of Stored Glycoconjugates

Not only may the glycoconjugate composition of the secretory cells of one airway, such as the trachea, differ between species, but the type and relative abundance of glycoconjugates can vary between airway levels in the same species. Computerized morphometric assessment of the volume of the secretory product per unit surface area demonstrates that there is considerable difference in the amount of glycoconjugate stored in airway epithelium of the rhesus monkey (Fig. 3-5) (Plopper *et al.,* 1988). Not only is there greater than a twofold difference in the amount of secretory material stored when comparing the trachea with terminal and respiratory bronchioles, but there is a difference in the sulfation of product being stored (Fig. 3-5a). The product is predominantly sulfated in proximal airways and acidic nonsulfated in more distal airways. The amount of product stored in submucosal glands of the conducting airways in the rhesus monkey shows the same degree of variability (Fig. 3-5b). However, our observations indicate the secretory product in the glands is predominantly a neutral glycoconjugate (Fig. 3-5b).

VI. SUMMARY

This comparative review of tracheobronchial epithelial composition emphasizes the high degree of diversity found in different species of mammals. Epithelial populations of the tracheobronchial tree vary in:

1. The same airway level in different species,
2. Different airway levels in the same species,
3. Types of cells present in all airways,
4. The overall abundance of secretory product and number of cells,
5. Carbohydrate content of secretory granule glycoconjugates,
6. The amount of stored secretory product, and
7. The ultrastructural composition of the same cell type in different species.

This wide diversity in epithelial populations of the tracheobronchial tree of different species challenges our ability to extrapolate findings from animal models to humans and care must be taken. This study emphasizes the need for careful selection of animal species for specific types of exposure and precise selection of the site for assessment of the lung's response. It should be emphasized that comparative pulmonary toxicologic studies, whether by inhalation or ingestion, are still needed to assess how this diversity will affect the reaction of the respiratory system of different species to the same environmental toxicant.

ACKNOWLEDGMENTS

This work was supported in part by NIH Grants HL28978, ES00628, and DRR00169.

REFERENCES

Breeze, R. G., and Wheeldon, E. B. (1977). The cells of the pulmonary airways. *Am. Rev. Respir. Dis.* **161,** 705–777.

Castleman, W. L., Dungworth, D. L., and Tyler, W. S. (1975). Intrapulmonary airway morphology in three species of monkeys: A correlated scanning and transmission electron microscopic study. *Am. J. Anat.* **152,** 107–122.

Chang, L.-Y., Mercer, R., and Crapo, J. D. (1986). Differential distribution of brush cells in rat lung. *Anat. Rec.* **216,** 49–54.

Clara, M. (1937). Zur Histobiologie des Bronchial-Epithels. *Z. Mikrosk. Anat. Forsch.* **41,** 321–347.

Emura, M., and Mohr, U. (1975). Morphological studies on the development of tracheal epithelium in the Syrian golden hamster. I. Light Microscopy. *Z. Versuchstierkd.* **17,** 14–26.

Freeman, G., Stephens, R., Coffin, D., and Stara, J. (1973). Changes in dog's lungs after long-term exposure to ozone. *Arch. Environ. Health* **26,** 209–216.

Hyde, D. M., and Plopper, C. G. (1984). Cell and organellar volumes of nonciliated bronchiolar cells of rabbits, cats and monkeys. *Anat. Rec.* **208,** 77A.

Hyde, D. M., Orthoefer, D., Dungworth, D., Tyler, W., Carter, R., and Lum, H. (1978). Morphometric and morphologic evaluation of pulmonary lesions in beagle dogs chronically exposed to high ambient levels of air pollutants. *Lab. Invest.* **38,** 455–469.

Hyde, D. M., Samuelson, D. A., Blakeney, W. H., and Kosch, P. C. (1979). A correlative light microscopy, transmission and scanning electron microscopy study of the ferret lung. *Scan. Electron Microsc.* **3,** 891–898.

Hyde, D. M., Plopper, C. G., Weir, A. J., Murnane, R. D., Warren, D. L., Last, J. A., and Pepelko, W. E. (1985). Peribronchiolar fibrosis in lungs of cats chronically exposed to diesel exhaust. *Lab. Invest.* **52,** 195–206.

Jeffery, P. K. (1977). Structure and function of mucus-secreting cells of cat and goose airway epithelium. *Cibra Found. Symp. Ser.* pp. 5–19.

Jeffery, P. K. (1983). Morphologic features of airway surface epithelial cells and glands *Am. Rev. Respir. Dis.* **128,** S14–S20.

McCarthy, C., and Reid, L. (1964). Acid mucopolysaccharide in the bronchial tree in mouse and rat (sialomucin and sulphate). *Q. J. Exp. Physiol.* **49,** 81–84.

Mariassy, A. T., and Plopper, C. G. (1983). Tracheobronchial epithelium of the sheep. I. Quantitative light microscopic study of epithelial cell abundance and distribution. *Anat. Rec.* **205,** 263–275.

Mariassy, A. T., and Plopper, C. G. (1984). Tracheobronchial epithelium of the sheep: II. Ultrastructural and morphometric analysis of the epithelial secretory cell types. *Anat. Rec.* **209,** 523–534.

Marsan, C., Cava, E., Lougnon, J., and Roujeau, J. (1978). Cytochemical and histochemical characterization of epithelial mucins in human bronchi. *Acta Cytol.* **22,** 562–565.

Pack, R. J., Al-Ugaily, L. H., and Morris, G. (1981). The cells of the tracheobronchial epithelium of the mouse: A quantitative light and electron microscopic study. *J. Anat.* **132,** 71–84.

Plopper, C. G. (1983). Comparative morphologic features of bronchiolar epithelial cells: The Clara cell. *Am. Rev. Respir. Dis.* **128,** S37–S41.

Plopper, C. G., Mariassy, A. T., and Hill, L. H. (1980a). Ultrastructure of the nonciliated bronchiolar epithelial (Clara) cell of mammalian lung. *Exp. Lung Res.* **1,** 139–154.

Plopper, C. G., Mariassy, A. T., and Hill, L. H. (1980b). Ultrastructure of the nonciliated bronchiolar epithelial (Clara) cell of mammalian lung. II. A comparison of horse, steer, sheep, dog, cat. *Exp. Lung Res.* **1,** 155–169.

Plopper, C. G., Hill, L. H., and Mariassy, A. T. (1980c). Comparative ultrastructure of the nonciliated bronchiolar epithelial (Clara) cell of the mammalian lung. III. A study of man with comparison of 15 mammalian species. *Exp. Lung Res.* **1,** 171–180.

Plopper, C. G., Halsebo, J. E., Berger, W. J., Sonstegard, K. S., and Nettesheim, P. K. (1983a). Distribution of nonciliated bronchiolar epithelial (Clara) cells in intra- and extra-pulmonary airways of the rabbit. *Exp. Lung. Res.* **5,** 79–98.

Plopper, C. G., Hyde, D. M., and Weir, A. J. (1983b). Centriacinar alterations in lungs of cats chronically exposed to diesel exhaust. *Lab. Invest.* **49,** 391–399.

Plopper, C. G., Mariassy, A. T., and Lollini, L. O. (1983c). Structure as revealed by airway dissection: A comparison of mammalian lungs. *Am. Rev. Respir. Dis.* **128,** S4–S7.

Plopper, C. G., Mariassy, A. T., Wilson, D. W., Alley, J. L., Nishio, S. J., and Nettesheim, P. (1983d). Comparison of nonciliated tracheal epithelial cells in six mammalian species: Ultrastructure and population densities. *Exp. Lung Res.* **5,** 281–294.

Plopper, C. G., St. George, J. A., Nishio, S. J., Etchison, J. R., and Nettesheim, P. (1984). Carbohydrate cytochemistry of tracheobronchial airway epithelium of the rabbit. *J. Histochem. Cytochem.* **32,** 209–218.

Plopper, C. G., Heidsiek, J., Weir, A., St. George, J., and Hyde, D. (1989). Tracheobronchial epithelium of the adult rhesus monkey: A quantitative histochemical and ultrastructural study. *Am. J. Anat.* **184,** 31–40.

Reid, L., and Jones, R. (1979). Bronchial mucosal cells. *Fed. Proc., Fed. Am. Soc. Exp. Biol.* **38,** 191–196.

Spicer, S. S., Chakrin, L., Wardell, Jr., J., and Kendrick, W. (1971). Histochemistry of mucosubstances in the canine and human respiratory tract. *Lab. Invest.* **25,** 483–490.

Spicer, S. S., Mochizuki, I., Setser, M. E., and Martinez, J. R. (1980). Complex carbohydrates of rat tracheobronchial surface epithelium visualized ultrastructurally. *Am. J. Anat.* **158,** 93–109.

Spicer, S. S., Schulte, B. A., and Thomopoulos, G. N. (1983). Histochemical properties of the respiratory tract epithelium in different species. *Am. Rev. Respir. Dis.* **128,** S20–S25.

St. George, J. A., Nishio, S. J., and Plopper, C. G. (1984). Carbohydrate cytochemistry of rhesus monkey tracheal epithelium. *Anat. Rec.* **210,** 293–302.

St. George, J. A., Nishio, S. J., Cranz, D. L., and Plopper, C. G. (1986). Carbohydrate cytochemistry of rhesus monkey submucosal glands. *Anat. Rec.* **216,** 60–67.

Tyler, N. K., and Plopper, C. G. (1985). Morphology of the distal conducting airways in rhesus monkey lungs. *Anat. Rec.* **211,** 295–303.

Tyler, W. S. (1983). Comparative subgross anatomy of lungs. *Am. Rev. Respir. Dis.* **128,** S32–S36.

Von Hayak, H. (1960). "The Human Lung." Hafner, New York.

Widdecombe, J. G., and Pack, R. J. (1982). The Clara cell. *Eur. J. Respir. Dis.* **63,** 202–220.

Wilson, D. W., Plopper, C. G., and Hyde, D. M. (1984). The tracheobronchial epithelium of the bonnet monkey: A quantitative ultrastructural study. *Am. J. Anat.* **171,** 25–40.

Chapter 4

Alternative Methods to Evaluate Species Differences in Upper Airway Structure – Function

D. F. Proctor
Environmental Health Sciences
Otolaryngology, Anesthesiology
Johns Hopkins Medical Institutions
Baltimore, Maryland 21205

I. INTRODUCTION

Lung dosimetry of inhaled materials is determined by the structure and function of the entire airway, a continuum from the nostrils or lips to alveoli (Brain *et al.,* 1977, 1979; Mygind *et al.,* 1983; Proctor and Andersen, 1982). While the tracheobronchial portion of the conducting airways can be studied with relative ease in living man, experimental animals, or models, the morphological complexity and multifunctional role of the upper airways make the investigation difficult in living man and extrapolation from research in experimental animals or models of dubious value. Nevertheless, investigations in animals or models are necessary approaches to many air pollution problems (Aharonson *et al.,* 1974; Barrow, 1984).

Webster's definition of extrapolation is: "To project by inference into an unexplored situation from observations in an explored field, on the assumption of continuity or correspondence . . ." Almost nothing is so tempting to the biological scientist as the possibility of extrapolating conclusions from readily demonstrable data that may be applicable to questions less susceptible to proven solutions. We take such action with caution, but in some instances where the use of living human subjects is reprehensible or impossible, we are forced to attempt to arrive at appropriate decisions through experiments in animals or in models. An example of the need for caution when we do so is the finding of nasal cancer in rodents when exposed to formaldehyde (Proctor and Chang, 1985). There is no convincing evidence that nasal cancer has been similarly associated with this agent in humans. Differences in ciliary clearance of surface secretions (Morgan *et al.,* 1986) or in contact between inhaled gas and nasal surfaces are two possible explanations for this discrepancy. Especially in considering the upper airways, we must be aware that toxicological investigations in experimental animals may lead to either an over- or underestimate of the hazards to humans in our ambient air.

Greater emphasis and effort should be spent on experiments in human volunteers. But I would be the last to deny the continuing necessity for nonhuman research. Experiments with sufficiently toxic substances cannot be done in man, and even relatively innocuous pollutant studies as well as some physiological investigations continue to be difficult or impossible in infants and children (Harding, 1986; Polgar and Weng, 1979).

Almost 100 years ago Goodale (1896) showed that accurate observations of nasal particle deposition are possible in humans. Over 25 years ago Pattle (1961) drew conclusions on nasal retention of particles and gases in the human nose, which experiments in succeeding years have chiefly confirmed. Such studies should remind us to resort to nonhuman experiments as little as is feasible.

We are in the paradoxical situation of having a special need to carry out investigations in the living human, but in many instances are faced with the impossibility of doing so. While we are forced to seek much of our information on the upper airways from nonhuman investigations, we must be especially cautious about extrapolating conclusions applicable to the health of man.

II. STRUCTURE AND FUNCTION IN RELATION TO AIRFLOW AND DEPOSITION

To understand how structure and function influences airflow and deposition, we must first examine what is now known about the morphology of the human upper airway (Proctor, 1977; Proctor and Andersen, 1982). The entrance to the nose provides the smallest total cross section in the entire respiratory tract (Haight and Cole, 1983). Just beyond this point the airway bends and becomes a narrow convoluted passage. The width of this passage is only 1–3 mm, but the introduction of measuring devices is difficult and inevitably interferes with airflow. At its posterior termination, there is a gradual bend of 90° to enter the pharynx, which is the most widely variable portion of the airways.

As a result of these anatomical features, there is a high velocity of the air stream at the nasal valve and the development of some turbulence in the main passage (Proctor, 1966, 1986, 1988). It is now known that deposition of particles in the anterior nares occurs where mucociliary clearance in humans carries them forward to a region from which they may be removed by nose blowing and cleaning (Andersen *et al.,* 1971; Hounam, 1975; Proctor *et al.,* 1977). It is unclear whether or not such anterior clearance and disposal occurs in animals. In the main passage, there is an exchange of temperature and water vapor, and an optimal opportunity for absorption of water soluble gases in surface secretions. Further deposition of particles may occur at the bend in the nasopharynx. The approximate main lines of inspiratory airflow in man, the baboon, barking deer, and rat are shown in Fig. 4-1. The significant differences and their probable effect on particle deposition are self evident.

After the bend in the nasopharynx, during quiet breathing, there is a more or less straight line to the larynx and trachea. But, when pulmonary ventilation demands airflows in excess of those possible through the nose, the lips are parted and oronasal breathing begins (Niinimaa *et al.,* 1981;

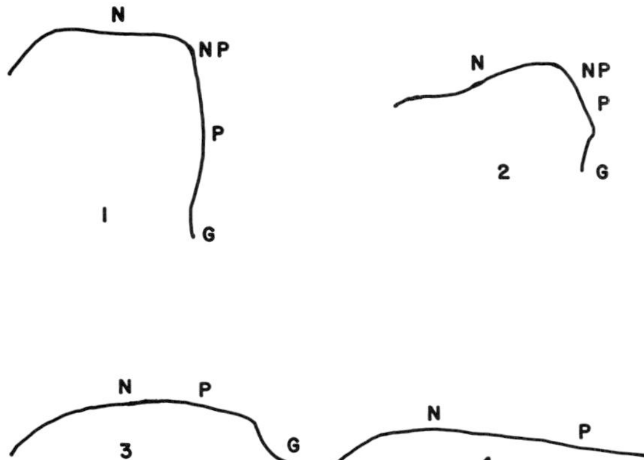

Fig. 4-1 Main lines of inspiratory airflow. Direction of flow is from left to right. (1) Man; (2) baboon; (3) barking deer; (4) rat. Nostrils are to the left; N, main nasal passage; NP, nasopharynx; P, pharynx; G, glottis. These lines are roughly redrawn from various sources in the literature.

Olson and Strohl, 1987; Proctor, 1983; Rodenstein and Stanescu, 1984, 1986; Strohl *et al.,* 1980, 1982). Thanks to the work of Rodenstein and Stanescu (1984, 1986) we know that positioning of the soft palate determines the division of airflow between the two parallel naso- and oropharyngeal airways. In most circumstances this is nearly an equal division. But, due to the multifunctional role of the pharynx, this is not always the case. During forced expiration (as in blowing up a balloon, playing a flute, or performing a forced vital capacity) the palate closes the velopharyngeal valve, and during deep inspirations, as in singing (Bates *et al.,* 1965) or yawning, the oropharyngeal airway opens widely, the tongue is depressed and drawn forward, and there is ready access of inspired air to the larynx. Oronasal breathing also takes place during ordinary conversation, but the equal division of inspiratory airflow between the two airways continues (Proctor, 1980). It has not yet been determined what occurs when the speaker must overcome loud ambient noise, such as in many industrial environments.

III. RATIONALE FOR CHOICES OF METHODOLOGY

How do these structure–function facts influence our choice of investigation methods? What are we seeking as the object of investigation? A major objective is to be able to determine what potentially hazardous materials in our ambient air may reach our lungs. In order to make such a prediction we try to find out what things may be taken in in the inspired air and what proportion of these will pass through those air passages intervening between the ambient air and the trachea. When making that determination we must consider not merely simple nasal breathing but also oronasal breathing and the variations which occur when breathing is used for nonrespiratory functions, such as speaking, singing, flute playing, yawning, sneezing, and coughing. Beyond all of that, some thought must be given to the possibility of damage to upper respiratory defenses resulting from the removal of injurious substances and a possible consequent increase of the burden placed on the lungs.

Owing to our unique upper airway configuration and to the many functions which we do not have in common with animals, it is difficult to reach a solution to the problem of air pollution and its health effects in humans. These objectives can only be partially realized by combining experiments in human volunteers with experiments in animals and ingeniously constructed models.

Airflow characteristics obviously affect the fate of materials carried from the ambient air into the inspired stream. It is presently impossible to make detailed airflow measurements in the nose of living humans. Using models in this case is helpful but they cannot simulate all of the dynamic changes in airway shape. Owing to the differences already referred to, experimental animal studies do not yield appropriate information (Proctor and Chang, 1985; Schreider, 1986). Therefore, we must resort to casts of human nasal passages and combine these findings with analysis of the result of oronasal breathing.

We should recognize the fact that in measuring the fate of inhaled materials during their inspiration through the upper airways, research in animals may lead to misleading information. Not only are the main lines of inspiratory airflow very different, but animals rarely employ oronasal breathing, do not indulge in conversation, and do not blow their noses. We should rely upon investigations in living humans where possible (Andersen *et al.,* 1971; Brain and Valberg, 1979; Hounam *et al.,* 1970; Hounam, 1975; Hounam and Morgan, 1977; Niinimaa *et al.,* 1981; Proctor *et al.,* 1977; Strohl *et al.,* 1980, 1982) and when necessary supplement this data with models of the human airways. Even in investigating toxic substances, it is possible to simulate their deposition characteristics by the use of physically similar but nontoxic materials.

When we come to the question of the multifunctional aspects of the naso-oro-pharyngo-laryngeal inspiratory passages, neither static models nor animal experiments can be reliably employed. Cineradiographic techniques are useful in examining the changing shape, especially of the oropharynx. These changes are important in evaluating what occurs during exercise (Olson and Strohl, 1987), phonation, deglutition, sneezing, coughing, yawning, and other actions in which this portion of the respiratory tract is involved. But the effects of these functions on the inspiratory fate of inhaled materials have not yet been fully elucidated.

IV. FLOW-LIMITING SEGMENTS AND AIRWAY DEFENSES

One factor that is of special importance is the sequence of inspiratory flow-limiting segments which compose the upper airways (Fig. 4-2) (Brancatisano *et al.,*, 1983; Fouke *et al.,* 1986; Proctor, 1983, 1988; Strohl *et al.,* 1980, 1982). These are the nasal valve (Haight and Cole, 1983), velopharyngeal valve, whole pharynx, and larynx. Fortunately, the caliber of the various portions of the upper airways is ordinarily stabilized by an inspiratory increase in tone of the muscles surrounding the region including the nasal valve and the pharynx (Proctor, 1983; Strohl *et al.,* 1980, 1982). In sleep apnea, there appears to be a failure of this reflex action, but we do not know to what degree a relative failure in that function may occur in seemingly normal persons and bring into play any one or all of the flow-limiting segments. Flow limitation at any point from nose to larynx will

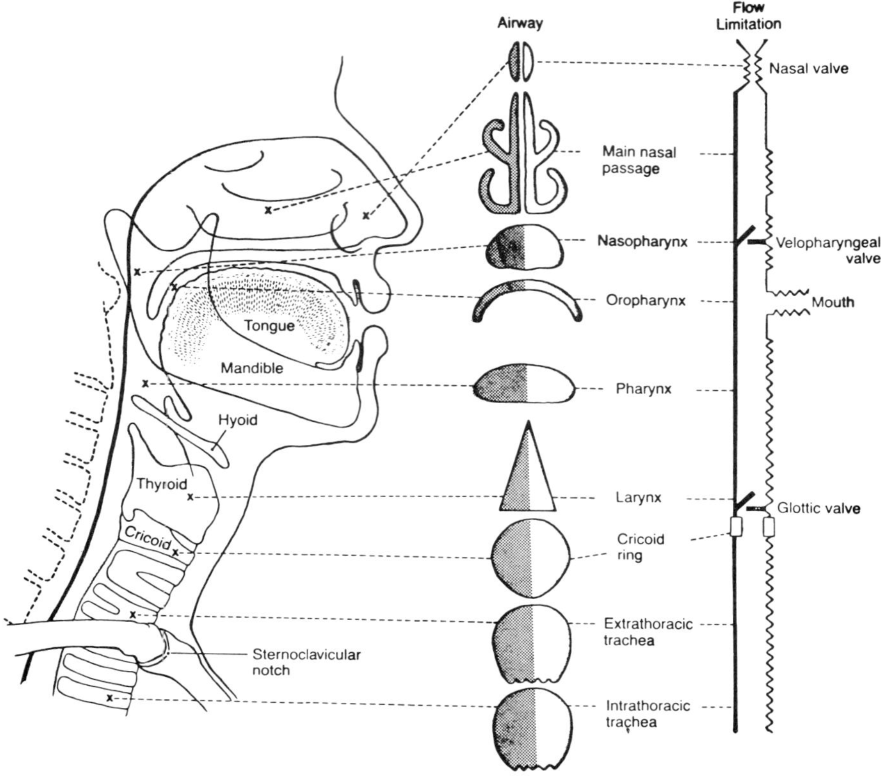

Fig. 4-2 Diagram illustrating the relationship in man between various segments of the upper airways and inspiratory flow-limiting segments. Left, anatomical structures; center, approximate cross sections; right, wavy lines indicate potential inspiratory flow limitations. Each bend in the airway and each flow limiting region may influence deposition of inhaled materials. (Reprinted from Proctor, 1988, with permission of the editors and publisher.)

result in the need for an increasingly negative inspiratory pressure at all subsequent points if the desired airflow is to be maintained. Such negative pressure may then narrow the involved airway segment.

Hounam and coworkers (1970, 1977) have shown that deposition characteristics of the nose are related to the airflow resistance at the nasal valve. If sufficiently narrowed, each of the other flow-limiting segments may also influence the fate of inhaled materials. Laryngeal function in relation to simple breathing probably has little effect on particle deposition or gas absorption. But, in spite of the excellent work in recent years, we still do not exactly know the results of either normal physiological function or malfunction of the larynx on the fate of inhaled particles and gases (Bartlett *et al.,* 1973; Bartlett, 1979; Brancatisano *et al.,* 1983; Proctor, 1986; Remmers and Bartlett, 1977).

Sometimes we overlook the possibility that upper airway defenses can be impaired by air pollutants and, as a consequence, the lungs may be increasingly exposed to injurious substances in the ambient air (Brain *et al.*, 1977; Proctor, 1977). Examples of such defense impairment are found in the studies of the effects of hardwood dust, sulfur dioxide, and formaldehyde on humans. Although animal experiments can provide useful information (Morgan *et al.*, 1986), cautiously planned human volunteer studies are the ideal approach to such problems.

V. CONCLUSIONS

Certainly, an understanding of lung dosimetry is impossible without a full knowledge of what transpires in the upper airways before air passes into the lungs. Considering all of the factors involved, we must conclude that a final decision on the control of any pollutant in our ambient air in relation to its likely effect upon human health should largely be based on results of studies of nasal and oronasal breathing in human volunteers, preferably in carefully controlled climate chambers. As a primary source of basic information, and the second most desirable alternative in the interpretation of toxicological studies, experiments, conducted with accurate models of the human respiratory tract must continue. Nevertheless, effective as the human volunteer approach is, there is no getting around the fact that some investigations will have to be done in experimental animals and more model experiments will be needed because it is difficult or impossible to carry out some studies in the living human. The usefulness of information gained will depend upon the accuracy of model construction, choice of appropriate experimental animals, and justifiable extrapolation from findings.

In summary, the use of both models and experimental animals is and will continue to be necessary to provide information which cannot be readily obtained through investigations in human volunteers. But extrapolation from such studies to decisions affecting human health must be done with extreme caution. The knowledge provided by research in humans is the only information upon which we can surely depend and from which final conclusions can be drawn.

REFERENCES

Aharonson, E. F., Menkes, H., Gurtner, G., Swift, D. L., and Proctor, D. F. (1974). The effect of respiratory airflow rate on the removal of soluble vapors by the nose. *J. Appl. Physiol.* **37,** 654–657.

Andersen, I., Lundqvist, G. R., and Proctor, D. F. (1971). Human nasal mucosal function in a controlled climate. *Arch. Environ. Health* **23,** 408–420.

Barrow, C. S. (1984). "Toxicology of the Nasal Passages." McGraw–Hill, New York.

Bartlett, D., Jr. (1979). Effects of hypercapnia and hypoxia on laryngeal resistance to airflow. *Respir. Physiol.* **37,** 293–302.

Bartlett, D., Jr., Remmers, J. E., and Gautier, H. (1973). Laryngeal regulation of respiratory airflow. *Respir. Physiol.* **18,** 194–204.

Bates, J. H., Potts, W. E., and Lewis, J. M. (1965). Epidemiology of primary tuberculosis in an industrial school. *N. Engl. J. Med.* **272,** 714–717.

Brain, J. D., and Valberg, P. A. (1979). Deposition of aerosol in the respiratory tract, state of the art. *Am. Rev. Respir. Dis.* **120,** 1325–1373.

Brain, J. D., Proctor, D. F., and Reid, L. M. (1977). "Respiratory Defense Mechanisms." Dekker, New York.

Brancatisano, T., Collett, P. W., and Engel, L. A. (1983). Respiratory movements of the vocal cords. *J. Appl. Physiol.* **54,** 1269–1276.

Fouke, J. M., Tetter, J. P., and Strohl, K. P. (1986). Pressure–volume behavior of the upper airways. *J. Appl. Physiol.* **61,** 912–918.

Goodale, J. L. (1896). An experimental study of the respiratory function of the nose. *Boston Med. Surg. J.* **135,** 457–460.

Haight, J. S. J., and Cole, P. (1983). The site and function of the nasal valve. *Laryngoscope* **93,** 49–55.

Harding, R. (1986). The upper respiratory tract in perinatal life. *In* "Reproductive and Perinatal Medicine. III. Respiratory Control and Lung Development in the Fetus and Newborn" (B. M. Johnston and P. D. Gluckman, eds.), pp. 331–376. Perinatology Press, Ithica, New York.

Hounam, R. F. (1975). The removal of particles from the nasopharyngeal (NP) compartment of the respiratory tract by nose blowing and swabbing. *Health Phys.* **28,** 743–750.

Hounam, R. F., and Morgan, A. (1977). Particle deposition. *In* "Respiratory Defense Mechanisms" (J. D. Brain, D. F. Proctor, and L. M. Reid, eds.), pp. 125–156. Dekker, New York.

Hounam, R. F., Black, A., and Walsh, M. (1970). The deposition of aerosol particles in the nasopharyngeal region of the human respiratory tract. *In* "Inhaled Particles III" (W. H. Walton, ed.), pp. 71–79. Unwin, Old Woking, Surrey.

Morgan, K. T., Patterson, D. L., and Gross, E. A. (1986). Responses of the nasal mucociliary apparatus of F-344 rats to formaldehyde gas. *Toxicol. Appl. Pharmacol.* **82,** 1–13.

Mygind, N., Rasmussen, F. V., and Molgaard, F. (1983). Preface to supplement. *Eur. J. Respir. Dis.* **64** (Suppl. 128, Part 1), U1.

Niinimaa, V., Cole, P., Mintz, S., and Shephard, R. J. (1981). Oronasal distribution of respiratory airflow. *Respir. Physiol.* **43,** 69–75.

Olson, L. G., and Strohl, K. P. (1987). The response of the nasal airway to exercise. *Am. Rev. Respir. Dis.* **135,** 356–359.

Pattle, R. E. (1961). The retention of gases and particles in the human nose. *In* "Inhaled Particles and Vapours" (C. N. Davies, ed.), pp. 302–309. Pergamon, Oxford.

Polgar, G., and Weng, T. R. (1979). The functional development of the respiratory system, state of the art. *Am. Rev. Respir. Dis.* **120,** 625–695.

Proctor, D. F. (1966). Airborne disease and the upper respiratory tract. *Bacteriol. Rev.* **30,** 498–513.

Proctor, D. F. (1977). The upper airways. I. Nasal physiology and defense of the lungs, state of the art. *Am. Rev. Respir. Dis.* **115,** 97–130.

Proctor, D. F. (1980). "Breathing Speech and Song." Springer-Verlag, Berlin and New York.

Proctor, D. F. (1983). The naso-oro-pharyngo-laryngeal airways. *Eur. J. Respir. Dis.* **64,** 89–96.

Proctor, D. F. (1986). The upper airways and larynx. *In* "Handbook of Physiology, Breathing Mechanics" (P. T. Macklem and J. Mead, eds.), pp. 57–61. American Physiological Society, Bethesda, Maryland.

Proctor, D. F. (1988). Disorders of the extrathoracic airways. *In* "Textbook of Respiratory Medicine" (J. F. Murray and J. A. Nadel, eds.), pp. 1153–1159. Saunders, Philadelphia.

Proctor, D. F., and Andersen, I. (1982). "The Nose, Upper Airway Physiology and the Atmospheric Environment." Elsevier, Amsterdam.

Proctor, D. F., and Chang, J. C. F. (1985). Comparative anatomy and physiology of the nasal cavity. *In* "Nasal Cavity Tumors in Animals and Man" (G. F. Reznik and S. F. Stinson, eds.), pp. 1–34. CRC Press, Boca Raton, Florida.

Proctor, D. F., Andersen, I., and Lundqvist, G. R. (1977). Human nasal mucosal function at controlled temperatures. *Respir. Physiol.* **30,** 109–124.

Remmers, J. E., and Bartlett, D., Jr. (1977). Reflex control of expiratory airflow and duration. *J. Appl. Physiol.* **42,** 80–87.

Rodenstein, D. O., and Stanescu, D. C. (1984). Soft palate and oro-nasal breathing in humans. *J. Appl. Physiol.* **57,** 651–657.

Rodenstein, D. O., and Stanescu, D. C. (1986). The soft palate and breathing, state of the art. *Am. Rev. Respir. Dis.* **134,** 311–325.

Schreider, J. P. (1986). Comparative anatomy and function of the nasal passages. *In* "Toxicology of the Nasal Passages" (C. S. Barrow, ed.), pp. 1–25. McGraw-Hill, New York.

Strohl, K. P., Hensley, M. J., Hallett, M., Saunders, N. A., and Ingram, R. H., Jr. (1980). Activation of upper airway muscles before onset of inspiration in normal humans. *J. Appl. Physiol.* **49,** 638–642.

Strohl, K. P., O'Cain, C. F., and Slutsky, A. S. (1982). Alae nasi activation and nasal resistance in healthy subjects. *J. Appl. Physiol.* **52,** 1432–1437.

Chapter 5

Preparation of Rat Nasal Airway Casts and Their Application to Studies of Nasal Airflow

K. T. Morgan
T. M. Monticello
Department of Experimental Pathology and Toxicology
Chemical Industry Institute of Toxicology
Research Triangle Park, North Carolina 27709

A. L. Patra
Health Effects Research
NSI Technology Services Corp.
Research Triangle Park, North Carolina 27709

A. Fleishman
Route 1, Box 203 K
Durham, North Carolina 27705

I. INTRODUCTION

Rats are used extensively in inhalation toxicology studies designed to assess human risks from exposure to environmental air pollutants. The nasal passages represent an important target site in these studies, for both neoplastic (Feron *et al.,* 1986) and nonneoplastic (Jiang *et al.,* 1986) responses. The anatomy of the nasal airways, which is complex in both laboratory animals and humans (Negus, 1958), has a major influence on nasal airflow characteristics (Proetz, 1951). Nasal airflow patterns play an important role in the deposition of inhaled materials in the nose (Proetz, 1951), and thus influence the location and severity of lesions induced by these mate-

Extrapolation of Dosimetric Relationships for Inhaled Particles and Gases

rials. While there is some information on nasal airflow in humans (Proetz, 1951; Masing, 1967; Girardin *et al.*, 1983), there is no data on airflow patterns in the nasal passages of rats.

Studies of nasal airflow, and determination of the relationship between airflow patterns and nasal lesion distribution, may provide a valuable point for direct comparison of rats with humans. The results of these studies will help clarify the value of the rat nose as a model for human risks associated with exposure to air pollutants. Much of our previous work has been directed towards determining site specificity of nasal lesions in rodents exposed by inhalation to a number of gaseous irritants (Buckley *et al.*, 1984; Morgan *et al.*, 1986; Gross *et al.*, 1987). For highly water-soluble gases, lesions were most severe in specific regions of the anterior nasal passages. This finding led us to investigate nasal airflow patterns in the rat in an attempt to determine whether this site specificity of irritant-induced lesions was attributable, at least in part, to nasal airflow patterns. The complexity and small size of the rat nasal passages necessitated the use of high quality airway casts.

The objectives of this chapter are to briefly review the literature on the application of nasal casting techniques, to describe the preparation of nasal casts and molds from rats, and to present initial results of airflow studies using these molds.

II. PREPARATION OF AIRWAY CASTS AND MOLDS

A. General Background

A cast is the positive or solid representation of the airway, while a mold, which is subsequently prepared from the cast, is a negative or hollow model of the airway (Collins, 1985). Casts are made of opaque materials, such as silicone rubber, wax, metals, or plastics. Silicone rubber permits close examination of complex casts due to its flexibility, but is not suitable for subsequent molding procedures involving small complex airways because it cannot be removed from the mold *(vide infra)*. Paraffin wax is used for casts because it can subsequently be removed from molds by heating or solvents. A number of metals have been used for casting anatomical specimens for about 300 years (McClenahan and Vogel, 1952). Cerralow 117, a low melting point metal alloy, yields high quality casts of biological structures as small as 25 μm diameter (McClenahan and Vogel, 1952). This material was used by Schreider (1977) to prepare casts and molds of nasal passages of guinea pigs. Molds are generally prepared from optically clear materials, such as acrylic plastic, to visualize the internal air passages. The objective of these procedures is to produce a faithful replica of the nasal airway. It is evident that defective casts or molds will result in artifacts which may yield misleading data.

A number of workers have prepared nasal casts and molds. However, for clear technical details of the complex and time consuming processes involved in the preparation of a high quality product, see the articles by Collins (1985) and Schreider (1977) for human and guinea pig nasal passages, respectively. If multiple copies of a single cast or mold are required, a sandwich mold technique may be applied (Patra *et al.,* 1986). The latter procedure is suitable for the large nasal airways of primates, but has not yet been successfully applied to the small complex nasal airways of rodents. For our ongoing studies of airflow in the nasal passages of rats, the procedure of Schreider (1977) was adopted with a number of modifications.

B. Preparation of Rat Nasal Casts

Healthy male F-344 (CDF[F-344]/Cr1Br) rats from Charles River Breeding Laboratories Inc., Kingston, New York, weighing 180–300 gm were killed by carbon dioxide narcosis and exsanguination via the abdominal aorta. After the head was detached, the lower jaw, skin, and excess muscle were removed, taking care to preserve the nares and adjacent skin intact. The calvarium was carefully dissected away, while taking special care not to split the nasal bones when opening the anterior portion of the cranial vault. The remaining tissue was then briefly air dried externally in a warm (about 60°C) airstream. Fine holes were pierced in the cribriform plate with a dissecting needle to permit drying of the nasal lumen and to avoid airlocks in the recesses of the ethmoid turbinates during casting. A short length (2 cm) of polyethylene tubing was fixed around the cribriform plate with hot glue. This tubing was used to flow casting material through the ethmoid region during the filling process. A second short length (4–5 cm) of polyethylene tubing was fixed into the nasopharynx or trachea, as appropriate, and sealed to provide a dead space to permit adequate filling of these airways. The nasal passages were then dried by aspirating warmed room air from a hair dryer using a vacuum line. Inadequate drying would produce defective casts, while excessive drying may lead to fragility of the tissues and subsequent leaking of casting material from the nasal lumen.

The nares were then carefully encased in the cut end of a short length (about 10 cm) of 1 cm inner diameter polyethylene tubing. Care was taken to avoid compressing or otherwise distorting the soft tissues of the nares, by carefully dissecting the surrounding skin and adnexa to provide a direct point for connecting the tubing to the nasal bones with a bridge of hot glue. The tube was then glued directly to the bones, while leaving the nares intact. This preparation, including all tubing and tissues, was completely embedded in hot glue to produce an airtight unit. Once the glue was set and completely hardened, the nose preparation was fixed in a clamp stand, with the nares uppermost. The upper tube was attached to a glass funnel, and the other end was attached to a vacuum line with a trap to collect excess casting material and allow the casting metal to pass into the nares and out through the ethmoid region.

The casting material selected was a low melting point metal alloy, Cerralow-117 (Cerro Metal Products, Bellefonte, Pennsylvania). During injection of the casting material, the tubes were closed as needed using hemostats, and the whole system was preheated in a 60°C incubator. The system was kept warm during injection of casting metal in a heated (about 60°C) airstream supplied by a large hairdryer to maintain free flow of the molten metal. The funnel was filled with preheated (60°C) Cerralow-117 and the upper tube closed. A vacuum of about 30 mm Hg was applied to the lower tube to produce a constant negative pressure in the nasal passages and tubing, and then metal flowed through the cast for about 10 sec. While metal was flowing, the lower tube was closed abruptly, while the upper tube was left connected to an open funnel containing excess metal to allow for shrinkage during overnight cooling at room temperature.

To obtain the nasal cast, extensive cleaning procedures were necessary. Initially, the plastic tubing and epoxy were removed by blunt dissection and immersion in methylene chloride. Tissues were then removed by repeatedly soaking (1–3 days) alternately in saturated, aqueous potassium hydroxide solution and distilled water, combined with very careful dissection. Removal of tissue from the ethmoid region, the nasal septum, and lateral scroll of the nasoturbinate required very careful and repeated cleaning with a waterpick. Casts were then stored at room temperature or in a refrigerator at 4°C, in a clean, airtight glass container.

C. Preparation of Transparent Molds from Nasal Casts

Nasal casts were attached to Swagelok fittings with Cerralow-117, using basic soldering techniques, to facilitate connections required for airflow determinations. Small boxes (about 2.5 × 10 × 1.5 cm) were then prepared from glass microscope slides and plexiglass which were precoated with polished wax as a releasing agent. The slides and plexiglass were connected to form the molding boxes with silicone adhesive, leaving the upper side open to permit filling with the molding plastic (Castolite, The

Castolite Company, Woodstock, Illinois). The nasal cast was mounted in the box which was then filled with molding plastic and closed with a glass slide coated with wax. During mold preparation, the mixture of plastic and hardener is critical as the polymerization procedure generates heat which, if excessive, may destroy the metal cast. The mold was polymerized under nitrogen gas for at least 2 days, and the glass sides were then removed to yield an optically clear surface (Fig. 5-1). The metal cast was subsequently removed from the mold by heating in boiling water, and flushing the lumen of the mold with warm (approximately 85°C) dilute acid (0.1 N HCl).

D. Determination of Cast and Mold Quality

Both casts and molds were examined grossly and by light microscopy for defects. Casts or molds with severe defects or artifacts in regions which would markedly influence airflow in the anterior nasal passages were not used for flow studies. Selected molds were cut into multiple cross-sections using a cooled rotary blade saw (Isomet Low Speed Buehler Saw, Buehler

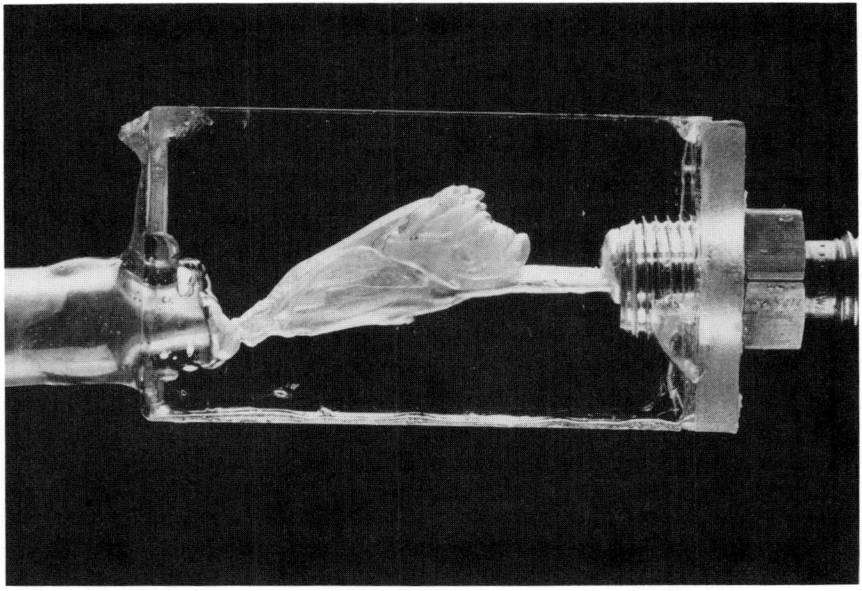

Fig. 5-1 Clear acrylic plastic mold prepared from a cast of the nasal passages of a rat.

Instruments, Lakebluff, Illinois). The sections were examined on a dissecting microscope. These cross-sections were compared with cross-sections of nasal passages of rats fixed in 10% neutral buffered formalin, and with histologic sections of the nasal passages available from previous studies in this laboratory (Morgan *et al.,* 1984; Jiang *et al.,* 1986; Randall *et al.,* 1987). Subjective comparisons demonstrated the absence of shrinkage or distortion during the casting–molding procedures, and indicated the production of a good facsimile of the original airways with excellent replication of major nasal structures, including the complex ethmoid turbinates.

E. Principal Technical Problems of Cast and Mold Preparation

The main difficulties of preparing the cast and mold were associated with adequately filling the nasal cavity with casting metal, especially in certain areas, including the complex ethmoid region. There was considerable variation between the heads for the degree of filling structures other than the nasal passages, such as the nasolacrimal ducts, nasal and cerebral vasculature, and clefts in interstitial connective tissues. Removing tissue debris from the casts also presented problems because small pieces of bone or cartilage were very difficult to extract from the lateral scroll of the nasoturbinates and the recesses of the ethmoid region. The waterpick markedly improved this cleaning process. Once the casts were clean, it was necessary to remove surplus adherent metal from the ethmoid region, and from points of leakage (flash marks), by carefully applying a heated dissecting needle. It was also difficult to clean some of the molds after the metal cast was removed. Residual tissue resulted in opacity, while metal in the maxillary recess and ethmoid region frequently required multiple rinses with the hot hydrochloric acid solution combined with vigorous shaking and vibration to dislodge droplets of metal in the smaller recesses of the mold.

F. Structure of Nasal Casts and Molds

The casting procedure yielded apparently faithful, and highly detailed, replica casts (Fig. 5-2 and 5-3) and molds (Fig. 5-1) of the nasal passages. Cross-sections of the molds demonstrated excellent reproduction of very delicate nasal structures, such as the naso-, maxillo-, and ethmoturbinates. Examination of these casts and molds was very instructive with respect to the anatomy of the nasal passages and their relationship to adjacent structures. It was evident that anterior nasal passages of the rat are intimately associated with the incisor teeth (Fig. 5-3), which may play a role in shaping these complex airways. The principal effect of the anatomical relationship between the incisor teeth and the anterior nasal airways appears to be the creation of a curved lateral recess on the dorso-lateral aspect

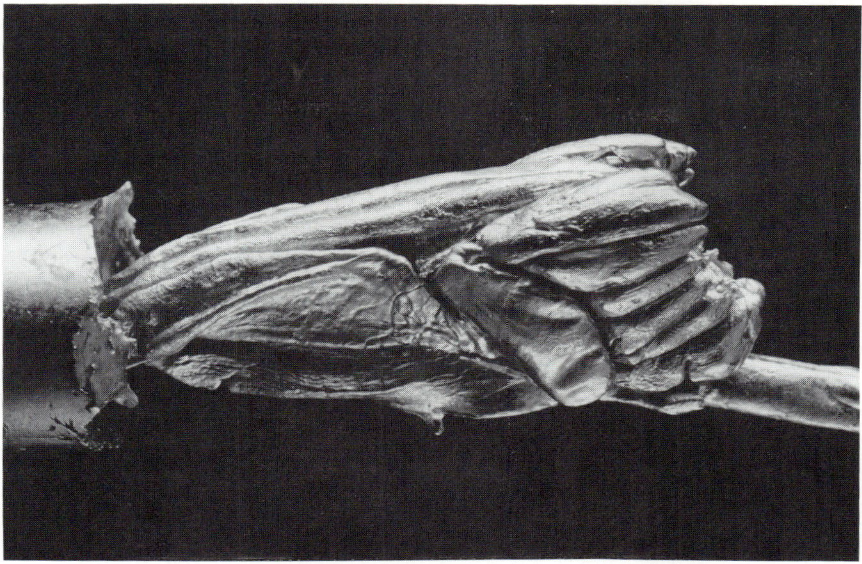

Fig. 5-2 Metal cast of rat nasal airways showing principal structures which are labelled in Fig. 5-3.

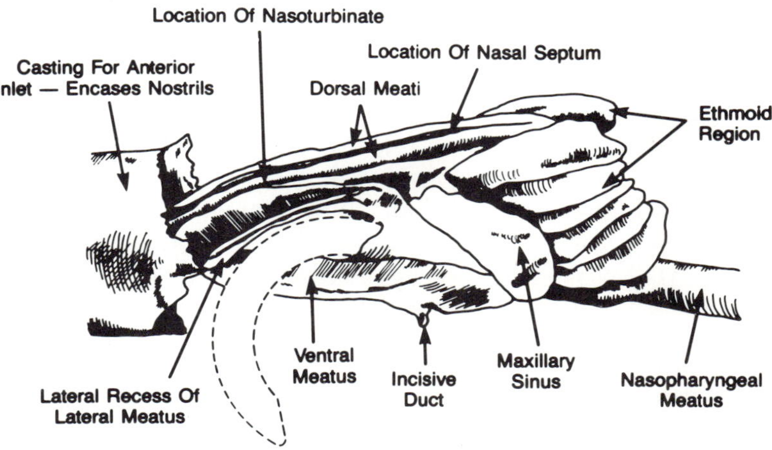

Fig. 5-3 Diagram of metal cast of rat nasal airways to show principal structures visible in the cast. The location of the incisor teeth is indicated by the broken line. Note the close anatomical association between the incisor tooth and the lateral recess of the lateral meatus.

of the lateral meatus (Fig. 5-1, 5-2, and 5-3). The form of the lateral meatus plays a role in determining airflow characteristics in the anterior nasal passages *(vide infra).*

Nasal casts also demonstrate the patency of the incisive ducts and the location of the eustachian tubes (Fig. 5-2 and 5-3). The eustachian tubes are separated from the main body of the nasal passages by the long, tubular nasopharyngeal meatus. These preparations also indicate the straight, smooth dorsal outline of the dorsal meati, the complexity of the ethmoid region, and the triangular profile of the maxillary sinus.

When using replicas of nasal airways, it is important to consider the effects of anesthesia, exsanguination, and other factors on the shape of the casts and molds. These preparations only provide a likeness of the original nasal airway, while they differ in many ways from the rat's nose which is a very dynamic structure. If the limitations of the technique are born in mind, these preparations provide a valuable means of studying nasal airflow characteristics.

III. DETERMINATION OF NASAL AIRFLOW CHARACTERISTICS

A. General Background

A number of techniques have been devised to study airflow, some of which have been applied directly to the nasal passages, while others use nasal molds or models. These techniques have included the use of transducers to determine pressure–flow relationships (Jaeger and Matthys, 1968, 1969), thermistor probes (Patra *et al.,* 1986) and hot-wire anemometry (Tans and Verberne, 1982) for studies of linear velocity, and laser anemometry for mapping velocity fields (Girardin *et al.,* 1983). These approaches provide valuable quantitative data on nasal airflow, but are not yet applicable to the very small nasal airways of rats. While these techniques are useful for determining the flow characteristics in specific sites, they are not suitable for detecting overall airflow patterns in complex airways, such as those found in the nose. Another approach to determine the airflow is to create visual images on cine-film of smoke laden air (Proetz, 1951) or dyes dissolved in water (Swift and Proctor, 1977; Patra *et al.,* 1986) passing through clear nasal models. Attempts to study airflow by passing dense tobacco smoke through rat nasal molds failed because the smoke was not visible within the mold. We therefore applied a water–dye system, similar to the one previously used for a baboon nasal mold in this laboratory (Patra *et al.,* 1986).

B. Determination of Water Flow Characteristics in Rat Nasal Molds

To detect water flow characteristics, the clear rat nasal molds were connected to a tap water supply using Swagelok fittings, with Teflon and polyethylene tubing. A Swagelok septum was fitted between the water supply and cast for the introduction of dye (1% aqueous Methylene Blue) by injection with a 10-ml hypodermic syringe and a 19-gauge steel needle. A glass and steel ball flow meter (Matheson, Morrow, Virginia), calibrated against tap water and connected between the water supply and the mold, was used to determine flow rate. The cast was placed on the viewing surface of a dissecting microscope, fitted with a Bolex 16-mm cinecamera with a motor drive to give a constant speed of 32 frames per sec. Images were recorded on high-speed color film (Kodak 7250 Tungsten Video News Film). The recorded images were examined and analyzed with an analytical projector (Athens, California), in both cine and frame-by-frame modes.

Water flow rates were selected on the basis of airflow rates determined from the estimated respiratory minute volume of 180 to 300 ml per min (Mauderly, 1986) and the sinusoidal pattern of inspiratory and expiratory airflow. Airflow rates to cover low, intermediate, and maximal inspiratory and expiratory flow rates (200–900 ml per min) were selected, and then water flow rates to mimic these airflow rates were determined using a Reynolds conversion factor, which simply adjusts for the differing dynamic viscosities of air and water at room temperature (Swift and Proctor, 1977). A conversion factor of 14.25, derived from physical characteristics of distilled water, gave water flow rates of 12 to 60 ml per min.

C. Airflow Characteristics Based on Water Flow in Rat Nasal Molds

Examining cinefilm of dye pulses in a continuous stream of water passing through the clear acrylic casts revealed numerous complex flow patterns, which were influenced by the rate of flow. Water (and thus air) entering the nose generally passed along one of four principal pathways (Fig. 5-4). These routes of flow, in order of significance, based on a visual assessment of the amount of dye passing along each route, were:

1. The lateral meatus
2. Middle meatus
3. Dorsal meatus
4. Ventral meatus.

Detailed analysis of these flow patterns is still in progress. However, there is evidently considerable turbulence in the region lateral to the nasoturbinate

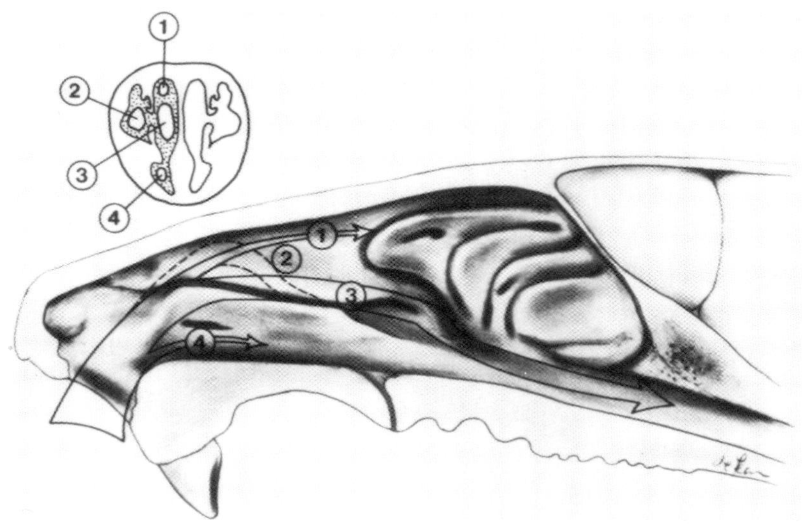

Fig. 5-4 Four main routes of airflow during inspiration in the rat, derived from studies of water–dye flow in rat nasal molds. The most prominent route of airflow during inspiration was along the lateral meatus in a region lateral to the nasoturbinate, indicated as a dotted line (airflow route 2); airflow in this region was associated with considerable turbulence. Another important route for inspired air was along the middle meatus (3), with much less air passing along the dorsal (1) and ventral meati (4).

where lesions are most severe following exposure to the highly water-soluble gaseous irritant, formaldehyde (Morgan *et al.*, 1986).

IV. DISCUSSION AND CONCLUSIONS

The nasal passages clean, humidify, and warm the inspired air (Proctor and Chang, 1983), and provide important protection for the more delicate lower airways. A number of studies have been conducted to determine airflow patterns in the nasal passages of humans (Proetz, 1951; Jaeger and Matthys, 1968, 1969; Girardin *et al.*, 1983; Swift and Proctor, 1977) and laboratory animals (Schreider, 1983a; Patra *et al.*, 1986). The present studies attempt for the first time to determine airflow patterns in the nasal passages of rats.

Rats are used extensively in inhalation toxicology studies directed towards assessing human risks associated with exposure to air pollutants. It is still uncertain whether the rat is a good model for man, because of differences in anatomy, physiology, biochemistry, and other factors. Until recently the nose was essentially ignored or incompletely examined in the majority of inhalation toxicology studies. Recent studies with a number of toxic gases, especially formaldehyde (Kerns *et al.,* 1983), have led to increased interest in the nose as a target organ. Inhaled chemicals induce a range of nasal lesions, including cytotoxic responses in the respiratory and olfactory epithelia, and in some cases neoplasia. For humans these changes may relate to risks of damage to airway clearance mechanisms, acute or chronic rhinitis, impairment of smell, and respiratory tract cancer. An important issue to be resolved is what relevance do nasal lesions induced in rodents by these chemicals have on man. The studies described here provide basic information on the nose of the rat, will improve assessment of important differences and similarities between rats and humans, and thus assist in estimating risk.

An important finding of this work was realizing the role played by the incisor teeth in shaping the anterior nasal passages of rats. This relationship, which in retrospect would be evident in serial or step sections of the nasal passages, was not noted until casts were examined during the tissue removal process. This observation highlights the value of casts for studies of anatomy. Incisor teeth of rats differ from those of primates because they have persistent pulps that grow throughout the life of the animal. The growing root of these teeth gives them an elongated, hook-like appearance, and their shape is reflected in the curved lateral recess of the lateral meatus of the anterior nasal passages. Examination of water–dye flow through molds of this region revealed that it was a major route of airflow during inspiration. It was evident that this curved recess directs air in a funnel-like manner on to the lateral aspect of the lateral scroll of the nasoturbinate. This site exhibits the severest lesions following acute exposures to gaseous irritants, such as formaldehyde (Morgan *et al.,* 1986a), and is a very common location for formaldehyde-induced nasal squamous cell carcinomas (Morgan *et al.,* 1986b). This finding suggests that nasal airflow patterns may play a major role in determining the location of these formaldehyde-induced responses.

It is of potential importance that the human nose differs in terms of anatomy and airflow from that of the rat. In the human nose most of the air flows along the middle meatus (Swift and Proctor, 1977), which probably accounts for the frequency of particle deposition and lesions on the anterior curvature of the middle turbinate in nickel workers (Torjussen, 1983). The nasal passages of rhesus monkeys anatomically resemble those of humans (Negus, 1958; Bridger and van Nostrand, 1978), and responses to acute formaldehyde exposure in rhesus monkeys are most severe on the middle turbinate (Monticello *et al.*,1989), suggesting a similarity of airflow between the two species. Thus the area of highest exposure, as based on airflow and lesion distribution, differs between rats, humans, and rhesus monkeys. It is also important to note that the target site in the rat (lateral meatus) is covered by a poorly ciliated (Morgan *et al.*, unpublished observations), cuboidal or low columnar epithelium with very few goblet cells (Monteiro-Riviere and Popp, 1984). The target site in the monkey (middle turbinate) is covered by densely ciliated, columnar epithelium, with fairly numerous goblet cells (Harkema *et al.*, 1987; T. M. Monticello, unpublished observations). Thus the airflow patterns result in exposure of very different cell populations between the rat and monkey. The two cell populations may differ in their sensitivity to the carcinogenic properties of formaldehyde. Therefore, species differences in regional gas deposition due to nasal airflow characteristics may represent an important issue for consideration when assessing human risk with data from studies in rats.

In conclusion, it is evident that much remains to be learned about the anatomy and airflow characteristics of the nasal passages of humans and laboratory animals. Casting and molding techniques can contribute in a major way to studies of these complex issues. However, it must be remembered that casts and molds represent a very poor imitation of the living nasal airway. Complex physiologic and anatomical changes, such as congestion of the nasal vasculature and movements of the nostrils, cannot yet be mimicked by rigid airway models. Studies which are cognizant of the limitations of these techniques, and their combination with *in vivo* procedures as available, should lead to valuable applications of nasal casts and molds.

ACKNOWLEDGMENTS

We wish to thank Dr. Jay P. Schreider for his helpful advice and comments during the initial stages of this work, Tim Chelmers for excellent photography, and Susan Sadler for artwork.

REFERENCES

Bridger, W. M., and van Nostrand, H. W. (1978). The nose and paranasal sinuses—applied surgical anatomy. *J. Otolaryngol.* **7,** 1–33.

Buckley, L. A., Jiang, X. Z., James, R. A., Morgan, K. T., and Barrow, C. S. (1984). Respiratory tract lesions by sensory irritants at the RD50 concentration. *Toxicol. Appl. Pharmacol.* **74,** 417–429.

Collins, M. P. (1985). A practical guide to the construction of a "cire perdue" model of the human nose. *Rhinology* **23,** 71–78.

Feron, V. J., Woutersen, R. A., and Spit, B. J. (1986). Pathology of chronic nasal toxic responses including cancer. *In* "Toxicology of the Nasal Passages" (C. S. Barrow, ed.), pp. 67–89. Hemisphere, New York.

Girardin, M., Bilgen, E., and Arbour, P. (1983). Experimental study of velocity fields in a human nasal fossa by laser anemometry. *Ann. Otol. Rhinol. Laryngol.* **92,** 231–236.

Gross, E. A., Patterson, D. L., and Morgan, K. T. (1987). Effects of acute and chronic dimethylamine exposure on the nasal mucociliary apparatus of F-344 rats. *Toxicol. Appl. Pharmacol.* **90,** 359–376.

Harkema, J. R., Plopper, C. G., Hyde, D. M., and St. George, J. A. (1987). Regional differences in quantities of histochemically detectable mucosubstances in nasal, paranasal and nasopharyngeal epithelium in the bonnet monkey. *J. Histochem. Cytochem.* **35,** 279–286.

Jaeger, M. J., and Matthys, H. (1968/1969). The pattern of flow in the upper human airways. *Respir. Physiol.* **6,** 113–117.

Jiang, X. Z., Morgan, K. T., and Beauchamp, R. O., Jr. (1986). Histopathology of acute and subacute nasal toxicity. *In* "Toxicology of the Nasal Passages" (C. S. Barrow, ed.), pp. 51–66. Hemisphere, New York.

Kerns, W. D., Pavkov, K. L., Donofrio, D. J., Gralla, E. J., and Swenberg, J. A. (1983). Carcinogenicity of formaldehyde in rats and mice after long-term inhalation exposure. *Cancer Res.* **43,** 4382–4392.

McClenahan, J. L., and Vogel, S. F. (1952). The use of fusible metal as a radiopaque contrast medium and in the preparation of anatomical castings. *Am J. Roentgenol. Radiat. Ther. Nucl. Med.* **68,** 406–412.

Masing, M. (1967). Patho-physiology of the nasal airflow. *Int. Rhinol.* **5,** 63–67.

Mauderly, J. L. (1986). Respiration of F-344 rats in nose-only inhalation exposure tubes. *J. Appl. Toxicol.* **6,** 25–30.

Monteiro-Riviere, N. A., and Popp, J. A. (1984). Ultrastructural characterization of the nasal respiratory epithelium in the rat. *Am. J. Anat.* **169,** 31–43.

Monticello, T. M., Morgan, K. T., Everitt, J. I., and Popp, J. A. (1989). Effects of formaldehyde gas on the respiratory tract of rhesus monkeys: Pathology and cell proliferation. *Am. J. Pathol.* **134,** 515–527.

Morgan, K. T., Jiang, X. Z., Patterson, D. L., and Gross, E. A. (1984). The nasal mucociliary apparatus: Correlation of structure and function in the rat. *Am. Rev. Respir. Dis.* **130,** 275–281.

Morgan, K. T., Gross, E. A., and Patterson, D. L. (1986a). Responses of the nasal mucociliary apparatus of F-344 rats to formaldehyde gas. *Toxicol. Appl. Pharmacol.* **82,** 1–13.

Morgan, K. T., Jiang, X. Z., Starr, T. B., and Kerns, W. D. (1986). More precise localization of nasal tumors associated with chronic exposure of F-344 rats to formaldehyde gas. *Toxicol. Appl. Pharmacol.* **82,** 264–271.

Morgan, M. S., and Frank, R. (1977). Uptake of pollutant gases by the respiratory system. *In* "Respiratory Defense Mechanisms" (J. D. Brain, D. F. Proctor, and L. M. Reid, eds.), pp. 157–189. Dekker, New York.

Negus, V. (1958). "The Comparative Anatomy and Physiology of the Nose and Paranasal Sinuses." Livingstone, Edinburgh.

Patra, A. L., Gooya, A., and Morgan, K. T. (1986). Airflow characteristics in a baboon nasal passage cast. *J. Appl. Physiol.* **61,** 1959–1966.

Proctor, D. F., and Chang, J. C. F. (1983). Comparative anatomy and physiology of the nasal cavity. *In* "Nasal Tumors in Animals and Man" (G. Reznik and S. F. Stinson, eds.), Vol. 1, pp. 1–33. CRC Press, Boca Raton, Florida.

Proetz, A. W. (1951). Air currents in the upper respiratory tract and their clinical importance. *Ann. Otol. Rhinol. Laryngol.* **60,** 439–467.

Randall, H. W., Bogdanffy, M. S., and Morgan, K. T. (1987). Enzyme histochemistry of the rat nasal mucosa embedded in cold glycol methacrylate. *Am. J. Anat.* **179,** 10–17.

Schreider, J. P. (1977). Lung anatomy and characteristics of aerosol retention of the guinea pig. Doctoral dissertation in the Department of Pharmacological and Physiological Sciences, University of Chicago.

Schreider, J. P. (1983a). Nasal airway and inhalation deposition in experimental animals and people. *In* "Nasal Tumors in Animals and Man" (G. Reznik and S. F. Stinson, eds.), Vol. 3, pp. 1–26. CRC Press, Boca Raton, Florida.

Swift, D. L., and Proctor, D. F. (1977). Access of air to the respiratory tract. *In* "Respiratory Defense Mechanisms" (J. D. Brain, D. F. Proctor, and L. M. Reid, eds.), pp. 63–93. Dekker, New York.

Tans, N., and Verberne, G. (1982). Flow measurements in the nose of unrestrained domestic rabbits. *Biotelemetry Patient Monit.* **9,** 89–97.

Torjussen, W. (1983). Nasal cancer in nickel workers. Histopathological findings and nickel concentrations in the nasal mucosa of nickel workers, and a short review of chromium and arsenic. *In* "Nasal Tumors in Animals and Man" (G. Reznik and S. F. Stinson, eds.), Vol. 2, pp. 33–53, CRC Press, Boca Raton, Florida.

Chapter 6

Age Related Morphometric Analysis of Human Lung Casts

JD Mortensen
L. Stout
B. Bagley
J. A. Burkart
R. N. Schaap
MIDMID and UBTL Laboratories
University of Utah Research Park
Salt Lake City, Utah 84108

For several years, under contract with the Environmental Protection Agency,[1] MIDMID/UBTL Laboratories have been conducting morphometric analysis of human respiratory tract airways to determine the relationships of postnatal growth and development parameters to specific morphometric characteristics of conducting airways.

The objective of this report is to describe the nature of our studies and to point out the statistical morphometric data base available to interested inhalation toxicologists, pulmonary dosimetry scientists, and other environmental pollution specialists.

I. DEVELOPMENT OF THE DATA BASE

Our data are derived from a detailed morphometric study of silicone rubber castings made by well-established *in situ* casting techniques (Phalen

[1] U.S. EPA contracts 68-01-4980 and 69-01-4062.

et al., 1973) which were applied to 36 human cadavers within 24 hours of death. The subjects ranged in age from 2 days to 26 years and included 17 females and 19 males. All died suddenly, usually of head/central nervous system (CNS) trauma or acute nonrespiratory disease. All of the lungs were normal clinically and at necropsy. While 36 castings were made, not all were fully suitable for all morphometric assessments because of inadequate filling of airway segments, breakage of the cast, and technically imperfect casts due to retained secretions, etc. Consequently, the number of cases utilized for specific morphometric data varied somewhat for specific parameters.

Growth parameters recorded for each subject were age, height, weight, chest circumference, and body surface area (which was computed from

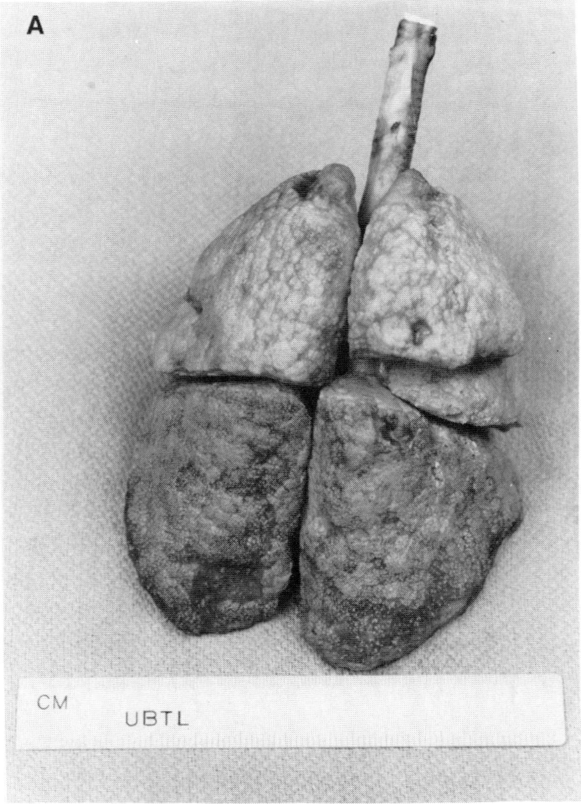

Fig. 6-1 Preparation of silicone rubber *in situ* casting of airways for morphometric study. A. Airway casting as it is removed from cadaver.

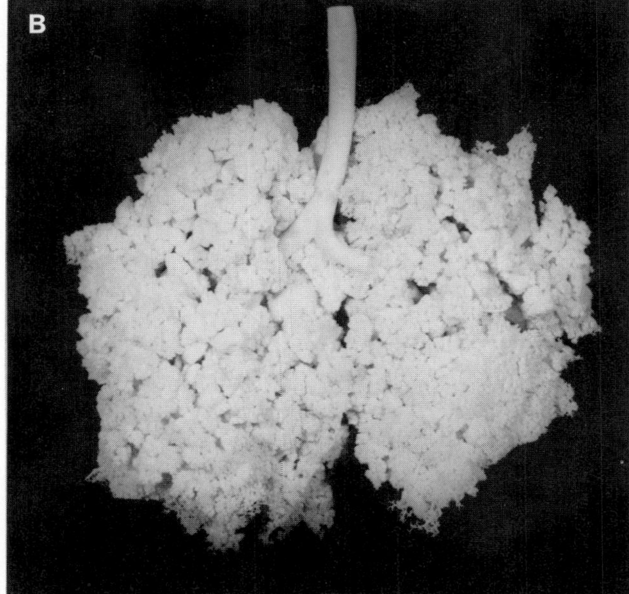

Fig. 6-1B Airway casting after pulmonary, vascular, bronchial, and parenchymal tissue has been digested off by means of sodium hydroxide bath. The casting at this stage represents both conducting and respiratory airways, the silicone rubber having filled the terminal bronchioles and alveoli as well as the trachea and bronchi.

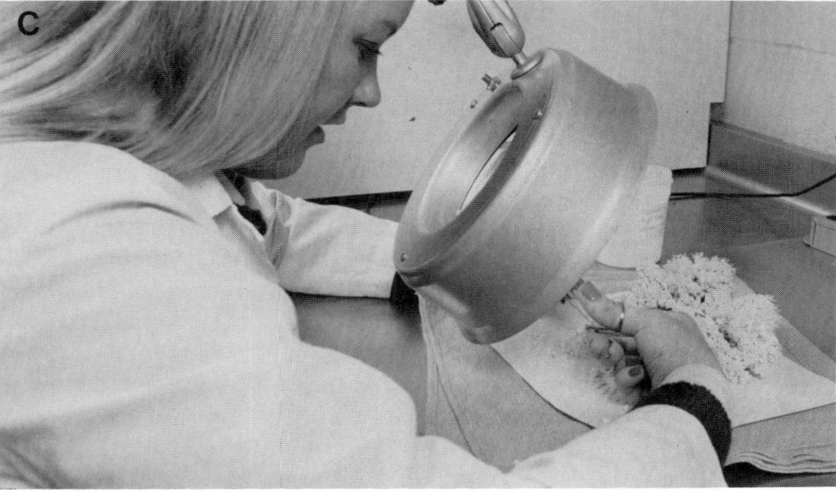

Fig. 6-1C Trimming the casting to remove the respiratory airways. These trimmings were saved and weighed to determine respiratory airway volume. *(Figure continues)*

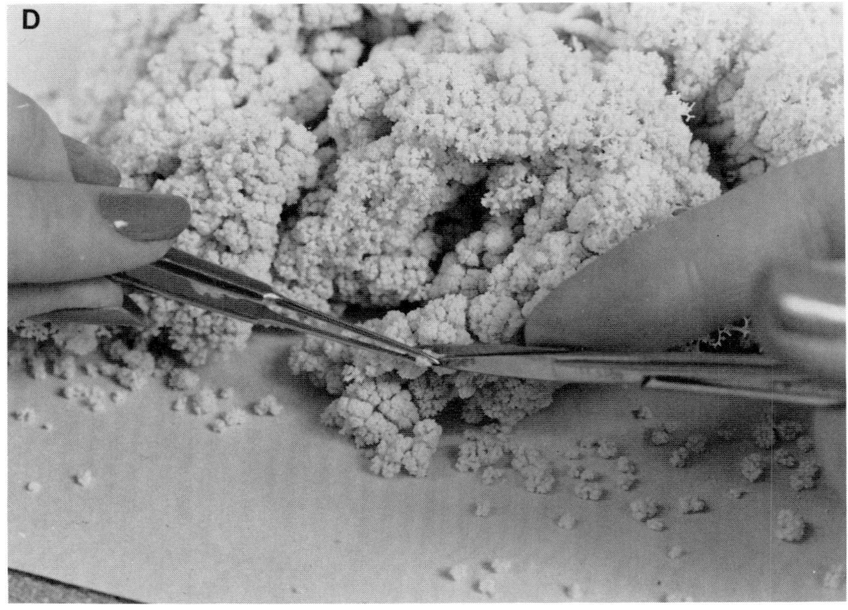

Fig. 6-1D Closer view of the trimming process.

age, height, and weight measurements). Our study was limited to subjects who were normal in size for their age, seemed normal with respect to their nutritional state and general health, and were free from known chronic illnesses that could influence pulmonary anatomy.

The airway castings were first carefully trimmed to remove the respiratory airways. The remaining conducting airways were then subjected to detailed morphometric assessments (Fig. 6-1, A–F).

Our morphometric analyses were limited to the conducting airways. Data were collected by direct measurements from each casting. Airway volumes were determined by weighing the specific airway trimmings, then multiplying this weight by the specific gravity of the casting material. By this method, the following were obtained: total airway volume, tracheal volume, conducting airway volume, and respiratory airway volume. From these measurements, seven different relevant airway volume-related ratios were calculated (Table 6-1).

Morphometric analyses also included determining two anatomic features of a representative airway path in each lobe from trachea to first

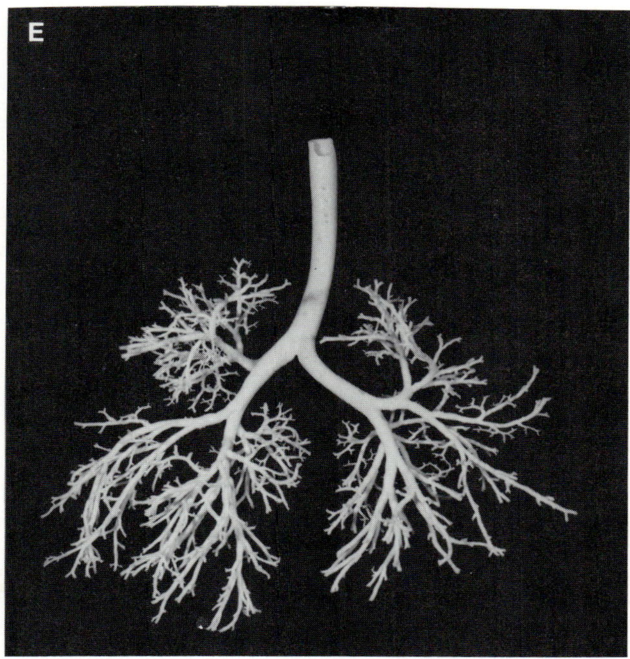

Fig. 6-1E Trimmed airways casting (6-1B) from a 4-month-old male subject. The casting now represents only the conducting airways. Most of our morphometric data come from castings of this type. *(Figure continues)*

alveolization, the number of branching generations and the total airway length.

Our most detailed morphometric analyses involved the proximal 10 generations of conducting airways. We first devised a unique airway classification system that permitted accurate, specific identification of each of the more than 1200 individual airway segments in the proximal 10 branching generations of every casting (Mortensen *et al.,* 1983). For each airway identified we measured and recorded its diameter proximally, at midpoint, and distally. Its length was also measured and recorded. Additionally, the ratio of diameter to length was calculated for each airway segment (Table 6-1, Fig. 6-2).

Spatial orientation of specific airway segments was assessed by suspending the casting with the trachea held in its proper relationship to gravity, then measuring the angle of the inclination (gravity angle) from a plumb

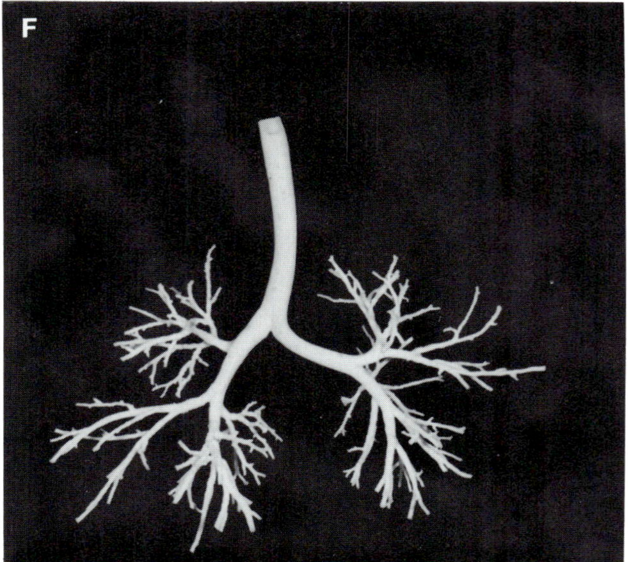

Fig. 6-1F Airway casting after a second trimming to remove conducting airways distal to the 10th branching generation.

line of as many specific airway segments as possible, usually the most proximal 4 or 5 generations. The tracheal gravity angle for proper spatial orientation was determined by an auxiliary unpublished study by the authors in which the tracheal gravity angle was determined from upright X rays of 170 normal adults and 50 normal children. This study indicated that in adults the trachea is inclined posteriorly an average of 24.5° and to the left an average of 9.1° from the vertical. In normal children ages 1 to 16 years, the average tracheal deviation is 14.6° posterially and 2.5° to the left. The angle at which each of the proximal 10 generations of airway segments branched or divided from its parent airway was measured and recorded as airway branching angle (Table 6-1, Fig. 6-2).

Finally, the branching pattern of conducting airways was assessed by recording the percentage incidence in each branching generation of the following branching patterns; regular dichotomy, irregular dichotomy, regular trichotomy, irregular trichotomy, and other or miscellaneous branching patterns (Table 6-1, Fig. 6-2).

All of these morphometric data points, which amount to 36 data books of more than 60 pages each, are available from EPA contract files (see Footnote[1], this chapter).

Table 6-1 Morphometric Characteristics Recorded from Casting of Normal Human Pulmonary Airways

Measured Airway Volumes	Calculated Airway Volume Ratios
Total Airway Volume (TAV)	$\frac{\text{RAV}}{\text{TAV}}$ $\frac{\text{TAV}}{\text{Body Weight}}$
Tracheal Volume (TV)	$\frac{\text{CAV}}{\text{TAV}}$ $\frac{\text{TAV}}{\text{Chest Circumference}}$
Conducting Airway Volume (CAV)	$\frac{\text{CAV}}{\text{RAV}}$ $\frac{\text{CAV}}{\text{Body Weight}}$
Respiratory Airway Volume (RAV)	$\frac{\text{CAV}}{\text{Chest Circumference}}$
Number of branching generations from trachea to first alveolization in each lobe	
Total length of a representative conducting airway from trachea to first alveolization in each lobe	
For the proximal 10 generations of conducting airways:	
Airway diameters (mm) proximal, mid, and distal diameter	
Airway lengths (mm) of each airway segment	
Ratio of diameter to length (D/L) of each airway segment	
Airway Angles	
Gravity angle of trachea (°)	
Gravity angle of each of the proximal 5 generations of conducting airways (°)	
Branching angle of each of the proximal 10 generations of conducting airways (°)	
Airway branching patterns in the proximal 10 generations of conducting airways (%, by airway generation):	
Incidence of regular dichotomy	
Incidence of irregular dichotomy	
Incidence of regular trichotomy	
Incidence of irregular trichotomy	
Incidence of other branching patterns	

II. CORRELATION OF GROWTH PARAMETERS WITH MORPHOMETRIC CHARACTERISTICS

To pursue the objective of determining the relationships of postnatal growth and development parameters to specific morphometric characteristics of conducting airways, the raw data were subjected to statistical analyses. Simple product-moment correlation coefficients were calculated for the relationships of each growth parameter to each morphometric characteristic (Fig. 6-3). The single best predictive growth parameter for each morphometric variable was chosen from these correlation coefficients and a simple linear regression equation was computed to describe the relationship. Similar plots, curves, and calculations were utilized to describe the relationship of each growth parameter to each morphometric variable studied.

MORPHOMETRY OF INDIVIDUAL AIRWAYS (by generation)

Generation	AIN	Diameter (mm) Prox	Mid	Dist	Length (mm)	D/L	Branching Angle	Gravity Angle	Branching Pattern: Reg Dichot #	%	Irreg Dichot #	%	Reg Trichot #	%	Irreg Trichot #	%	Other #	%
1	1	12.4	9.2	10.9	26.4	0.35	—	15.1°										
2	11	7.9	7.5	8.2	14.9	0.50	41°	33°	1									
	12	7.3	6.2	7.6	28.6	0.22	47°	37°	1									
	Entire Generation		6.9		21.8	0.31	88° *		2	100%	0	0	0	0	0	0	0	0
3	111	6.4	5.4	5.6	8.4	0.64	67°	82°			1							
	110	6.9	5.5	5.5	13.7	0.40	19°	23°			1							
	121	5.6	5.7	6.3	6.9	0.83	41°	77°	1									
	126	5.2	5.0	5.1	5.7	0.88	34°	31°	1									
	Entire Generation		5.4		8.7	0.62			2	50%	2	50%	0	0	0	0	0	0
4	111-1	2.8	2.5	2.4	4.4	0.57	52°	162°					1					
	111-2	3.4	3.1	2.7	5.5	0.56	74°	74°					1					
	111-5	2.6	3.0	2.7	5.8	0.52	10°	83°					1					
	110-2	4.2	3.9	3.6	7.8	0.50	19°	24°			1							
	110-0	5.2	5.0	5.3	3.3	1.52	15°	29°			1							
	121-1	4.0	3.8	4.3	5.2	0.73	38°	118°	1									
	121-6	3.0	2.7	2.9	7.8	0.35	36°	66°	1									
	126-0	4.7	4.9	5.7	10.6	0.46	0	24°			1							
	126-5	3.1	3.2	2.8	5.6	0.57	88°	95°			1							
	Entire Generation		3.6		6.2	0.57			2	22%	4	44%	3	33%	0	0	0	0
5	111-12	2.4	2.4	2.5	4.2	0.57	23°	168°	1									
	111-15	1.9	2.2	2.0	4.0	0.55	11°	166°	1									
	111-23	2.8	2.7	2.7	4.1	0.66	42°	61°	1									

Fig. 6-2 A representative page from a raw data book demonstrating type of raw morphometric data collected from each airway casting. Approximately 60 pages of data of this type were collected for each casting.

III. SUMMARY AND CONCLUSIONS

Statistical analysis of all the morphometric data collected and correlation of these data with the specific growth and development parameters ascertained for each subject produced the following summary statements and conclusions:

1. Measured airway volumes bear direct relationships to age, height, weight, chest circumference, and body surface area. Of these growth and development parameters, chest circumference relates most specifically and reproducibly to total airway volume, producing a correlation coefficient of 0.9136.
2. The average number of branching generations of conducting airways from trachea to terminal bronchioles (first alveolization) ranges from 15 to 23. The relationship of this morphometric feature to the various growth parameters is not very pronounced; correlation coefficients range from 0.6798 (for chest circumference) to 0.5863 (for age). As the normal postnatal human grows from infancy to adulthood, the average number of branching generations of airways from trachea to first alveolization appears to increase from about 17 to 19.

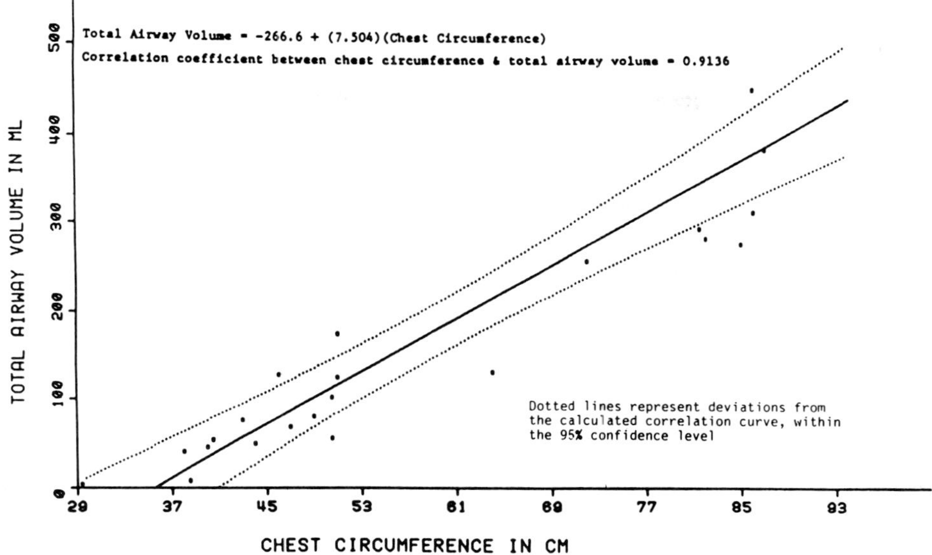

Fig. 6-3 Plot demonstrating the relationship of total airway volume to chest circumference, and method of calculating the numerical correlation coefficient for these two variables.

3. The total airway length from trachea to first alveolization is directly related to all growth and development parameters, producing correlation coefficients from 0.9753 (for chest circumference) to 0.9496 (for height).
4. The total length of airways comprising the proximal 10 generations of conducting airways from the trachea is directly related to all growth

Table 6-2 Computer Calculations of Correlation Coefficients for Total Airway Volume Related to Each of the Recorded Growth Parameters[a]

Predictor Variables[b]	Correlation with Dependent Variable[b]
X_1 − Age (years)	0.9118
X_2 − Height (cm)	0.8868
X_3 − Weight (kg)	0.9061
X_4 − Chest circumference (cm)	0.9136
X_5 − Body surface (m^2)	0.9022

[a] Y = Dependent variable: total airway volume (ml); $N = 23$ subjects. Similar tables were developed demonstrating the correlation coefficients for the relationships of each growth parameter to each morphometric variable studied.

[b] Best linear relationship, $Y = -266.6 + 7.504\ X_4$; Correlation coefficient = 0.9136, $p < .001$. Best multiple variable relationship, $Y = 195.8 - 8.781\ X_4 + .1287\ X_4^2$; multiple correlation coefficient = 0.9376, $p < .001$.

parameters studied, with correlation coefficients from 0.9757 (for chest circumference) to 0.9519 (for body weight).

5. Neither the total number of airway segments in the proximal 10 branching generations from the trachea nor the incidence of various branching patterns for conducting airways appears to be significantly influenced by any of the growth parameters studied.
6. The branching angles at which airways arise from proximal parent airways in the proximal 10 branching generations are not significantly influenced by any of the growth parameters studied.
7. The gravity angle of the trachea in a normal human in the upright position increases in its posterior deviation and in its lateral deviation as the subject matures from infancy to childhood, adolescence, and adulthood.
8. Of the five growth parameters studied (age, height, weight, body surface area, and chest circumference), the latter (chest circumference) is consistently the most closely correlated with airway morphometric variables, followed by age of the subject. Since all five growth parameters are closely interrelated to each other, there appears to be little value in combining growth parameters in attempts to devise a better predictor of pulmonary airway morphometric characteristics than utilization of the chest circumference.

REFERENCES

Mortensen, JD, Young, J. D., Stout, L., Stout, A., Bagley, B., and Schaap, R. N. (1983). A numerical identification system for airways in the lung. *Anat. Rec.* **206,** 103–114.

Phalen, R. F., Yeh, R. C., Raabe, O. C., and Velasquez, D. J. (1973). Casting of the lungs *in situ. Anat. Rec.* **177,** 255.

Chapter 7

Anatomical Modeling of Microdosimetry of Inhaled Particles and Gases in the Lung

R. R. Mercer
J. D. Crapo
Departments of Medicine and Pathology
Duke University, Durham, North Carolina 27710

I. INTRODUCTION

Determining the dose delivered to specific sites in the lungs is a critical first step in modeling the potential toxic effects of airborne pollutants. An important recent development in estimating site specific dosimetry has been combining sophisticated analytical models to determine the delivery of inhaled pollutants to the lungs with relatively simplistic anatomic models (Miller *et al.,* 1978). These methods are particularly important in extrapolating experimental results in laboratory animals to predict effects in humans. The goal in such modeling efforts is to characterize the average response of the lungs. However, in evaluating lung pathology following low level, chronic exposures to reactive gases, one quickly becomes aware that the lesions are far from uniform. The presence of such patchy lesions resulting from chronic exposures of laboratory animals or lifetime exposures of humans to various environmental pollutants is a frequent finding. The observation of a diseased area of the lung immediately adjacent to a normal healthy region is so common that it almost stifles the question of why it occurs. Discovering why such punctate lesions can be found next to healthy regions of apparently similar anatomic structure, will require dosimetric calculations on a much finer scale than previously carried out, combined with a critical consideration of the anatomic factors involved.

Extrapolation of Dosimetric Relationships for Inhaled Particles and Gases

For example, in considering the dosimetry of a reactive gas in the lungs the bronchiole–alveolar duct junction is one of the major sites for injury. Thus, a critical factor is the volume of pollutant laden inspired gas delivered through the bronchiole–alveolar duct junction to the gas exchange unit. If one begins with the simple assumption of a square wave front of inspired gas, the volume of fresh inspired gas delivered through each bronchiole–alveolar duct junction is proportional to the percentage of the inspiratory cycle of that gas exchange unit that receives fresh inspired gas. To compare the reactive delivered dose to the gas exchange units of a region, one must determine the relative transit times to the bronchiole–alveolar duct junctions of the gas exchange units. Three-dimensional anatomical reconstructions of the lung have been used to determine the relative transit times of all gas exchange units in a specific lung region. These calculations provide a means to determine the site specific dosimetry in a region of the lungs and suggest one reasonable cause for the patchy distribution of lesions commonly found following chronic, low level exposures to reactive gases.

II. METHODS

A. Fixation

For this study, lungs of male Sprague Dawley, specific pathogen-free rats [CD(SD)BR, Charles River, Wilmington, Massachusetts] with a mean body weight of 283 gm were perfusion fixed. This was done by methods similar to those previously reported by Mercer and Crapo (1987). For perfusion fixation, the animals were anesthetized with sodium pentobarbital (Nembutal, 50 mg/kg) by intraperitoneal injection. A tracheal cannula was inserted and a laparotomy was made. The inferior vena cava was then exposed and a 1.4-mm i.d. polyethylene perfusion cannula was inserted into the vena cava and threaded forward until it passed the diaphragm. In quick succession the tracheal cannula was connected to a constant pressure source, the clearing solution was perfused through the inferior vena cava, and the abdominal aorta was severed. During the perfusion of clearing solution, the lungs were inflated with air to 6 cm H_2O pressure. After perfusion of 25 ml of clearing solution, the perfusate was switched to a fixative solution containing 2% glutaraldehyde and 1% formaldehyde in cacodylate buffer adjusted to pH 7.4. After 5 min of fixative perfusion, the lungs were dehydrated by perfusion (Oldmixon *et al.*, 1985) using graded (20, 50, 70, 90, 95, and 100%) alcohol solutions. Approximately 2 ml were perfused at each of the alcohol concentrations until the final perfusion with 100% alcohol which was carried out with approximately 50 ml.

B. Embedding and Serial Sectioning

At the end of the perfusion dehydration, the lungs were removed from the chest cavity, and the heart and mediastinal tissue were dissected from the lungs. The fixed lung volume was determined by volume displacement in ethanol, after which the left lung was taken and passed through several transitional solutions of water free glycol methacrylate and embedded. After polymerization of the block, the left lung was serially sectioned at 10-μm spacing utilizing a Leitz 1512 rotary microtome with "D" profile steel knives. These sections were then stained with Lee's Methylene Blue–basic fuchsin, cover slipped, photographed on a Hasselblad 500-ELX with a 120-mm macro lens, and printed on 11 $\times$ 14-in paper at a final magnification of $\times$14.

C. Computer Entry and Analysis

For this study a ventilatory unit is defined as all of the gas exchange region found distal to one bronchiole–alveolar duct junction. Three-dimensional reconstructions have demonstrated that in the rat these are discrete units with no overlap between adjacent units (Mercer and Crapo, 1987). The number of bronchiole–alveolar duct junctions and thus the number of ventilatory units were determined for the left lung by examining each series of prints and marking every bronchiole–alveolar duct junction with a unique identifying number.

To determine the size distribution of gas exchange units along a single path, a 1-mm branch off the lobar bronchus of the left lung was identified in the series. The area occupied by this branch was then photographed with a Leitz Orthoplan microscope using a $\times$6.3 plan-apochromat objective. These negatives were then printed on 11 $\times$ 14-in paper at final magnification of $\times$75. The 400 pictures, at a spacing of 10 μm, from this region were then used to determine the dead space distribution to each of the uniquely identified bronchiole–alveolar duct junctions as well as the volume of the gas exchange unit distal to the bronchiole–alveolar junction.

The volume of the ventilatory units in this series was determined by a modification of the methods previously described to determine the volume of alveoli and alveolar ducts distal to the bronchiole–alveolar duct junction. This was done by placing an overlay grid containing 40 points over approximately 50 equally spaced regions throughout the series of prints. The points falling on the airspace of the gas exchange region were then identified. For each point, the series of pictures was used to follow the region of interest up and down until it was determined which gas exchange unit the point had landed on. A tally was made of the number of points falling on each ventilatory unit, and the absolute volume of each ventilatory unit was calculated by taking the ratio of the points for that unit and dividing by the total points in the volume throughout which the point counting was carried out.

A complete three-dimensional reconstruction of the airways was completed using the same series in which ventilatory volumes were determined. To do this, each airway profile within the series of prints was identified. The prints throughout this series were aligned by placing reference marks in the lower left and upper right regions of the middle print of the series. A light box was used to overlay the next print of the series and orient it correctly relative to the print with the reference marks on it. The reference marks were transferred on this overlayed print and the process was repeated for all prints, in both the forward and reverse directions from the middle print.

The dead space distribution to each of the bronchiole–alveolar duct junctions was determined by forming cylindrical approximations for the airways (Fig. 7-1). This was done by reducing the outlines to their appropriate centers of mass throughout the airway reconstruction. A computer

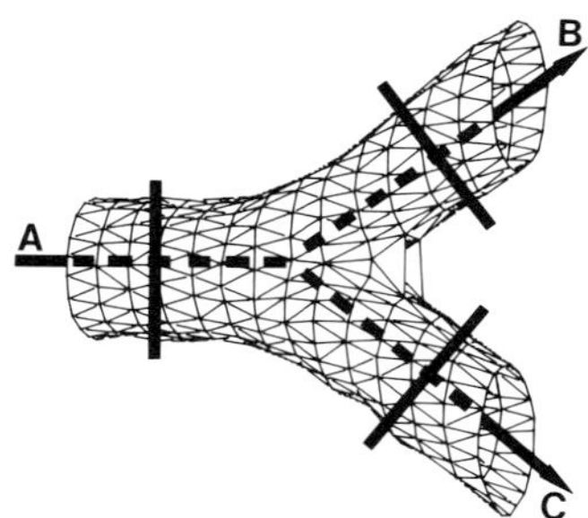

Fig. 7-1 Illustration of the method for determining the central axis and mean diameter of reconstructed airways. The original data are entered as a series of points on each sectioning plane. By stacking these correctly in three dimensions, the points can be connected to form a series of triangles which provide an image of the object's surface. To locate the center line at any point along the tube's surface, we find the minimum perimeter around the surface at that point. The center line of the tube and the tube's mean diameter can then be determined from this new polygon. To form the center line approximation for a branching network of airways this process needs to be repeated at equally spaced points throughout the reconstruction. Center axes of airways A, B, and C are shown, with the heavy solid lines indicating approximate orientation of the sites of analysis used to determine the axis and the diameter.

program was used to link together these centers of mass points throughout the airways. The dead space volume for each path was calculated by summing the distance along the centers of mass points times the perimeter of each airway segment.

A computer program was written to determine the transit time distribution to the gas exchange units as well as the relative delivered dose to the gas exchange units. This consisted basically of carrying out the computations given by Horsfield and Cumming (1968) on a linked list data structure, where each item on the list represented an airway generation or gas exchange unit. A marker indicated both the proximal and distal airway branches to which each airway generation was connected. The program then determined the transit time through each generation by taking the flow at the entry to that generation and dividing it by the dead space volume of that unit. Flows in the daughter generations were determined under the assumption that the flow at a branch is split in proportion to the distal alveolar gas volume of the daughter branches. Thus, the flow for the next generation was computed as the fraction of the distal ventilatory unit's volume(s) of the daughter branches times the flow in the parent branch.

To simplify the calculations the pollutant gas or particle was assumed to move forward as a square wave function and to react or be removed during the inspiratory portion of the respiratory cycle. Thus, the concentration of the pollutant would be negligible in the expired gas, and a simplified estimate of the relative delivered dose to each ventilatory unit can be computed by taking the tidal volume of each ventilatory unit times the fraction of time during inspiration that the ventilatory unit would receive fresh, pollutant laden air.

III. RESULTS

By using the glycol methacrylate technique to serially section large blocks, we carried out three-dimensional reconstructions of the complete airway branching pattern in the lung from trachea to the bronchiole–alveolar duct junctions. A hidden line view of one such reconstruction of the rat lung is shown in Fig. 7-2. Using the airway reconstructions, it was possible to determine the dead space for pathways from the trachea to the bronchiole–alveolar duct junctions. Figure 7-3 shows the distribution of dead space from trachea to bronchiole–alveolar duct junction for each lobe of the rat lung. It is evident that the monopodial branching pattern present in the large airways of the rat lung (Phalen *et al.*, 1978, Yeh *et al.*, 1979) causes the dead space of a given pathway to be predominantly determined by the lobe it is in. This is quite unlike the distributions seen in

Fig. 7-2 Hidden line view of three-dimensional reconstruction of the airways of a rat lung from the trachea to the bronchiole–alveolar duct junction.

dog lungs where Ross (1957) showed that there were large variations in pathway dead space distributions which overlapped significantly between lobes.

To compare dosimetric calculations which use the distal number of ventilatory units as a flow partitioning rule at branch points, as opposed to using the distal alveolar volume at a branch point, we reconstructed an airway and determined the volumes of the ventilatory units distal to it. The airway reconstruction is illustrated in Fig. 7-4 with the corresponding relative delivered dose of fresh pollutant laden air given adjacent to the bronchiole–alveolar duct junction of each ventilatory unit. The average volume of the ventilatory units in this reconstruction was 0.49×10^{-3} ml and was not significantly different from that determined in our previous study (Mercer and Crapo, 1987).

The relative transit times to the bronchiole–alveolar duct junctions could be determined based on either flow partitioning rule. The results in Fig. 7-5 show that the distribution of transit times given by these two methods have similar means. However, the distribution calculated by flow

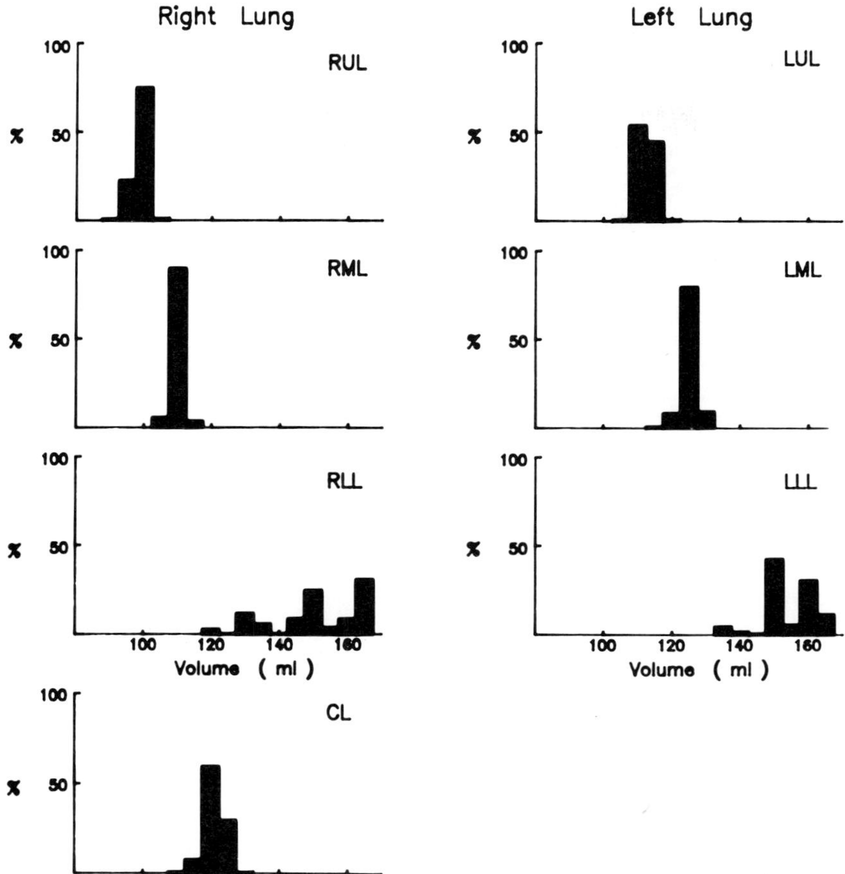

Fig. 7-3 Distribution of dead space from the trachea to the bronchiole–alveolar duct junction for each lobe of the rat lung.

partition based on distal alveolar volume was more dispersed than was the distribution calculated using the distal number of ventilatory units method. The delivered dose was computed based on the relative transit times to the different units and the volumes of the units. When these calculations were made for the reconstruction shown in Fig. 7-4, it was found that ventilatory unit volume was a significant factor in determining the calculated variation in delivered dose. Approximately 5% of the ventilatory units were proximally located. In these cases the combined effects of large ventilatory unit size and low dead space to the unit resulted in a threefold greater delivery of fresh inspired gas to these units compared to the average ventilatory unit. Two of these high-dose ventilatory units are indicated by arrows in Fig. 7-4.

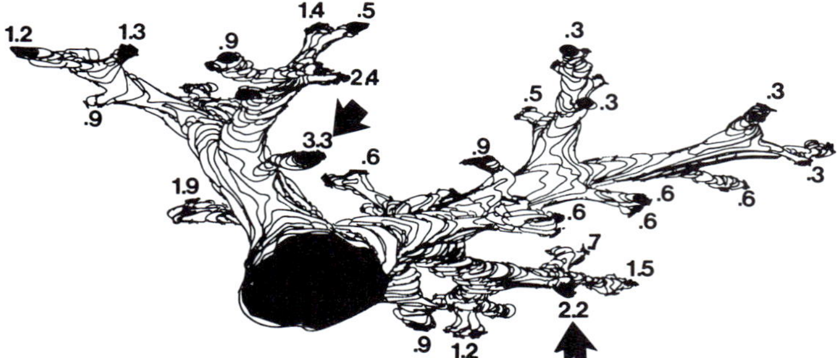

Fig. 7-4 Hidden line view of a three-dimensional reconstruction of the airways for a region of a rat lung. Shown is the reconstruction of the airways which connect to a series of ventilatory units for which individual ventilatory unit volume has been determined. The large blackened region at the center of the figure was the point at which the airway joined the lobar bronchus. The diameter at this point was approximately 0.7 mm. The junctions between the ventilatory units and the airways are darkened. The ventilatory units supplied by these airways accounted for approximately 4% of the left lung volume. The numbers placed adjacent to each bronchiole–alveolar duct junction are the relative delivered dose of fresh inspired gas according to the calculations explained in the Methods section. The arrows point to ventilatory units which, due to their size and proximal location, would receive a two- or threefold greater dose of fresh inspired gas compared to the average for all ventilatory units.

IV. DISCUSSION

The results of this study, although based on relatively small samples, suggest that the dead space distribution to the gas exchange units and the variations in gas exchange unit volume are critical in determining the delivery of inhaled pollutants. Models for dosimetry have typically made the assumption that all gas exchange units at the end of the airways are equal. But the results of three-dimensional reconstructions have indicated that this is not true. The number of gas exchange units at the end of the terminal bronchiole vary from two to six (Mercer and Crapo, 1987) and the volume of ventilatory units varies by two- to threefold as well. Thus our study suggests that not all gas exchange units have the same volume. This is similar to the qualitative observations given by Boyden (1971, 1974) who suggested that different sizes of gas exchange units in the adult were related to variations in growth rate during development.

Finally, because the distributions of dead space volume in different lobes are significantly overlapped in larger species, it is suspected that in all lobes there are pathways which receive a significantly greater portion of the inhaled gas than others. Thus, we suspect that, particularly in larger species, the critically exposed pathways may appear to be diffusely located throughout the lungs. These critical areas could be entirely overlooked in

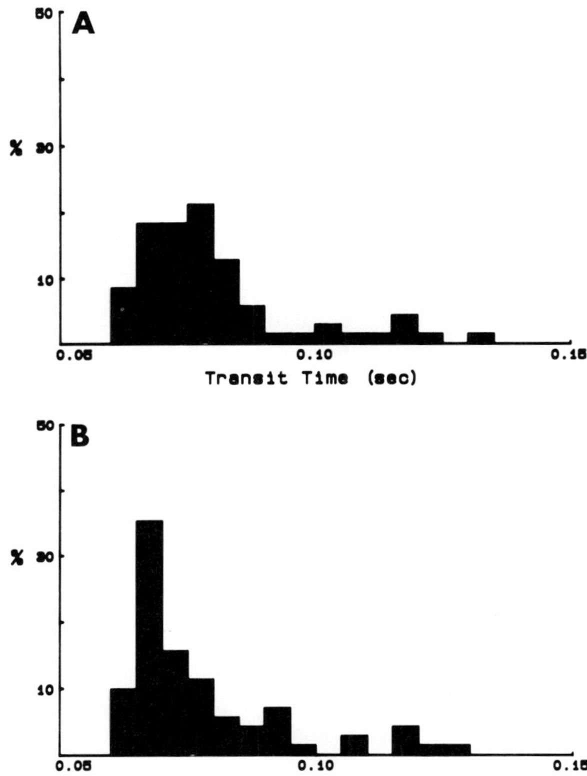

Fig. 7-5 Comparison of methods for computing the transit time for inspired gas to the bronchiole–alveolar duct junctions for the ventilatory units shown in Fig. 7-4. A, the distribution is given based on flow partitioning due to the distal alveolar volume at a branch; B is based on flow partitioning due to the distal number of bronchiole–alveolar duct junctions.

studies which only consider lobar or central–peripheral variation in pollutant exposures.

The presence of large, proximally located ventilatory units and the significant dispersion of the dead space distributions could account for punctate-like lesions in chronic, low-level exposures to pollutants. While these results are preliminary, they do point out the need for studies comparing the predicted microdosimetry in regions of the lungs and the actual damage resulting from exposure.

ACKNOWLEDGMENTS

This work was supported by U.S. Environmental Protection Agency Cooperative Agreement Number CR813113, Department of Energy grant DE-FG05-88ER60654, a grant from RJR

Nabisco, Inc., and a grant from Southern California Edison Company. Although the research described in this article has been funded in part by the E.P.A., it has not been subjected to its required peer and policy review and therefore does not necessarily reflect the views of the Agency and no official endorsement should be inferred.

REFERENCES

Boyden, E. A. (1971). The structure of the pulmonary acinus in a child of six years and eight months. *Am. J. Anat.* **132,** 275–300.

Boyden, E. A. (1974). The mode of origin of pulmonary acini and respiratory bronchioles in the fetal lung. *Am. J. Anat.* **141,** 317–328.

Horsfield K., and Cumming, G. (1968). Functional consequences of airway morphology. *J. Appl. Physiol.* **24,** 384–390.

Mercer, R. R., and Crapo, J. D. (1987). Three-dimensional reconstruction of the rat acinus. *J. Appl. Physiol.* **63,** 785–794.

Miller, F. J., Menzel, D. B., and Coffin, D. L. (1978). Similarity between man and laboratory animals in regional pulmonary deposition of ozone. *Environ. Res.* **17,** 84–101.

Oldmixon, E. H., Suzuki, S., Butler, J. P., and Hoppin, F. C. (1985). Perfusion dehydration fixes elastin and preserves lung air space dimensions. *J. Appl. Physiol.* **58,** 105–113.

Phalen, R. F., Yeh, H. C., Schum, G. M., and Raabe, O. G. (1978). Application of an idealized model to morphometry of the mammalian tracheobronchial tree. *Anat. Rec.* **190,** 167–176.

Ross, B. B. (1957). Influence of bronchial tree structure on ventilation in the dog's lung as inferred from measurements of a plastic cast. *J. Appl. Physiol.* **10,** 1–14.

Yeh, H. C., Schum, G. M., and Duggan, M. T. (1979). Anatomic models of the tracheobronchial and pulmonary regions of the rat. *Anat. Rec.* **195,** 483–492.

Part Three

Experimental Dosimetry of Inhaled Particles and Gases

Chapter 8

Experimental Dosimetry: Introduction

O. G. Raabe
Veterinary Pharmacology and Toxicology, Civil Engineering,
and Institute for Environmental Health Research
University of California
Davis, California 95616

I. INTRODUCTION

Although there are various models of deposition and uptake of inhaled aerosols, vapors, and gases in the respiratory tract, most of these oversimplify the complex anatomical, physiological, aerodynamic, and biochemical phenomena associated with inhalation exposure to toxic materials. Experimental dosimetry provides data under controlled and measured conditions that describe the dosimetric effects of actual inhalation exposures. Experiments that are designed to quantitatively determine the dose to target tissues are essential to improving our understanding of the consequences of exposures and the relationships that describe interspecies scaling or extrapolation.

To the toxicological purist, *dose* refers specifically to the total quantity of chemical, biochemical, or radiative energy or specific structural and/or biochemical alteration transferred to or effected in a known mass of sensitive tissue. The *dose rate* of this transfer or alteration is essentially proportional to the concentration of toxicant in the tissue; hence, the concentration of toxic agent associated with lung tissue is a surrogate parameter for dose rate. The time integral of dose rate yields the total dose delivered to the tissue. Dose rate and total dose may vary markedly in various regions of the respiratory tract depending on physical factors (deposition of particles or uptake of vapor/gas) and temporal factors (clearance rate and

Extrapolation of Dosimetric Relationships for Inhaled Particles and Gases

consequent retention pattern). The dose rate and cumulative dose to sensitive tissue associated with inhalation exposures must be determined to describe dose–response relationships, scale among mammalian species, and extrapolate for risk assessment purposes.

The broader use of the term dose traditionally associated with medicines and drugs can lead to confusion and is best avoided in inhalation toxicology. More descriptive terminology that facilitates clarity is preferred. For example, the aerosol or gas exposure concentration (mg/m^3 or ppm) and the intake rate (mg/day) should not be called the dose.

Most of the chapters in this section concern experimental dosimetry associated with the inhalation exposure of aerosols. Aerosols are relatively stable suspensions of finely divided liquid droplets or solid particles in a gaseous medium, usually in air. If inhaled, aerosol particles may be deposited by contact upon the various surfaces of the respiratory tract leading to potential injury, receptor responses, biochemical alterations, desirable therapeutic action, or planned diagnostic behavior depending on the particular properties of the particles. Inhalable aerosols affecting the lung are those consisting primarily of particles smaller than 10 μm in aerodynamic equivalent diameter since larger particles tend to deposit in the head airways. Because of the high permeability of the lung and the copious blood flow, soluble particles depositing in the lung can readily enter the blood for action throughout the body, while other less soluble particles may directly influence the airway epithelium and effect responses.

The behavior of inhaled airborne particles in the respiratory airways and their alternative fates of either deposition in the various airway regions (nasal–pharyngeal airways, mouth, oral pharynx, larynx, tracheobronchial airways, and pulmonary parenchyma) or exhalation depend upon the aerodynamic behavior of the aerosol particles and physiological and anatomical factors. Particle aerodynamic behavior depends on particle physical density, shape, size distribution, electrostatic charge, and hygroscopicity or deliquescence. Physiological factors include breathing patterns, breathing rate, tidal volume (TV), functional residual capacity (FRC), air flow dynamics in the airways, and variations in relative humidity and temperature within the respiratory tract. The exact anatomy of the airways from nose or mouth to the lung parenchyma including the diameters, lengths, and branching angles of airway segments also influences aerosol particle deposition (Lippmann, 1977; Raabe, 1982, 1984a).

Utilizing the specific details of these controlling factors and simplifying assumptions, theoretical models of regional deposition have been developed to predict the fate of inhaled particles of various types. Carefully collected data from experiments described in this section provide the basis for testing these predictions.

II. AEROSOL PROPERTIES

The aerodynamic properties of aerosol particles depend upon several physical characteristics including size, shape, and physical densities of the particles. Two important properties of aerosol particles are the inertial properties, which apply most appropriately to particles larger than 0.5 μm in diameter (related to the settling speed in air under the influence of the earth's gravity), and the diffusional properties which are most important for particles smaller than 0.5 μm in diameter (related to the particle diffusivity; Raabe, 1980). The inertial properties are described in terms of an equivalent aerodynamic diameter; particles with identical inertial properties have the same aerodynamic diameter irrespective of their actual sizes, shapes, or densities. The aerodynamic diameter most generally used is the aerodynamic equivalent diameter (D_{ae}) defined by Hatch and Gross (1964) as "the diameter of a unit density sphere having the same settling speed (under gravity) as the particle in question of whatever shape and density." Raabe (1976) has recommended the use of an "aerodynamic resistance diameter" (D_{ar}) to describe particle mechanics for aerosol sampler calibrations. These two definitions are about equivalent with the aerodynamic resistance diameter being about 0.08 μm larger than the aerodynamic equivalent diameter. The aerodynamic diameter of a particle is indicative of its rate of settling under the influence of the earth's gravity, or its inertia that controls particle traversal of airflow streamlines. Both are involved in inhalation deposition. Experiments in this section involving particles larger than about 1 μm are dependent upon aerodynamic diameter and associated inertial properties. In those cases, the particle deposition tends to be focal at airway branches and turns, and shows special accumulation in the larger airways.

Bombardment of airborne particles by gas molecules in air produces random zigzag (Brownian) motion which causes very small particles (smaller than 0.5 μm) to move and mix even under tranquil conditions. This property of small particles can be described in terms of a diffusive diameter, D_{df}, defined by Raabe (1980) as "equal to the geometric diameter of an ideal spherical particle with the same diffusivity as the actual particle under identical conditions." The diffusive diameter is indicative of deposition of small particles in small airways at low flow rates. Experiments in this section involving particles smaller than about 0.5 μm depend upon diffusional properties and tend to show diffuse deposition throughout the airways, especially in the lung parenchyma.

Aerosols are comprised of particles or droplets having different physical and aerodynamic sizes, so they must be described in terms of size distribution parameters. Referring to aerosols as if they consist of one size of particle can be misleading when, in fact, a range of sizes are mixed together. It has become customary to assume that most experimental aerosols are lognormally distributed with respect to size, and to describe them in terms of their respective median size and associated geometric standard deviation (σ_g; Raabe, 1971).

If the particle aerodynamic or diffusive diameter had been determined rather than the physical diameter, the median would have been the count median aerodynamic diameter (CMAD) or count median diffusive diameter (CMDD), respectively. Half of the particles are smaller than the CMD (or CMAD or CMDD) and half are larger. If the mass of an aerosol is under study, which is usually more important than particle number, then the median that is used is the mass median diameter (MMD), mass median aerodynamic diameter (MMAD) or mass median diffusive diameter (MMDD). Half the mass of an aerosol is associated with particles smaller than the mass median and half with larger particles. If a particular component of an aerosol, such as a particular toxicant, fluorescent label, or radiolabel is of interest, then the activity median diameter (AMD), activity median aerodynamic diameter (AMAD), or activity median diffusive diameter (AMDD) is used. Potentially toxic aerosols are usually described in terms of their mass median aerodynamic diameter (MMAD) and associated geometric standard deviation (Raabe, 1980).

III. INHALATION DEPOSITION OF AEROSOLS

The collection of inhaled airborne particles in the respiratory tract and the initial regional pattern of these collected particles is called *deposition.* Inhaled particles that pass the head airways region (HAR) of the respiratory tract during inhalation may deposit in the tracheobronchial region (TBR) or the gas exchange region (GER) of the lung. The magnitude of this deposition will depend upon aerodynamic and diffusive particle sizes, respiratory mechanics, and anatomical relationships.

All particles that come in contact with the moist wall of the airways are deposited. Most aerosols are electrostatically charged by the aerosol generation processes, and these charged particles may be attracted to the wall of the airway by the image–charge effect. However, the overall influence of charge on the deposition of most aerosols is probably small. Likewise, the noninertial incidental contact of a particle with the wall of the airways leads to deposition by interception, but this process is primarily important for elongated particle shapes such as fibrous aerosols. The most important physical processes associated with inhalation deposition of aerosols are impaction, gravitational setting, and Brownian diffusion.

Deposition by gravitational settling occurs throughout the respiratory tract due to the influence of the earth's gravity on small particles suspended in air. The settling rate for small particles increases proportionally to the square of the aerodynamic diameter, D^2_{ar} (or approximately D^2_{ae}). Gravitational deposition is especially important in the distal regions of the bronchial airways and in portions of the gas exchange region.

Inertial impaction is the dominant mechanism of deposition of particles larger than 3 μm in aerodynamic diameter which occurs primarily in the head airways or tracheobronchial airway regions. In this process the airborne particles, because of their inertia, do not follow changes in direction or speed of air streamlines, and they may collide with the wall of the airway. For example, if air is directed toward an airway surface (such as a branch carina), but the forward velocity is suddenly reduced because of the change in flow direction caused by the obstruction of the surface, inertial momentum may carry larger particles across the air streamlines and into the moist surface of the tract where they are deposited. Aerodynamic separation of this type may be characterized in terms of an impaction parameter given by $D^2_{ar}Q$ (or approximately by $D^2_{ae}Q$) where Q is the average inspiratory flow rate and D_{ar} or D_{ae} is the aerodynamic diameter.

Unlike impaction and settling that becomes greater with larger particle size, deposition by Brownian diffusion increases with decreasing size and depends on the diffusive rather than the aerodynamic diameter. Diffusional deposition in the gas exchange region of the lung is the predominant collection mechanism for particles smaller than 0.5 μm. Deposition by diffusion can also occur in the nose, mouth, and pharyngeal airways for very small particles (smaller than 0.01 μm, Cheng *et al.*, 1989).

Even with carefully developed theoretical models with the support of reliable deposition data, it is impossible to predict exactly the quantitative regional deposition of particles of a given inhaled aerosol in a particular person. Considerable variability in respiratory parameters may occur among individuals, particularly when healthy adults are contrasted with children, or women are compared with men. Biological variability between individuals, differences in health, confounding factors such as cigarette smoking, differences that relate to age or breathing styles, and inherent variations in airway sizes can alter the fraction of an inhaled aerosol that may deposit in the airways and in the quantity of particulate mass that is actually breathed. For the purposes of aerosol planning, reasonable predictions must be made using the available models and data employing certain simplifying assumptions concerning biological factors. However, experimental studies are needed to challenge, test, and/or confirm these models.

The most widely used models of regional deposition vs particle size were developed by the International Commission on Radiological Protection (ICRP) Task Group on Lung Dynamics under the chairmanship of P. E. Morrow. Morrow *et al.* (1964) used representative values for normal respiratory parameters described for "reference man" (Snyder, 1975). Assumed values for a typical adult include a body weight of 70 kg, height of 175 cm, and body surface area of 1.8 m^2. The ICRP Task Group used the anatomical model and general methods of Findeisen (1935) and Landahl (1950) for calculating deposition in the tracheobronchial and pulmonary regions. Particles were assumed to be insoluble, stable, and spherical with physical densities of 1 g/cm^3. Regional deposition was calculated for a breathing rate of 15 breaths/min (BPM) via the nose for three tidal volumes (TV):

1. TV 750 ml, modest activity
2. TV 1450 ml, moderate activity
3. TV 2150 ml, strenuous activity.

Their result for inhaling via the nose at a TV 750 ml is summarized in Fig. 8-1. Not shown in the figure is the nasal deposition of ultrafine particles (0.01 μm; Cheng *et al.*, 1989).

Morrow *et al.* (1964) calculated the theoretically expected deposition of particles in the airways of the lung during breathing, but they did not theoretically determine head airway deposition. Their nasal–pharyngeal

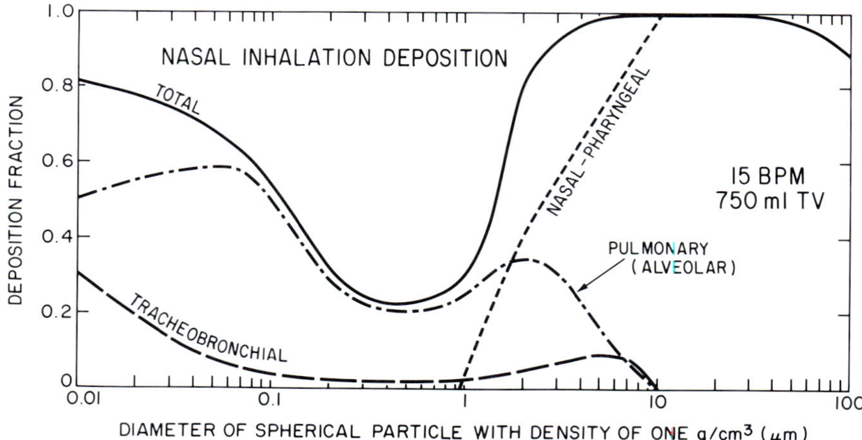

Fig. 8-1 Total and regional deposition fractions for aerosols entering the nose for various sizes of inhaled airborne spherical particles with physical density of 1 g per cm^3 in the human respiratory tract as calculated by the International Commission on Radiological Protection (ICRP) Task Group on Lung Dynamics (Morrow *et al.*, 1964) for nasal breathing at a rate of 15 breaths per min (BPM) and tidal volume (TV) of 750 ml.

(NP) deposition calculations were based upon the empirical log–linear equation of Pattle (1961) (deposition fraction $= -0.62 + 0.475 \log (D_{ar}^2 Q)$ where log is the base 10 logarithm) for NP deposition during inspiration. They assumed that the log–linear Pattle equation also applied to the exhalation process to calculate the total nasal–pharyngeal deposition. Raabe (1984b) used available data to develop a log–linear function (deposition $= -1.4 + 0.6 \log(D_{ar}^2 Q)$ with log the base 10 logarithm) to describe head airway deposition associated with inhalation via the mouth. With this relationship, the ICRP Task Group results can be modified to estimate gas exchange region (GER) deposition and tracheobronchial region (TBR) deposition during mouth breathing. The resulting regional deposition of inhaled particle for mouth breathing is shown in Fig. 8-2.

Clearly, thoracic deposition in the tracheobronchial airways and deep lung is emphasized during oral breathing. The deposition of inhaled aerosol particles in the nasal–pharyngeal region during nasal breathing is much higher for all sizes than the deposition that occurs in the oral–pharyngeal head airways during mouth breathing. Specifically, much larger particles that normally are completely collected in the nasal–pharyngeal portion of the head airways during breathing via the nose may pass the epiglottis and larynx and deposit in the lung during breathing via the mouth. The inhalation of deliquescent or hygroscopic aerosols will alter the deposition char-

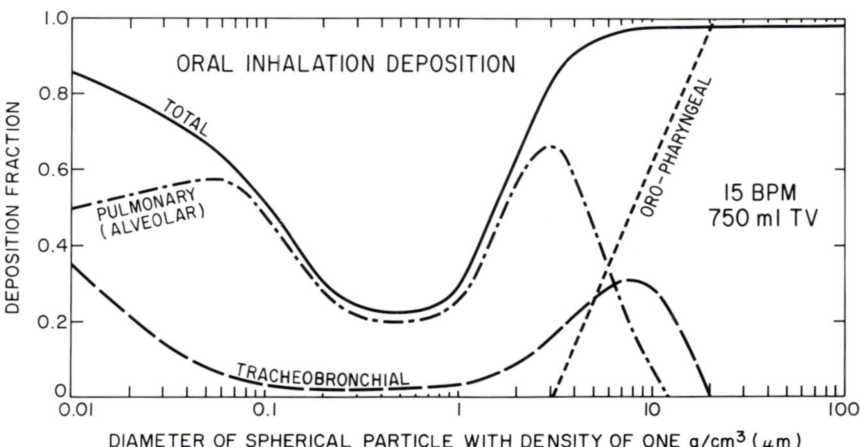

Fig. 8-2 Total and regional deposition fractions for aerosols entering the mouth for various sizes of inhaled airborne spherical particles with physical density of 1 g per cm^3 in the human respiratory tract as calculated by the International Commission on Radiological Protection (ICRP) Task Group on Lung Dynamics (Morrow *et al.*, 1964) with the head airway deposition function given by Raabe (1984b) for oral breathing at a rate of 15 breaths per min (BPM) and tidal volume (TV) of 750 ml.

acteristics since these particles will tend to grow rapidly in the humid environment of the respiratory tract.

After initial deposition of particles associated with inhaled aerosols, these particles are subjected to various biological, physical, and chemical processes including dissolution into body fluids with absorption by the blood, uptake by cells by phagocytosis or pinocytosis, and movement with mucus and body fluids. The term *clearance* is used to describe the translocation, transformation, and removal of deposited particles from the various regions of the respiratory tract. The temporal distribution of uncleared deposited particles or their resultant transformation products is called *retention.*

Relatively insoluble particles deposited in the ciliated region of the tracheobronchial airways are moved with mucus flow towards the epiglottis where they are swallowed or expectorated. This process is relatively efficient in that most particles deposited in the TBR are probably cleared within a few hours and all by one day postexposure. Insoluble particles deposited in the nonciliated bronchioles or the alveoli of the gas exchange region (GER), are not rapidly cleared from the lung. Most are engulfed by and phagocytized by scavenger pulmonary alveolar macrophage cells recruited in the GER, and some of these cells may ultimately enter the TBR mucus flow. But this does not seem to be a rapid process in human lungs and requires up to 1 yr to effectively clear half of GER deposited particles. Some of the macrophages may enter the pulmonary lymph circulation and be transported to tracheobronchial lymph nodes, but this is a slow process requiring up to 1 yr for clearance of half of the deposited particles from the GER. Consequently, very insoluble particles are tenaciously retained in the GER. Clearance by dissolution becomes an important process for sparingly soluble particles (e.g., particles of glass; Raabe, 1982).

IV. INHALATION UPTAKE OF GASES AND VAPORS

One of the studies in this section involves the inhalation of a potentially toxic gas. There are significant differences in the dosimetric factors associated with the inhalation of gases or vapors as contrasted to those with aerosols. The transfer of gas molecules from inhaled air to the epithelium surfaces of the airways depends upon convective diffusion, which in turn depends upon breathing rates, volumes, and molecular diffusivity. It cannot be assumed that all molecules that hit the surface of the airways adhere to or are taken up into the tissues, as in the case of particles. Some adsorbed or absorbed gas or vapor molecules may be released back into the air and exhaled. The uptake process is affected by the factors controlling the rate of transfer of these molecules through the liquid–gas interface and epithelial membranes including the air-to-tissue and air-to-blood partition coefficients which depend upon solubility in body fluids. The metabolic fate of the absorbed gas or vapor molecules will determine the net concentration in lung tissue and the consequent dose rate to those tissues (Miller and Menzel, 1984).

During inspiration, the air entering the alveolar (gas exchange) region of the lung includes air remaining in the conductive airways of the respiratory tract from the previous inhalation. These conductive airways represent a respiratory dead space, since gas exchange or vapor uptake may or may not occur in them. The volume of this respiratory dead space in people is estimated at 30% of the resting tidal volume, or about 150 ml (Fiserova-Bergerova, 1983). The common assumption of complete mixing in the alveolar space during breathing is probably not valid. The air in the lungs is not well mixed, and the passage of gas or vapor from the incoming air to the walls of the alveoli is a function of vapor diffusivity, which is the proportionality constant describing the rate of flow of gases from regions of high to low concentration.

The use of simulation or pharmacokinetic models of the processes associated with inhalation uptake, distribution, metabolism, and excretion of vapors and gases provides a convenient basis for predicting the uptake of a vapor or gas and establishing the expected concentration in the blood, liver, kidneys, and other body organs. It also provides a basis for understanding the results of experimental measurements (Fiserova-Bergerova, 1983).

REFERENCES

Cheng, Y. S., Yanada, Y., and Yeh, H. C. (1988). Diffusional deposition of ultrafine aerosols in a human nose cast. *J. Aerosol Sci.*, **19**, 741–751.

Findeisen, W. (1935). Uber das Absetzen kleiner, in der Luft suspendierter Teiche in der menschlichen Lunge bei der Atmung. *Pflugers Arch. Ges. Physiol.* **236**, 367–379.

Fiserova-Bergerova, V. (1983). "Modeling of Inhalation Exposure to Vapors: Uptake, Distribution, and Elimination." CRC Press, Boca Raton, Florida.

Hatch, T. E., and Gross, P. (1964) "Pulmonary Deposition and Retention of Inhaled Aerosols." Academic Press, New York.

Landahl, H. D. (1950). On the removal of airborne droplets by the human respiratory tract. I. The lung. *Bull. Math. Biophys.* **12**, 43–56.

Lippmann, M. (1977). Regional deposition of particles in the human respiratory tract. *In* "Handbook of Physiology, Section 9: Reactions to Environmental Agents" (D. H. K. Lee, H. L. Falk, and S. D. Murphy, eds.), pp. 213–232. American Physiological Society, Bethesda, Maryland.

Miller, F. J., and Menzel, D. B. (1984). "Extrapolation Modeling of Inhaled Toxicants: Ozone and Nitrogen Dioxide." Hemisphere, Washington, D.C.

Morrow, P. E., Bates, D. V., Fish, B. R., Hatch, T. F., and Mercer, T. T. (1964). Deposition and retention models for internal dosimetry of the human respiratory tract. (Report of the

International Commission on Radiological Protection (ICRP) Task Group on Lung Dynamics). *Health Phys.* **12,** 173–207.

Pattle, R. E. (1961). The retention of gases and particles in the human nose. *In* "Inhaled Particles and Vapours" (C. N. Davies, ed.), pp. 302–309. Pergamon, Oxford.

Raabe, O. G. (1971). Particle size analysis utilizing grouped data and the log-normal distribution. *J. Aerosol Sci.* **2,** 289–303.

Raabe, O. G. (1976). Aerosol aerodynamic size conventions for inertial sampler calibration. *J. Air Pollut. Control Assoc.* **26,** 856–860.

Raabe, O. G. (1980). Physical properties of aerosols affecting inhalation toxicology. *In* "Pulmonary Toxicology of Respirable Particles, Conference—791002" (C. L. Sanders, F. T. Cross, G. E. Dagle, and J. A. Mahaffey, eds.), pp. 1–28. U. S. Dept. of Energy, National Technical Information Service, Springfield, Virginia.

Raabe, O. G. (1982). Deposition and clearance of inhaled aerosol. *In* "Mechanisms in Respiratory Toxicology" (H. R. Witschi and P. Nettesheim, eds.), Chap. 2, pp. 27–76. CRC Press, Boca Raton, Florida.

Raabe, O. G. (1984a). Deposition and clearance of inhaled particles. *In* "Occupational Lung Diseases" (J. B. L. Gee, W. K. C. Morgan, and S. M. Brooks, eds.), pp. 1–37. Raven, New York.

Raabe, O. G. (1984b). Size-selective sampling criteria for thoracic and respirable mass fraction. *Ann. Am. Conf. Ind. Hyg.* **11,** 53–85.

Snyder, W. S. (1975). "Report of the Task Group on Reference Man." International Commission on Radiological Protection (ICRP) Task Group on Reference Man. Pergamon, Oxford.

Chapter 9

Particle Deposition at the Alveolar Duct Bifurcations

A. R. Brody
Laboratory of Pulmonary Pathobiology
National Institute of Environmental Health Sciences
Research Triangle Park, North Carolina 27709

C. P. Yu
Department of Mechanical and Aerospace Engineering
State University of New York at Buffalo
Buffalo, New York 14260

I. INTRODUCTION

We are conducting studies to elucidate the initial biochemical and cellular alterations which lead to particle-induced interstitial lung disease (Brody, 1986; Warheit *et al.,* 1985). It is highly probable that the initial pattern of particle deposition at the alveolar level plays a major role in the nature of the cellular response (Brody and Hill, 1982; Warheit *et al.,* 1985, 1986). Thus, we have carried out studies on the initial deposition pattern of a variety of particles small enough to reach the alveolar surfaces (Brody *et al.,* 1981, 1982; Brody and Roe, 1983). In addition, we present a theoretical explanation for the deposition pattern observed (Cai and Yu, 1988).

II. MATERIALS AND METHODS

Caged male white (CD) rats were acutely exposed in open chambers to the aerosols chrysotile and crocidolite asbestos, fiberglass, silica, and ash from Mount St. Helens volcano (Brody and Roe, 1983). Mice (strain B10, D2/nSn) were exposed to chrysotile asbestos. Exposure time ranged from 1 to 5 hr, and the animals were sacrificed immediately (4 min) after cessation of exposure (Brody *et al.*, 1981). Lung fixation was carried out by perfusing the lungs *in situ* with Karnovsky's fixture through the pulmonary artery as previously described (Brody *et al.*, 1981; Brody and Hill, 1982). Fixed lungs were removed from the chest cavity and dissected to reveal the major airways and their bronchioles. Tissue blocks measuring approximately 1 cm^3 were cut from the lungs and critical point dried (Brody *et al.*, 1981; Warheit *et al.*, 1984). Then specific dissection techniques were used (Brody *et al.*, 1981; Warheit *et al.*, 1984). Critical point dried tissue blocks were attached to carbon disks with carbon paint and the block surface was thinly sliced away with a razor blade so that terminal bronchioles and their alveolar ducts could be clearly observed under a dissecting microscope (Brody *et al.*, 1981; Brody and Roe, 1983; Warheit *et al.*, 1984). These freshly dissected surfaces then are gold-coated (Brody *et al.*, 1981; Brody and Roe, 1983; Warheit *et al.*, 1984) and studied by scanning electron microscopy (SEM) (Fig. 9-1). SEM allows a high resolution view of airway and alveolar surfaces where the inhaled particles were counted (Brody and Roe, 1983).

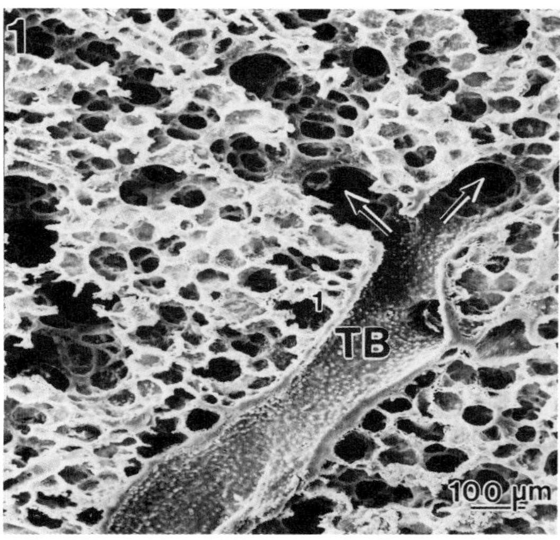

Fig. 9-1 Scanning electron micrograph of the lung parenchyma of a rat prepared by critical point drying and dissection. Terminal bronchioles (TB) and alveolar ducts (arrows) are clear.

III. RESULTS AND DISCUSSION

A. Observed Deposition Pattern

We found that the majority of particles which passed through the terminal bronchioles were deposited at the bifurcations of alveolar ducts (Fig. 9-2). There were more particles on the first duct bifurcation than on the second, and more on the second than third. Few particles of any type were found in alveolar spaces or on duct surfaces between bifurcations (Brody and Roe, 1983). SEM clearly showed the accumulation of fibers and particulates at the edges of the bifurcations (Fig. 9-2). This deposition pattern is not what one would expect based upon theoretical calculations of airflow velocity at the alveolar level. We believed we would find an even distribution of particles in alveolar spaces because airflow past terminal bronchioles is so low that inhaled particles would be expected to settle after random diffusion. This does not appear to be the case in rats and mice. In the following section we present a theoretical scheme which could explain the mechanism leading to the deposition pattern observed. It is important to point out that this pattern is the one most likely to occur *in vivo* since the initial cellular responses and a consequent interstitial lesion are induced at the first alveolar duct bifurcations after asbestos inhalation (Brody, 1986; Brody and Hill, 1982; Chang *et al.*, 1988; Warheit *et al.*, 1984, 1986).

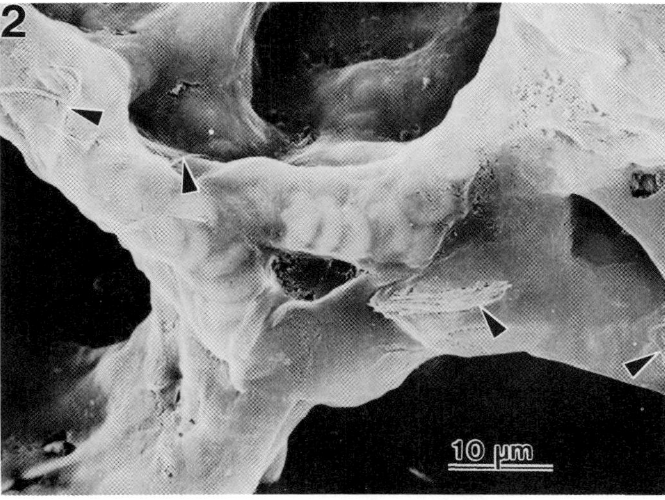

Fig. 9-2 Higher magnification scanning electron microscopy showing the distribution of chrysotile asbestos fibers on a first alveolar duct bifurcation.

B. Theoretical Deposition Patterns

The observed enhanced deposition at alveolar duct bifurcations can be explained qualitatively on a theoretical basis. In a bifurcation airway, particle deposition can be controlled by five independent mechanisms: impaction, interception, diffusion, sedimentation, and electrostatic attraction. The relative importance of each mechanism, however, depends on airway geometry and airflow velocity as well as on a particle's mass, charge, and size. It is known that deposition contribution by electrostatic attraction becomes significant only when the particle charge level exceeds a threshold limit (Yu, 1985). We shall neglect this effect in the present analysis and will only consider deposition during the inspiratory phase of a respiratory cycle because the majority of deposition occurs in this phase.

In the rat lung, the airflow velocity in the terminal bronchioles is calculated to be about 6.4 cm/sec based upon the whole lung model of Schum and Yeh (1980). This leads to a Reynolds number of about 0.85 for the flow. Thus, at the exit of a terminal bronchiole and in its alveolar ducts, the airflow will have a parabolic velocity profile. The deposition efficiencies for various mechanisms in bifurcating airways with parabolic velocity profiles have been derived from other studies (Cai and Yu, 1988; Ingham, 1975; Pich, 1972). The formulas for these efficiencies are summarized as follows:

1. Impaction for spherical particles:

$$\eta = G(\alpha, R/R_0)St \tag{1}$$

where η is the deposition efficiency, α is the bifurcation angle, R and R_0 are, respectively, the radii of the alveolar duct and terminal bronchiole, and St is the particle Stokes number, given by

$$St = \frac{C\rho_p d_p^2 v_o}{36\mu R_o} \tag{2}$$

in which C is the slip correction factor, ρ_p is the particle mass density, d_p is the particle diameter, v_o is the airflow velocity in the terminal bronchiole, and μ is the viscosity of the air. The function G in equation (1) has the form

$$G(\alpha, R/R_0) = \frac{8 \sin \alpha f_1(\alpha, R/R_0)}{(R/R_0) f_0(\alpha, R/R_0)} \tag{3}$$

where

$$f_0(\alpha, R/R_0) = \pi\left[1 - \frac{1}{4}\left(\frac{R}{R_0}\right)^2\right] - \frac{4}{3}\left(\frac{15}{16}\pi - 2\right)\left(\frac{R}{R_0}\right)^2 \cos^2 \alpha \tag{4}$$

and

$$f_1(\alpha, R/R_0) = 1 - \frac{1}{3}\left(\frac{R}{R_0}\right)^2 + \left(\pi - \frac{11}{3}\right)\left(\frac{R}{R_0}\right)^2 \cos^2\alpha - \frac{1}{3}\left(\frac{R}{R_0}\right)^2 \sin\alpha$$
$$+ \left(\frac{2}{3} - \frac{\pi}{8}\right)\left(\frac{R}{R_0}\right)^4 \cos^2\alpha + \frac{1}{5}\left(\frac{R}{R_0}\right)^4 \sin^2\alpha$$
$$+ \left(6 - \frac{15}{8}\pi\right)\left(\frac{R}{R_0}\right)^4 \cos^4\alpha$$
$$+ \left(\frac{7}{15} - \frac{\pi}{8}\right)\left(\frac{R}{R_0}\right)^4 \sin^2\alpha \cos^2\alpha. \quad (5)$$

2. Impaction for fibers in random orientation:

$$\eta = G(\alpha, R/R_0)(St)_f \quad (6)$$

where $(St)_f$ is the Stokes number of a fiber, defined as

$$(St)_f = \frac{C\rho_p d_f^3 \beta v_0}{36\mu\left[\left(1 - \frac{\pi^2}{16}\right) d_{f\parallel} + \frac{\pi^2}{16} d_{f\perp}\right] R_0}, \quad (7)$$

in which d_f is the fiber diameter, β is the aspect ratio, and $d_{f\parallel}$ and $d_{f\perp}$ are, respectively, the equivalent Stokes diameters of the fiber for the cases when the fiber moves with its axis parallel and perpendicular to the direction of motion. The expressions for $d_{f\parallel}$ and $d_{f\perp}$ are

$$d_{f\parallel} = \frac{\frac{4}{3}(\beta^2 - 1)d_f}{\frac{2\beta^2 - 1}{\sqrt{\beta^2 - 1}} \ln(\beta + \sqrt{\beta^2 - 1}) - \beta} \quad (8)$$

and

$$d_{f\perp} = \frac{\frac{8}{3}(\beta^2 - 1)d_f}{\frac{2\beta^2 - 1}{\sqrt{\beta^2 - 1}} \ln(\beta + \sqrt{\beta^2 - 1}) + \beta}. \quad (9)$$

3. Interception for spherical particles:

$$\eta = 2I(\alpha, R/R_0)d_p/R \quad (10)$$

4. Interception for fibers in random orientation:

$$\eta = I(\alpha, R/R_0)l_f/R \quad (11)$$

where

$$I(\alpha, R/R_0) = \frac{\left[\frac{1}{2} - \frac{1}{6}\left(\frac{R}{R_0}\right)^2 + \left(\frac{5}{6} - \frac{\pi}{4}\right)\left(\frac{R}{R_0}\right)^2 \cos^2 \alpha\right]}{f_0\left(\alpha, \frac{R}{R_0}\right)}. \tag{12}$$

5. Sedimentation:

$$\eta = \frac{2}{\pi}\left(2\epsilon\sqrt{1-\epsilon^{2/3}} - \epsilon^{1/3}\sqrt{1-\epsilon^{2/3}} + \arcsin \epsilon^{1/3}\right) \tag{13}$$

where

$$\epsilon = \frac{3\pi v_s z}{32 R v} \tag{14}$$

in which v_s is the particle settling velocity, R and z are, respectively, the airway radius and length. Differentiating equation (13) with respect to z, we obtain the following expression for deposition per unit length of the alveolar duct at any location z as

$$\frac{d\eta}{dz} = \left(\frac{d\eta}{d\epsilon}\right)\left(\frac{d\epsilon}{dz}\right) = \frac{v_s}{2Rv}\sqrt{1-\epsilon^{2/3}}. \tag{15}$$

6. Diffusion:

$$\begin{aligned}\eta = 1 &- 0.819 \exp(-14.63\Delta) - 0.0976 \exp(-89.22\Delta)\\ &- 0.0325 \exp(-228.0\Delta) - 0.0509 \exp(-125.9\Delta^{2/3})\end{aligned} \tag{16}$$

where

$$\Delta = \frac{Dz}{4vR^2} \tag{17}$$

and D is the diffusion coefficient. At any location z in the alveolar duct, deposition per unit length is found by differentiating equation (16) with z. The result is

$$\begin{aligned}\frac{d\eta}{dz} = \frac{d\eta}{d\Delta} \cdot \frac{d\Delta}{dz} = \frac{D}{4vR^2}&[11.982 \exp(-14.63\Delta) + 8.7079 \exp(-89.22\Delta)\\ &+ 7.41 \exp(-228.0\Delta) + 4.2722\Delta^{-1/3} \exp(-125.9\Delta^{2/3})].\end{aligned} \tag{18}$$

Using the above formulas, deposition patterns have been calculated for quartz particles and chrysotile asbestos fibers in the alveolar ducts immediately following a terminal bronchiole in the rat lung. We used a spherical particle model for quartz particles and an ellipsoidal particle model for chrysotile asbestos fibers and assumed a random fiber orientation in the flow. The particle size and mass density used for calculation are 1.4 μm diameter and 2.65 g/cm^3 mass density for quartz particles, and 0.28 μm diameter, 10 μm length and 3 g/cm^3 mass density for fibers, assuming that both particles are monodisperse.

Figures 9-3 and 9-4 show the deposition patterns of these particles with respect to the nondimensional distance along the alveolar duct. The deposition fractions by impaction and interception are assumed to take place in the beginning one tenth of the duct length. The total deposition fraction due to all mechanisms in this beginning section was found to be over 15 times greater than those in the other sections of the duct for quartz particles, and over 53 times for asbestos fibers.

The high deposition efficiencies for impaction and interception shown in Fig. 9-3 and 9-4 are not surprising. Because of their high breathing frequency and small airway size, for rats, the air velocity in the terminal bronchioles is significantly larger than in humans, resulting in larger deposition by impaction and interception. Deposition patterns calculated for humans using the same particle characteristics in a Weibel lung model (Weibel, 1963) suggest that the degree of enhancement of deposition may not occur at the first alveolar duct bifurcations, although it is clear that human asbestotic lesions are most prominent at bronchiolar–alveolar regions.

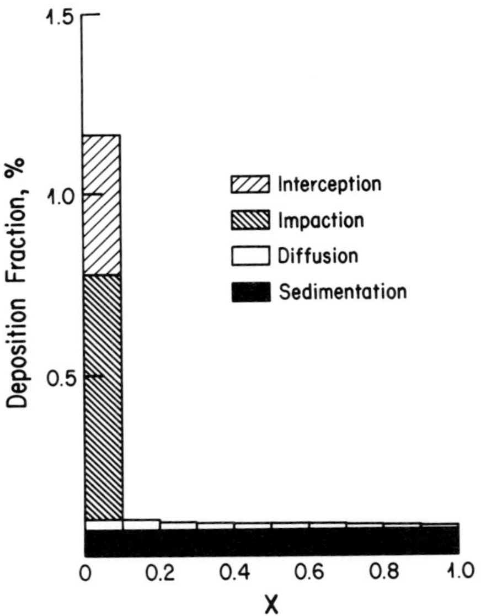

Fig. 9-3 Calculated deposition pattern of quartz particles.

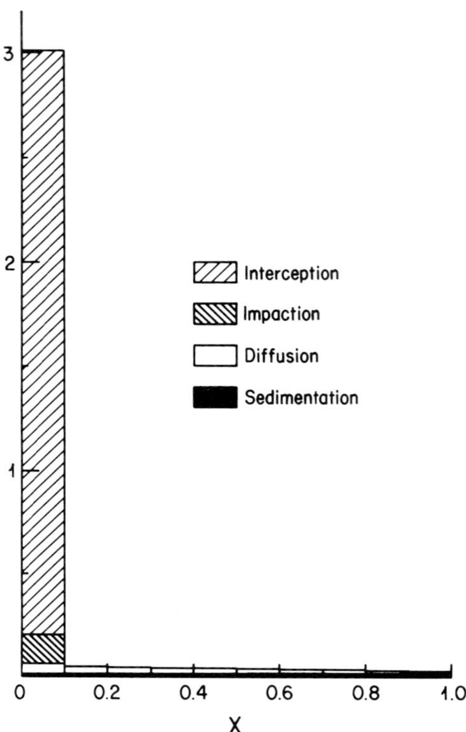

Fig. 9-4 Calculated deposition pattern of chrysotile asbestos fibers.

IV. SUMMARY

Inhaled particles small enough to pass through the terminal bronchioles are deposited initially at the alveolar duct bifurcations in the lungs of rats and mice. This deposition pattern can be attributed theoretically to the unusually large deposition by impaction and interception in the alveolar ducts of these animals because of their high breathing frequency and small airway size. Calculated deposition patterns from inspiratory air support this view. However, the enhanced deposition at alveolar duct bifurcations may not occur in all species. It is clear that early pneumoconiotic lesions are found at the bronchoalveolar junctions in the lungs of humans. It has not yet been determined whether the deposition pattern seen in rats and mice occurs in humans with well developed respiratory bronchioles.

ACKNOWLEDGMENTS

The authors thank Helena Bonner and Lila Overby for preparation of the manuscript and figures. C. P. Yu's work is supported in part by NIH Grant HL 38503.

REFERENCES

Brody, A. R. (1986). Pulmonary cell interactions with asbestos fibers *in vivo* and *in vitro*. *Chest* **89,** 155–159.

Brody, A. R., and Hill, L. H. (1982). Interstitial accumulation of inhaled chrysotile asbestos fibers and consequent formation of microcalcifications. *Am. J. Pathol.* **109,** 107–114.

Brody, A. R., and Roe, M. W. (1983). Deposition pattern of inorganic particles at the alveolar level in the lungs of rats and mice. *Am. Rev. Respir. Dis.* **128,** 724–729.

Brody, A. R., Hill, L. H., Adkins, Jr., B., and O'Connor, R. W. (1981). Chrysotile asbestos inhalation in rats: Deposition pattern and reaction of alveolar epithelium and pulmonary macrophages. *Am. Rev. Respir. Dis.* **123,** 670–679.

Brody, A. R., Roe, M. W., Evans, J. N., and Davis, G. S. (1982). Deposition and translocation of inhaled silica in rats: Quantification of particle distribution, macrophage participation and function. *Lab. Invest.* **47,** 533–542.

Cai, F.-S., and Yu, C. P. (1988). Inertial and interceptional deposition of spherical particles and fibers in bifurcating airways. *J. Aerosol. Sci.* **19,** 679–688

Chang, L. Y., Overby, L. H., Brody, A. R., and Crapo, J. D. (1988). Progressive lung cell reactions and extracellular matrix production after a brief exposure to asbestos. *Am. J. Pathol.* **131,** 156–170.

Ingham, D. B. (1975). Diffusion of aerosols from a stream flowing through a cylindrical tube. *J. Aerosol Sci.* **6,** 125–132.

Pich, J. (1972). Theory of gravitational deposition of particles from laminar flows in channels. *J. Aerosol Sci.* **3,** 351–361.

Schum, M., and Yeh, H.-C. (1980). Theoretical evaluation of aerosol deposition in anatomical models of mammalian lung airways. *Bull. Math. Biol.* **42,** 1–15.

Warheit, D. B., Chang, L. Y., Hill, L. H., Hook, G. E., Crapo, J. D., and Brody, A. R. (1984). Pulmonary macrophage accumulation and asbestos-induced lesions as sites of fiber deposition. *Am. Rev. Respir. Dis.* **129,** 301–310.

Warheit, D. B., George, G., Hill, L. H., Snyderman, R., and Brody, A. R. (1985). Inhaled asbestos activates a complement-dependent chemoattractant for macrophages. *Lab. Invest.* **52,** 505–514.

Warheit, D. B., Hill, L. H., George, G., and Brody, A. R. (1986). Time course of chemotactic factor generation and the corresponding macrophage response to asbestos inhalation. *Am. Rev. Respir. Dis.* **134,** 128–133.

Weibel, E. R. (1963). "Morphometry of the Human Lung." Academic Press, New York.

Yu, C. P. (1985). Theories of electrostatic lung deposition of inhaled aerosols. *Ann. Occup. Hyg.* **29,** 219–227.

Chapter 10

Effects of Airway Branch Angle, Branch Point Number, and Gravity on Particle Deposition and Retention

K. E. Pinkerton
C. G. Plopper
Department of Anatomy, School of Veterinary Medicine, and Institute for Environmental Health Research
University of California, Davis
Davis, California 95616

C. P. Yu
Department of Mechanical and Aerospace Engineering
State University of New York at Buffalo
Buffalo, New York 14260

I. INTRODUCTION

It is well recognized that inhalation of asbestos fibers constitutes a significant human health risk. Lung injury associated with inhalation of asbestos is typically a patchy diffuse form of fibrosis which tends to be more extensive in the basilar regions of the lungs, particularly in workers occupationally exposed to asbestos dust. The cause for this pattern of fibrosis is assumed to be due to a greater fiber burden in the lower portions of the lung. This deposition pattern is thought to be a result of gravitational effects of particles having a discrete mass. In addition, the more linear path to the lung parenchyma which intrapulmonary airways take to reach the basilar lung regions may be the preferential route for passage of particles carried in the airstream.

A number of investigators have examined the distribution of fibers in the lungs of workers occupationally exposed to asbestos (Ashcroft and Hepleston, 1973; Davis *et al.,* 1978; Sebastien *et al.,* 1977; Churg, 1980; Churg and Warnock, 1980; Morgan and Holmes, 1984). Wide variations were found in fiber concentration, even within adjacent tissue blocks. However, in all of these studies the importance of branch angle, path length, and branch point number of the pulmonary airways was not examined, and in most instances, no correlation between fiber concentration and parenchymal tissue response was made.

The purpose of this study was to address these specific questions using an animal model of asbestosis. The primary focus was to understand airway characteristics which may influence the deposition and subsequent retention of asbestos fibers within the lung parenchyma. A second objective was to correlate asbestos fiber concentration to the response of parenchymal tissues within each isolated lung region. A third interest was to compare lung regions which must be reached by moving with or against gravity to determine whether or not gravity influences asbestos fiber size distribution. An earlier published study from our laboratories examined a number of these parameters (Pinkerton *et al.*, 1986). This study extends that work to examine fiber concentrations and alveolar tissue response in regions arising from pulmonary airways of identical path length.

II. MATERIALS AND METHODS

Specific pathogen-free male Fischer 344 rats were exposed to chrysotile asbestos by inhalation for 7 hr per day, 5 days per week for a total of 12 months. Exposures were done in 5.0 m^3-capacity stainless steel chambers. A Timbrell generator was used to aerosolize asbestos. Chamber fiber mass was monitored daily to facilitate the generation of a constant exposure regimen (Pinkerton *et al.*, 1983). The mean chamber concentration during hours of exposure for the study period was 11.36 ± 2.18 mg/m^3. Following the end of fiber exposure, animals were kept in filtered air conditions for one week prior to sacrifice. Animals were anesthetized by intraperitoneal injection of pentobarbital. The lungs were inflation-fixed by intratracheal instillation of 2% glutaraldehyde at a hydrostatic pressure of 20 cm of fixative.

The parenchymal regions identified in Fig. 10-1 were isolated by microdissection of the airways using a pair of fine scissors, forceps, and razor blades with the aid of a dissecting microscope and fiberoptic illumination (Plopper *et al.*, 1983). To facilitate identification of the airway location during the dissection, a corrosion cast of the bronchopulmonary tree (Fig. 10-2) was used as a guide. Due to the monopodial branching system of the rat (Phalen *et al.*, 1978), a binary classification of airway generations based on airway size can be used to identify and follow the same pathways in different animals. Major and minor airway daughter branches arising from

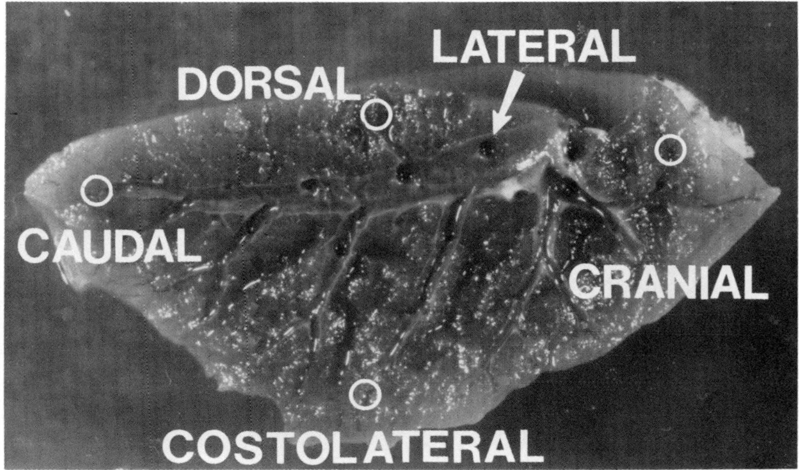

Fig. 10-1 Airway microdissection of the left lung of a Fischer 344 rat. The parenchymal regions isolated for this study are labeled.

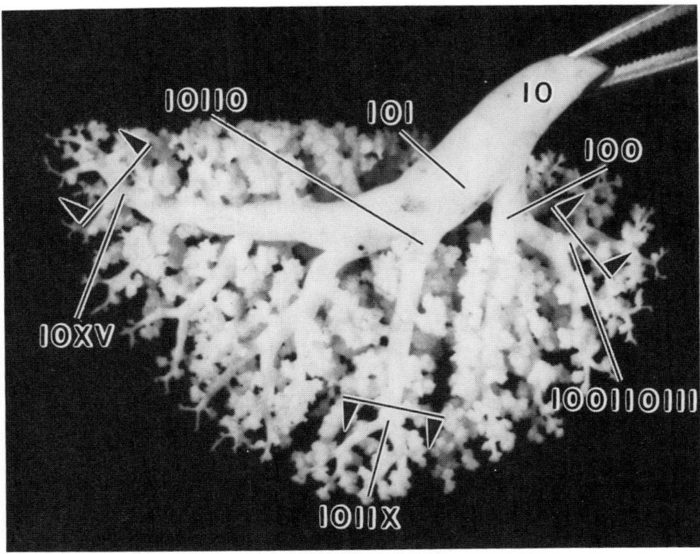

Fig. 10-2 Corrosion cast of the airways of the left lung from a Fischer 344 rat. The binary classification of the airways is shown beginning at the left pulmonary bronchus labeled 10. For each new airway generation followed, the number 0 or 1 is added to denote a minor or major airway respectively. The binary airway classifications to the cranial, costolateral, and caudal regions are shown.

each parent airway are classified as 1 and 0, respectively. Beginning with the trachea, a new number is added for each new airway generation, thus providing the unique branching history of each pathway followed. Using these methods, identical airways were dissected to within 2 to 4 generations of the terminal bronchiole in a total of four animals. Slabs of tissue $1.5 \times 1.5 \times 0.4$ mm cut in a perpendicular orientation to the axis of the last dissected airway were used to measure asbestos fiber concentration and alveolar tissue density in parenchymal tissues arising from each isolated pathway.

Of particular interest in this study was the analysis of parenchymal tissues arising from airways of similar path length, but with different branch angles, branch point numbers, and anatomical locations in the lungs. Three of the five pathways isolated meeting these criteria were the cranial, dorsal, and lateral regions. The costolateral and caudal regions were also included in the analyses performed for purposes of comparison. Table 10-1 summarizes a number of characteristics for each of these airways.

Asbestos fiber concentration was determined by methods previously described (Roggli and Brody, 1984; Pinkerton *et al.*, 1986). Tissue was dissolved in 5.25% sodium hypochlorite, rinsed in distilled water and the fibers were recovered by passing the digested solution through a 0.2 μm Nuclepore filter. The fibers and filter were sputter-coated with a thin layer of gold and viewed under a scanning electron microscope (Fig. 10-3). Randomly selected fields were examined, and fibers having their midpoint within the field were counted until a total of 200 fibers or 100 fields had been examined. The length and diameter of 50 fibers per filter were also measured. From these measurements fiber concentration and fiber size

Table 10-1 Characteristics of Pulmonary Airways Isolated from the Left Lung[a]

Region of Airway Termination	Path Length from Carina[b] (mm)	Cumulative Branch Angle Relative to the Carina[b]	Gravitational Orientation
Dorsal	13.6 ± 1.4	113.3 ± 15.3	Against gravity
Lateral	13.0 ± 0.8	126.7 ± 27.5	None[c]
Cranial	14.0 ± 1.2	225.0 ± 21.8	With gravity
Costolateral	18.3 ± 0.9	68.3 ± 1.4	With gravity
Caudal	22.3 ± 0.3	30.0 ± 8.6	With gravity

[a] Modified from Table 1 and Figures 4 and 5, Pinkerton *et al.*, 1986.

[b] Values are mean $\pm$SD.

[c] This airway path has neither a dorsal nor a ventral orientation in the left lung as do the other microdissected airways. Therefore, this region has not been classified as moving either with or against gravity.

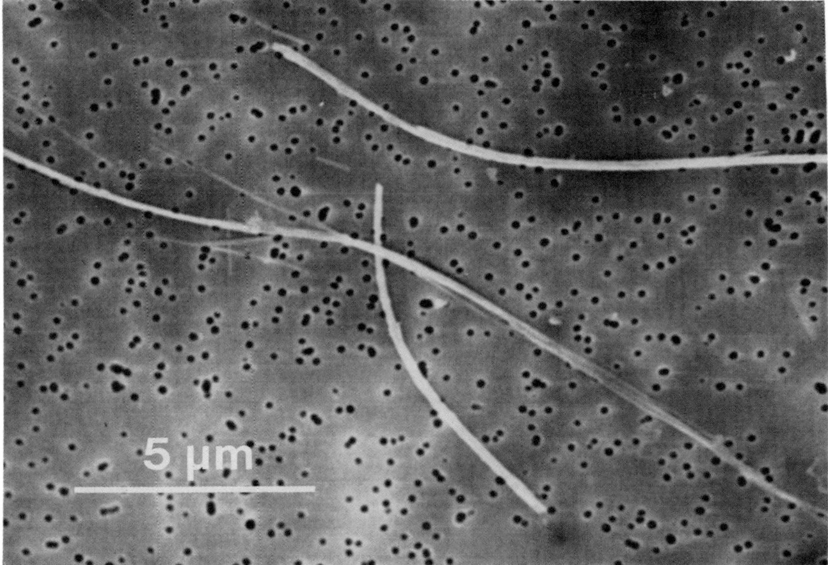

Fig. 10-3 Chrysotile asbestos fibers recovered by lung tissue digestion on a 0.2-μm pore size Nuclepore filter. The diverse size of fibers recovered from the lungs is evident.

distribution could be determined (Roggli and Brody, 1984; Pinkerton *et al.,* 1986).

An adjacent tissue block from each region was dehydrated in a graded series of alcohol and embedded in Epoxy resin. Orientation of the tissue in the open end of a conical beem capsule permitted sectioning of the entire tissue block face. Sections 0.5 μm thick were placed on glass slides and stained with Toluidine Blue. Volume density of the parenchyma was determined with the aid of an eyepiece etched with a 21 line Weibel test lattice (Weibel, 1979). A minimum of 1000 points spaced in an equidistant manner over the entire tissue section were tallied as falling on alveolar air, alveolar tissue, capillary bed, or nonparenchyma (i.e., airways and blood vessels greater than 30 μm in diameter). The ratio of points falling on alveolar tissue to the total number of points falling on the parenchymal regions of the section is directly proportional to alveolar tissue density (Weibel, 1979).

The calculated deposition pattern of asbestos fibers in the rat were based on a deposition model for fibers (Asgharian and Yu, 1987). For this model the airway structure and dimensions of the rat lung were assumed to follow the data of the whole lung model of Schum and Yeh (1980). Although this model does not contain sufficient details for the airway geometry at the lobular level, it does provide an adequate base for deposition calculations from which a better understanding can be derived of the relationship of deposition, airway size, and location.

III. RESULTS

For each of the isolated parenchymal regions of the lung, asbestos fiber number per unit volume of fixed lung tissue and the effect of branch point number on the actual and calculated fiber deposition within parenchymal tissues arising from each airway are shown in Fig. 10-4. The distribution of fibers according to their length in each region is illustrated in Fig. 10-5. As the length is increased, the percentage of fibers longer than the stated length decreases. However, the percentage of fibers greater than 20 μm in length is significantly greater in the cranial region compared to the other lung regions examined.

Alveolar tissue volume density for each region is shown in Fig. 10-6 as a percentage change from control values. The alveolar tissue volume density of the cranial region was significantly different from the lateral region, and the lateral region was significantly different from the dorsal, costolateral, and caudal regions. Due to the greater fiber concentration of the dorsal region, compared to the cranial and lateral regions, the pattern of alveolar tissue volume density changes in these regions was unexpected. Alveolar septal tissue changes of the cranial, lateral, and dorsal regions are illustrated and compared to control lung in Fig. 10-7.

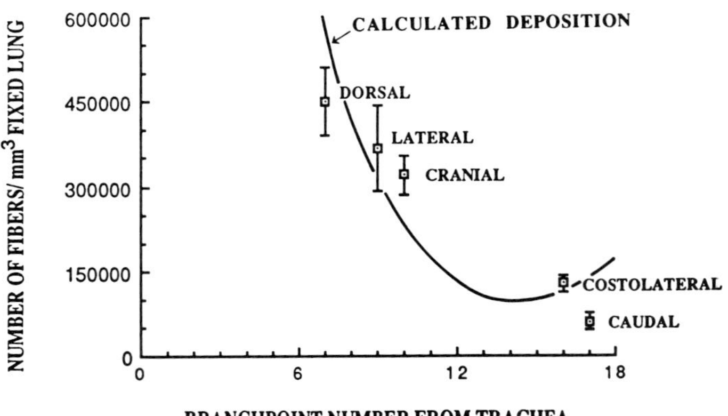

Fig. 10-4 The effect of airway branch point number on asbestos fiber concentration in the lung parenchyma. All observed values are given as means ± 1 SD. The theoretical calculated fiber deposition is given as a line.

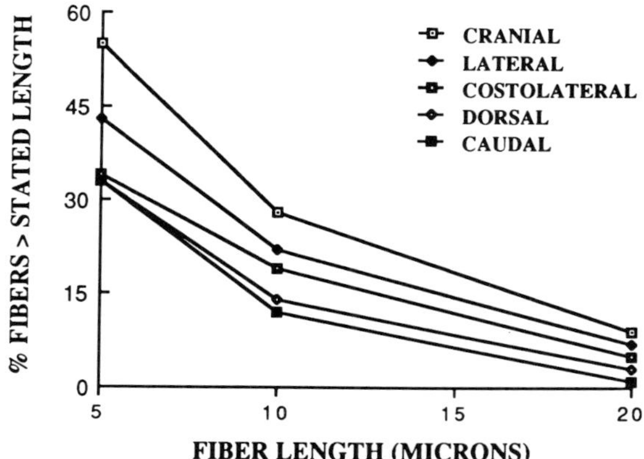

Fig. 10-5 Chrysotile asbestos fiber length distribution within each isolated region of the left lung. Points on each line represent the percentage of fibers greater than the stated length on the x axis. The proportion of fibers greater than 20 μm in length in the cranial region is significantly greater than that found in the remaining isolated regions.

IV. DISCUSSION

Earlier studies in our laboratories have demonstrated that characteristics of the intrapulmonary airways play an important role in the deposition and retention of asbestos fibers in the lungs (Pinkerton *et al.*, 1986). Of the

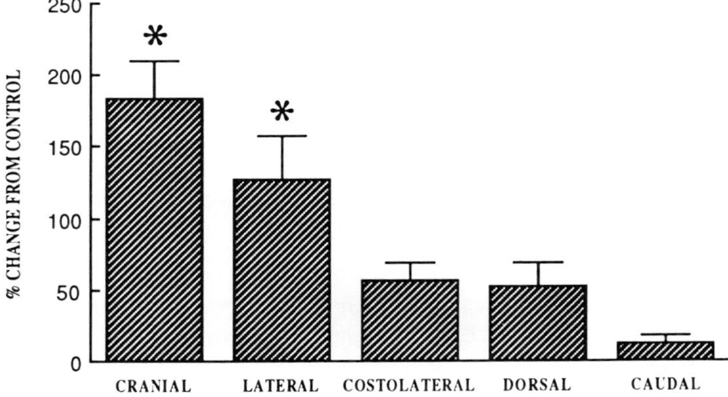

Fig. 10-6 Alveolar tissue density in the lungs of animals chronically exposed to chrysotile asbestos. For each isolated lung region, tissue density is expressed as a percentage change from control (± 1 SD.). Alveolar tissue density is significantly greater in the cranial region compared to the lateral region and the cranial and lateral regions are significantly different (*) from the costolateral, dorsal, and caudal regions.

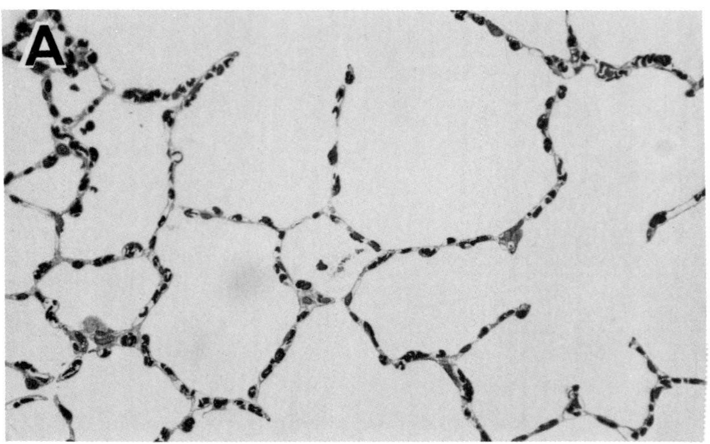

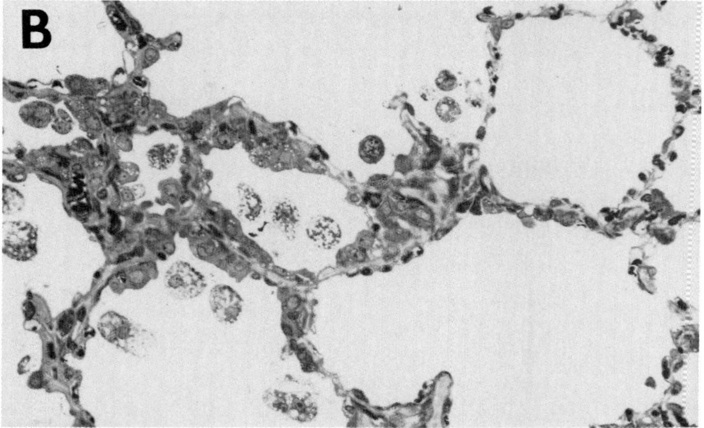

Fig. 10-7 Alveolar septal changes following chronic inhalation of chrysotile asbestos. A, control lung; B, the cranial region; C, the lateral region; and D, the dorsal region. A more diffuse septal thickening is present in the cranial region compared to the lateral and dorsal regions. The scale bar is equal to 100 μm. *(Figure continues)*

various parameters measured, airway path length and branch point number correlated best with asbestos fiber concentration in alveolar tissues arising from anatomically distinct pathways of the bronchopulmonary tree. In this study, these patterns of fiber deposition are compared to theoretical deposition patterns. It was found that the observed fiber concentration in each region was inversely proportional to the number of airway branch points passed to reach the lung parenchyma. Because inhaled fibers deposit on surfaces, particularly at bifurcations as the fibers

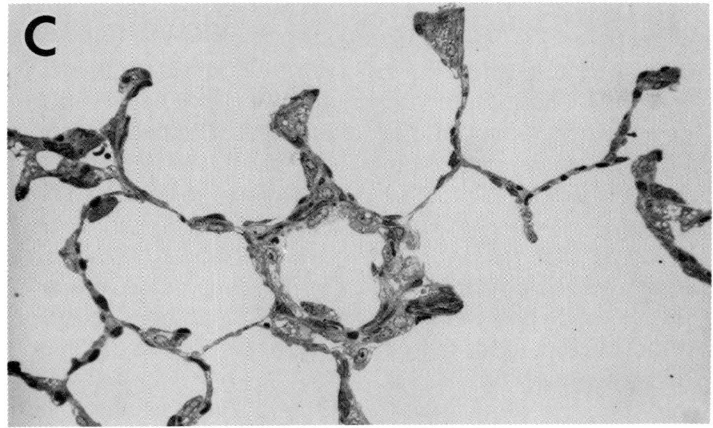

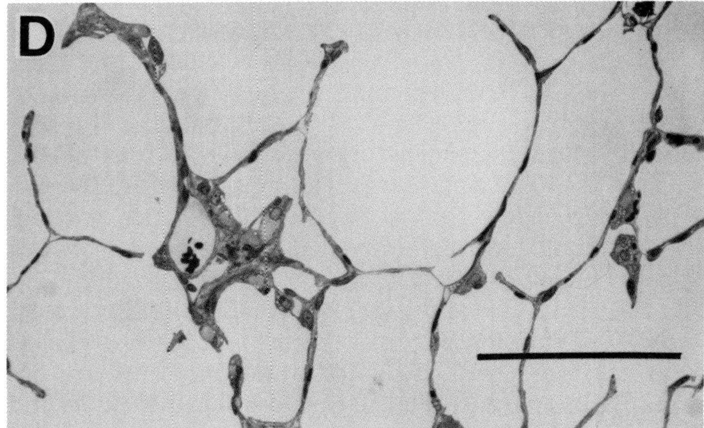

Fig. 10-7 *(continued)*

travel along the airways, and as the surface area increases with increasing branch point number, the deposition of fibers in parenchymal tissues arising from these airways will rapidly decrease. As the branch point number reaches 13, the amount of deposition levels off and then slightly rises due to increases in sedimentation and interception (Asgharian and Yu, 1988). This change is reflected in the calculated fiber deposition line shown in Fig. 10-4. The calculated deposition pattern compares favorably with the observed fiber deposition per unit volume of lung tissue as illustrated in Fig. 10-4.

In our earlier study, a gravitational effect for fiber deposition in small animals was not clearly established (Pinkerton *et al.,* 1986). In this study these same animals were re-evaluated for gravitational effects on particle deposition by analyzing fiber length distribution and alveolar tissue response within three regions: dorsal, cranial, and lateral. The distance from the trachea to the alveolar tissues of each region was similar (Table 10-1), but the distribution of fibers by length (Fig. 10-5) and the alveolar tissue volume density (Fig. 10-6) were significantly different for each region. Due to similarities in airway path length from the tracheal carina, these three regions presented a unique opportunity to examine directly the effects of airway branch angle, branch point number, and gravity on fiber deposition and response. As a further analysis, these three regions were also compared to the costolateral and caudal regions.

The branch point number along airways leading to alveolar tissues plays an important role in parenchymal fiber deposition. The theoretical calculation of fiber deposition by Asgharian and Yu (1988) based on the lung model of Schum and Yeh (1980) suggests that deposition efficiency of fibrous particles is due to a variety of conditions including diffusion, sedimentation, impaction, and interception. They found that impaction of fibers at bifurcations is greatest between airway generations 3 to 13 (Asgharian and Yu, 1988). For parenchymal tissues arising from airways of short path length (and therefore a smaller surface area) and fewer branch point numbers, it would follow that greater numbers of parenchymal asbestos fibers will be found compared to regions arising from airways of longer path length and branch point number.

An unexpected finding was the less severe degree of change in alveolar tissues of the dorsal region compared to the cranial and lateral regions (Fig. 10-6). Asbestos fiber number was greatest in the dorsal region compared to the other regions analyzed (Fig. 10-4). However, the cranial and lateral regions contained significantly greater proportions of fibers greater than 10 μm in length compared to the dorsal region (Fig. 10-5). Fibers in excess of 10 to 20 μm in length are thought to be more injurious than shorter fibers (Vorwald *et al.,* 1951). This model of asbestosis in the rat conforms to this hypothesis.

It is unclear what causes variations in fiber size distribution for different regions of the lung. Of the five regions examined, the dorsal region arises from an airway with the fewest branch point numbers and represents a net movement upward, against gravity, with its dorsal branching from the left main bronchus. It is not apparent that the airway branch point number plays a significant role in the distribution of fibers of given lengths. Parenchymal tissues arising from airways with a larger branch point number, such as the cranial and lateral regions, have greater proportions of long fibers (Fig. 10-6) which are more likely to impact along airway surfaces or be intercepted at branch points along the airways. These findings from the dorsal region of the rat may have important implications to explain the human pattern of asbestosis. Sebastien and colleagues found in a number of workers occupationally exposed to asbestos that greater numbers of asbestos fibers were present within the apical regions of the lungs, while the more extensive parenchymal lesions were evident in the basilar regions (Sebastien *et al.,* 1977). Such variations may be due strictly to a difference in fiber size. Although fiber size distribution was not discussed in the study by Sebastien *et al.,* a selective process for deposition of fibers of various lengths in different regions of the lungs may play an important role in the observed changes of asbestos-induced lung disease in humans.

An important implication of this study is the need to exercise care in analyzing parenchymal lung changes which may be strongly influenced by the deposition of respired particles and/or gases due to the characteristics of the intrapulmonary airways.

ACKNOWLEDGMENTS

The authors thank Dr. James D. Crapo for helpful comments and suggestions and Patricia Tom and Melissa Chechowitz for technical assistance. This work is supported in part by NIH Grants ES 04338, ES 00628, HL 38503, and a grant from the American Lung Association.

REFERENCES

Asgharian, B., and Yu, C. P. (1987). Deposition of fibrous particles in the human lung. Symposium on Extrapolation Modeling of Inhaled Particles and Gases: Lung Dosimetry, Duke University (Abstract).

Asgharian, B., and Yu, C. P. (1988). Deposition of inhaled fibrous particles in the human lung. *J. Med. Aerosol,* **1,** 37–50.

Ashcroft, T., and Heppleston, A. G. (1973). The optical and electron microscopic determination of pulmonary asbestos fibre concentration and its relation to the human pathological reaction. *J. Clin. Pathol.* **26,** 224–234.

Churg, A. (1980). Asbestos fiber content of the lungs in patients with and without asbestos airway disease. *Am. Rev. Respir. Dis.* **127,** 470–473.

Churg, A., and Warnock, M. L. (1980). Asbestos fibers in the general population. *Am. Rev. Resp. Dis.* **122,** 669–678.

Davis, J. M. G., Beckett, S. T., Bolton, R. E., Collings, P., and Middleton, A. P. (1978). Mass and number of fibres in the pathogenesis of asbestos-related lung disease in rats. *Br. J. Cancer* **37,** 673–688.

Morgan, A., and Holmes, A. (1984). The distribution and characteristics of asbestos fibers in the lungs of Finnish anthophyllite mine-workers. *Environ. Res.* **33,** 62–75.

Phalen, R. F., Yeh, H. C., Schum, G. M., and Raabe, O. G. (1978). Application of an idealized model to morphometry of the mammalian tracheobronchial tree. *Anat. Rec.* **190,** 167–176.

Pinkerton, K. E., Brody, A. R., McLaurin, D. A., Adkins, B., Jr., O'Connor, R. W., Pratt, P. C., and Crapo, J. D. (1983). Characterization of three types of chrysotile asbestos after aerosolization. *Environ. Res.* **31,** 32–53.

Pinkerton, K. E., Plopper, C. G., Mercer, R. R., Roggli, V. L., Patra, A. L., Brody, A. R., and Crapo, J. D. (1986). Airway branching patterns influence asbestos fiber location and the extent of tissue injury in the pulmonary parenchyma. *Lab. Invest.* **55,** 688–695.

Plopper, C. G., Mariassy, A. T., and Lollini, L. O. (1983). Structure as revealed by airway dissection: A comparison of mammalian lungs. *Am. Rev. Respir. Dis.* **138,** S4–S7.

Roggli, V. L., and Brody, A. R. (1984). Changes in numbers and dimensions of chrysotile asbestos fibers in lungs of rats following short-term exposure. *Exp. Lung Res.* **7,** 133–147.

Schum, M., and Yeh, H. C. (1980). Theoretical evaluation of aerosol deposition in anatomical models of mammalian lung airways. *Bull. Math. Biol.* **42,** 1–15.

Sebastien, P., Fondimare, A., Bignon, J., Monchaux, G., Desbordes, J., and Bonnaud, G. (1977). Topographic distribution in relation to occupational and non-occupational exposure. *In* "Inhaled Particles IV" (W. H. Walton, ed.), pp. 435–444. Pergamon, Oxford.

Spencer, H. (1977). "Pathology of the Lung." Saunders, Philadelphia.

Vorwald, A. J., Durkan, T. M., and Pratt, P. C. (1951). Experimental studies of asbestosis. *Arch. Ind. Hyg. Occup. Med.* **3,** 1–43.

Weibel, E. R. (1979). "Stereological Methods, Volume 1." Academic Press, New York.

Chapter 11

Regional Deposition of Inhaled Particulates in Conducting Airways and Lung Segments

W. M. Foster
Pulmonary Disease Division
Department of Medicine, School of Medicine
State University of New York, Stony Brook, New York 11794

I. INTRODUCTION

The membrane surfaces of the respiratory tract represent the largest interface between humans and the environment. These surfaces are constantly exposed to a host of particles suspended in the inspired air. The fraction of respired aerosol which deposits within oronasal, pharyngeal, and lower respiratory tract regions has been evaluated and modeled for a broad range of particle sizes and ventilatory patterns (Heyder *et al.,* 1975; Chan and Lippmann, 1980; Martonen, 1983). However, the regional distribution of particles which have deposited within the lower respiratory tract and their potential to interact with, and/or overburden, local defense mechanisms are not well characterized.

Inertial impaction, sedimentation, and diffusion are the primary mechanisms which govern particle deposition in the human lung. Secondary processes include interception, which usually applies to fibers, and electrostatic precipitation for charged particles (Lippmann, 1977). During particle exposures, ventilatory parameters (Smaldone and Messina, 1985) known to affect regional lung deposition include: the depth of the tidal volume which determines aerosol penetrance into small airways; mean inspiratory airflow which is an index of forces governing aerosol impaction onto airway walls; and the duration of the breath cycle which determines respiratory residence time and is influential to sedimentation and diffusion of particles.

Investigations are described which evaluated the effects of exercise, chronic obstructive airway disease, and expiratory airflow limitation on the regional distribution of aerosol deposition. Techniques applied to humans utilized radiolabeled particles and noninvasively assessed their planar distributions within the thorax postinhalation and when applicable, their retention 24 hr postinhalation, as an index of penetration into terminal airways and lung parenchyma.

II. MATERIALS AND METHODS

Indices to Characterize Regional Distribution of Deposited Aerosol

Aerosol composition, solubility, sizing, and labeling techniques have been described (Foster *et al.*, 1980, 1982; Smaldone *et al.*, 1983). The nonpermeant aerosol used in each protocol was radiolabeled with ^{99m}Tc and the deposition within the lung at FRC with subjects in a seated position was assessed immediately following inhalation periods by a posteriorly positioned Anger camera. Subjects were re-imaged 24 hr postinhalation to record residual retention of deposited aerosol. The camera was set with a 15% window around the peak energy of 140 keV and shielded by a parallel hole collimator. For each of the described studies, tidal airflows were measured at the mouth by a Fleisch pneumotachograph and the flow signal was integrated to provide a continuous measure of tidal volume. Initial distributions of deposited aerosol were characterized by computer analysis of the chest images acquired immediately after aerosol inhalation. A region of interest, based on equilibrium scan during ventilation but before aerosol inhalation with radioactive xenon gas (^{133}Xe) was identified covering the posteroanterior image of the lung fields. Using this region, two indices have been developed to quantitate distribution of deposited aerosol.

A central to peripheral index (C/P) (Smaldone and Messina, 1985) divides the region of interest into a central component which contains large central airways and parenchyma (approximately the inner 33% of the region area that surrounds the hilus) and a peripheral component which contains terminal airways and parenchyma (outer 67% of region area). The distribution of counts central and peripheral of the aerosol scan is quantitated as a C/P ratio and is divided by the C/P ratio of equilibrium gas scan counts to normalize for regional volume (Smaldone and Messina, 1985). Thus, for homogeneous deposition of aerosol equivalent to the distribution of regional volume of the central and peripheral regions, the index has a value of 1.0 and when particles preferentially deposit centrally, e.g., in lobar and segmental bronchi over that of smaller airways in the periphery, the value of the index increases above 1.0.

A central fraction index (CF) (Foster *et al.*, 1980) divides the region of interest into four concentric zones surrounding the hilus; each zone encompassed approximately 25% of the region area. An activity profile of counts vs region area was constructed to quantitate the distribution of deposited aerosol within the region of interest. This plot was characterized by the percentage counts contained in the inner 50% zone, referred to as central fraction of the lung deposition. If the distribution of deposited aerosol is increased within central airways, the activity profile has an increased slope and the value of the central fraction index increases.

The 24-hr image was considered to record the maximal clearance of particles from airways with mucociliary function in normal subjects, because particles retained after this time are cleared at such slow rates (20-fold slower) which suggests that their deposition is at other sites (Lippmann *et al.*, 1971; Foster *et al.*, 1976). The amount of counts (corrected for isotope decay) at 24 hr within the lung region of interest, expressed as a percentage of the initial counts, is thus considered to reflect aerosol which penetrated to the lung periphery and deposited within slowly clearing terminal airways and lung parenchyma.

III. EFFECT OF EXERCISE

The distribution of aerosols deposited during steady state breathing at rest and exercise was assessed in healthy nonsmokers (Bennett *et al.*, 1985). The group, comprised of three males and two females had a mean age of 23.4 ± 1.5 (SD) years and pulmonary function within predicted normal values. Subjects breathed via the mouth a 2.6-μm equivalent aerodynamic diameter (EAD) monodisperse aerosol (sigma g <1.1), at rest and while exercising on a bicycle ergometer at 60–70% predicted maximum heart rate (corresponding to a work rate of 80–120 W for these subjects); rest and exercise exposures were accomplished on separate days. Aerosol was inhaled for 5–10-min periods after development of steady spontaneous breathing patterns at rest and during exercise.

During exercise subjects breathed aerosol with increased tidal volume and frequency, such that minute ventilation was raised four- to fivefold above ventilation at rest. Tidal volume more than doubled from an average

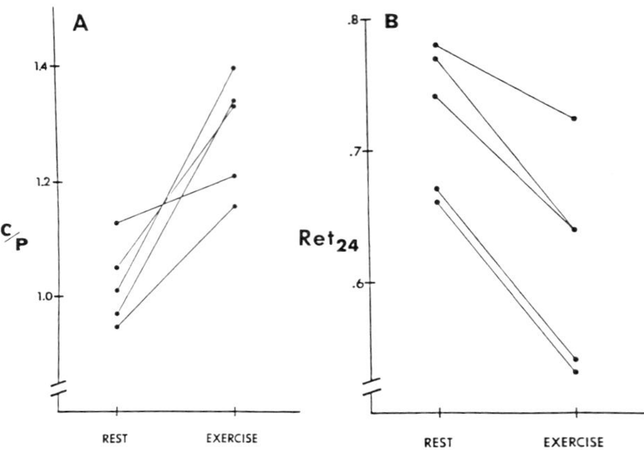

Fig. 11-1 A, Central-to-peripheral distribution of aerosol deposition (C/P) following exposures at rest (group mean $1.02 \pm .07$ SD) and exercise (group mean $1.29 \pm .10$ SD). $p < .01$ by paired analysis. B, lung retention at 24 hr (R_{24}) following exposures at rest (group mean $0.72 \pm .06$ SD) and exercise (group mean $0.61 \pm .08$ SD). $p < .005$ by paired analysis. Reproduced from Bennett *et al.* (1985) with permission of The American Physiological Society.

of 614 to 1582 ml, as did frequency from 13.4 to 31.4/min. Figure 11-1 demonstrates the effect of exercise on C/P and 24-hr retention indices which were utilized to quantify the distributions of deposited aerosol. The C/P index, largely homogeneous at rest (group mean $1.02 \pm .07$ [SD] value) became elevated during exercise, which reflected a shift of particulate deposition into bronchial airways. Simultaneous measure of fractional aerosol deposition (see Bennett *et al.,* 1985) indicated that the fraction of inhaled aerosol deposited per breath during rest was not influenced by exercise, i.e., less aerosol deposited peripherally during exercise, but a coincident increase in deposition occurred centrally. The percentage of particles retained at 24 hr decreased during exercise, suggesting that despite increases in tidal volume, a smaller fraction of aerosol may have penetrated to the lung periphery because of increased flows and impaction of particles on inspiration. By intrasubject analysis, the differences between rest and exercise of C/P and 24-hr indices were significant.

IV. EFFECT OF CHRONIC OBSTRUCTIVE AIRWAY DISEASE

In chronic obstructive lung disease when patients are exposed to test aerosols, lung scans are noted with focal, hyperdeposition of particles. Often these aerosol patterns occur in central lung regions which contain large airways (Foster *et al.,* 1982; Itoh *et al.,* 1981). In progressive airway disease, this effect appears to be independent of aerosol diameter. Ten subjects, similar in age but who were different because of the presence of chronic airway disease, were evaluated for variations in aerosol distribution (Smaldone and Messina, 1985). The group, comprised of five normal subjects at a mean age of 63.8 years with VC, FRC, and DL_{CO} that were on average >90% predicted and a mean $FEV_{1.0}$, expressed as a percentage of FVC of 72.0 ± 6.5 (SD) % and 5 subjects with airway obstruction, mean age of 56.8 years with mean VC, FRC, and DL_{CO} of 73.4 ± 8 (SD) %, 178.0 ± 33 %, and 50.8 ± 6 % predicted, respectively, and $FEV_{1.0}$, expressed as a percentage of FVC of 28 ± 10.4 (SD) %. Tracings of maximal flow–volume and tidal loops are shown in Fig. 11-2. The two groups were separated by the presence or absence of airway flow limitation on expiration, i.e., superimposition of tidal loops on the maximum expiratory flow–volume (MEFV) envelope (see Fig. 11-2); this distinction was associated with the presence of chronic airway disease, which also separated the groups. Most of the obstructed subjects had impairment due to pulmonary emphysema illustrated by low diffusing capacities, e.g., mean DL_{CO} was 50.8 ± 6.4 (SD) % of predicted values. Subjects inhaled by mouth a 1.5-μm EAD monodisperse radiolabeled aerosol (sigma g <1.1) during quiet tidal breathing.

Breathing parameters known to affect aerosol penetration in the lung were comparable for the two subject groups. For example, aerosols were inhaled on average by the normal subjects with 510 ± 139 (SD) ml breaths versus 576 ± 169 ml for flow-limited subjects, and peak inspiratory flows were similar, 1.09 ± .39 (SD) liter/sec and 1.37 ± .25 liter/sec, for normals

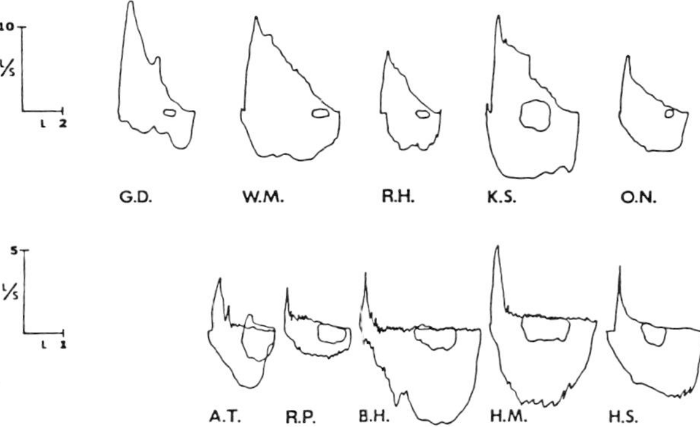

Fig. 11-2 Maximal flow–volume and tidal loops for normal (top) and flow-limited subjects (bottom). Reproduced from Smaldone and Messina (1985) with permission of The American Physiological Society.

Table 11-1 Central-to-Peripheral (C/P) Ratios for Normal and Flow-Limited Subjects

Normal		Flow-Limited	
Subject	C/P ratio	Subject	C/P ratio
GD	1.10	AT	1.81
WM	1.06	RP	2.37
RH	1.06, 0.94[a]	BH	1.69
KS	0.94	HM	2.46
ON	1.02	HS	0.92
Mean	1.02		1.85
±SD	0.07		0.62

[a] Subject studied twice. Reproduced from Smaldone and Messina (1985) with permission of The American Physiological Society.

and flow-limited subjects, respectively. Inspiratory and expiratory residence times were not significantly different between the subject groups.

The C/P index of each subject is presented in Table 11-1. The mean value of C/P index for the normal subjects was near unity, i.e., 1.02 ± .07 (SD), indicative of the homogeneous distribution of deposited aerosol; four of the flow-limited subjects had values which ranged between 1.69 and 2.46, with a resultant mean for the five subjects of 1.85 ± .13, significantly different than the C/P indices of the normal subjects with $p < .05$. Similarity between the groups in factors (aerosol diameter, tidal volume, peak inspiratory flow, and inspiratory and expiratory residence times) capable of shifting sites of preferential aerosol deposition, suggested that centralization of aerosol in the obstructed subjects may have resulted from some aspect of airway disease which separated the two groups, e.g., dynamic collapse and airway flow limitation by major bronchi on expiration.

V. EFFECT OF AIRWAY FLOW LIMITATION

To investigate further the association between airway flow limitation and distribution of particle deposition within central bronchi, healthy subjects inhaled aerosol with maneuvers designed to induce compression of large bronchi and limit airflow on expiration (Foster *et al.,* 1988). The group, comprised of six males and three females with a mean age of 25.1 ± 3 (SD) years had pulmonary function within predicted normal values. Subjects in haled by mouth for 3 to 5 min a monodisperse aerosol between 4 and 6 μm EAD (sigma g < 1.14) on three separate occasions: tidal volume breathing with peak inspiratory, expiratory flow rates <1.0 liter/sec; inspiratory capacity breathing with peak inspiratory, expiratory flow rates <1.0 liter/sec; and inspiratory capacity breathing with peak inspiratory flow rates <1.0 liter/sec, followed by maximal expiratory effort and airflow limitation on expiration (superimposition of tidal expiratory curve onto MEFV curve).

In Table 11-2, data from aerosol exposures with the three breath patterns are presented and include aerosol diameters and breathing volumes for each subject's exposure. Between test differences were not significant in particle diameters inhaled with the three breath patterns, i.e., particles had mean diameters for the three tests of 4.96, 4.94, and 5.06 μm EAD. Differences between volumes (expressed as a percentage of the subject's IC) inhaled with tidal and IC volumes were significant, $p < .005$, with mean values of 29% for the tidal pattern and 69% and 70% for the two IC patterns. Figure 11-3 graphically compares the central fraction indices obtained during tidal and IC breathing at similar inspiratory and expiratory flow rates, i.e., peak flows <1.0 liter/sec. In seven of the eight subjects studied, IC breathing shifted distributions of aerosol deposition peripherally (central fraction index decreased); as the subjects increased their inhalational volume, on average from 29 ± 8 (SD) % to 69 ± 16% of the IC, the distribution of aerosol deposited centrally decreased significantly, on average from 86 ± 5% to 71 ± 7%. Therefore when aerosol was inhaled with

Table 11-2 Inhalation and Deposition Parameters[a]

	Diameter (μm)			Inhalation Volume (% IC)		
Subject	T	IC	IC↑E	T	IC	IC↑E
S1	4.05	4.21	4.44	24	73	70
S2	4.79	4.82	5.12	40	88	58
S3	4.94	4.75	5.08	19	57	95
S4	5.52	5.51	—	38	79	70
S5	5.78	5.86	6.02	35	52	60
S6	—	4.96	4.88	20	80	75
S7		4.72	4.79		40	50
S8	5.34	5.31	5.55	30	85	95
S9	4.32	4.30	4.62	27	65	55
Avg	4.96	4.94	5.06	29	69	70
±SD	.6	.5	.5	8	16	17

[a] Parameters listed are for tidal (T), inspiratory capacity (IC), and inspiratory capacity with forced exhalation (IC↑E) breathing. Diameters of particles are listed as equivalent aerodynamic diameter (sigma g <1.14); inhalation volumes of respective tests are presented as a percentage of the subject's inspiratory capacity.

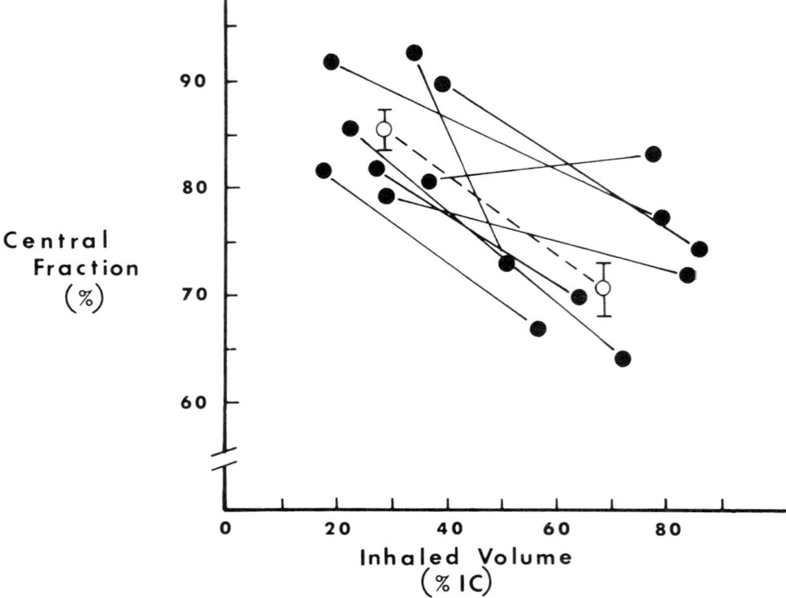

Fig. 11-3 Relationship between inhaled volume and central fraction index. Filled circles represent individual central fraction values (expressed as a percentage of aerosol deposited) for particles inhaled with tidal and inspiratory capacity, volume breathing. Solid lines connect central fraction values observed in an individual at the indicated inhalation volumes (expressed as a percentage of the subject's IC). Open circles represent the mean central fraction ±SEM, for tidal and IC breathing of all subjects; the mean values are connected by a dashed line. Differences between values of central fraction for two breathing volumes were significant, $p < .01$.

larger volumes but without increased flows and frequency (breaths/min) which occur during exercise (see Section III), particle penetration to the lung periphery increased and distributions of deposited aerosol shifted peripherally into small airways and lung parenchyma.

Figure 11–4 compares activity profiles for a single subject who inhaled aerosol with low flows and isovolumes (IC breaths) but exhaled with either low flows, <1.0 liter/sec, or maximal expiratory effort. On the forced exhalation test day, the curve of the activity profile shifted up and to the left (with an almost three-fold increase in slope) indicative of a more central pattern of aerosol distribution. This is demonstrated in the superimposed aerosol scans for forced exhalation in which hyperdeposition of aerosol is visualized within hilar regions of the lung that contain lobar and segmental bronchi. Central fraction indices for deposition patterns when aerosol was expired with either low flows or maximal expiratory effort are compared in Fig. 11–5. All nine subjects had redistribution of aerosol into

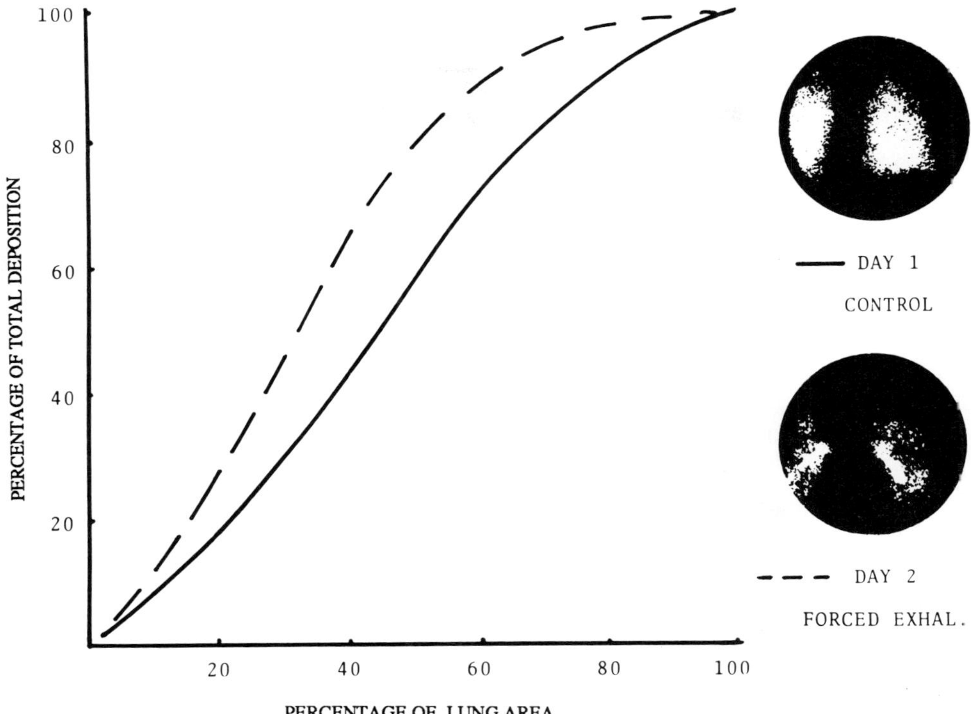

Fig. 11-4 Influence of maximal expiratory effort on cumulative activity profile. Relationship of percentage deposition to percentage area of region of interest on two study days: subject inhaled particles with normal flows and iso-volumes, but exhaled on day 1 (solid line, control profile) with normal effort, and on day 2 (dashed curve, forced exhalation profile) with maximal effort. Shift of dashed profile up and to left is indicative of more centralized distribution of aerosol deposition. Deposition images, acquired posteriorly, of right lung fields visually display hyperdeposition of aerosol, centrally distributed about the hilus on day 2 compared to the homogeneous, diffuse distribution of day 1.

hilar regions (central fraction index increased) during expiratory flow limitation as compared to aerosol inhaled with near identical volume (listed in Table 11–2) and flow but exhaled at low flow. This change was significant, $p < .01$.

VI. CONCLUSIONS

The distribution of aerosol deposited within the lung during mouth breathing was influenced by the depth of the inhalational volume, flow rates, obstructive airway disease, and expiratory flow maneuvers known to

exceeded by deposition of aerosol within central bronchi during the expiratory limb of the breath cycle. For exposures with expiratory maneuvers, deposition appears enhanced in regions which contain airways of the second, third, and fourth generations of the bronchial tree (see aerosol scans in Fig. 11-4); these same airways are known to undergo compression and limit maximal flow achieved over lung volumes from 75 to 25% of the VC during maximal expiratory flow (Macklem and Wilson, 1965) or cough (Leith, 1985) maneuvers. Increased deposition of particles at sites of flow limitation or downstream (toward the trachea) by impaction would explain the observed shift in particle distributions. In animal models where locations of flow limitation could be measured directly, particle deposition was enhanced during forced exhalation at these sites (Smaldone *et al.*, 1979).

The significance of these relationships is that increased ventilation during physical activity and expiratory airflow limitation (which occurs during cough, a primary response to respiratory irritants), events common to daily living, and the prevalence of obstructive lung disease (70,000 Americans succumbed to chronic obstructive airway disease in 1985) may predispose a given individual to increased risk during exposure to an airborne hazard. A persistent question related to regional aerosol deposition is whether a causal relationship exists between inhalational hazard and regional lung injury and/or airway disease. Chronic re-exposure to respiratory irritants and carcinogens with repetitive, preferential deposition within central bronchi may contribute to the incidence and predilection of certain carcinomas for tracheobronchial airways (Moersch and McDonald, 1953). The viewpoint that centralization of particulate deposition may be beneficial, because it would serve to deposit hazardous materials onto mucosal surfaces of central airways which have rapid clearance times as compared to respiratory bronchioles and alveolar ducts, is uncertain. Recent investigations of bronchial mucociliary clearance suggest that airflow limitation, as well as airway pathology of chronic obstructive lung disease, uncouple continuous mucus flow through central airways and prolong particulate retention times within these same airways (Foster *et al.*, 1982, 1988).

ACKNOWLEDGMENTS

The views expressed in this chapter are the author's; they do, however, reflect discussions with colleagues, W. D. Bennett and G. C. Smaldone during and after the symposium. Research was supported by the Veterans Administration and NIH Grant No. HL31429, Washington, D.C.

REFERENCES

Bennett, W. D., Messina, M. S., and Smaldone, G. C. (1985). Effect of exercise on deposition and subsequent retention of inhaled particles. *J. Appl. Physiol.* **59,** 1046–1054.

Chan, T. L., and Lippmann, M. (1980). Experimental measurements and empirical modelling of the regional deposition of inhaled particles in humans. *Am. Ind. Hyg. Assoc. J.* **41,** 399–409.

Foster, W. M., Bergofsky, E. H., Bohning, D. E., Lippmann, M., and Albert, R. E. (1976). Effect of adrenergic agents and their mode of action on mucociliary clearance in man. *J. Appl. Physiol.* **41,** 146–152.

Foster, W. M., Langenback, E. G., and Bergofsky, E. H. (1980). Measurement of tracheal and bronchial mucus velocities in man: Relation to lung clearance. *J. Appl. Physiol.* **41,** 146–152.

Foster, W. M., Langenback, E. G., and Bergofsky, E. H. (1982). Lung mucociliary function in man: Interdependence of bronchial and tracheal mucus transport velocities with lung clearance in bronchial asthma and healthy subjects. *Ann. Occup. Hyg.* **26,** 227–244.

Foster, W. M., Langenback, E. G., Smaldone, G. C., and Bergofsky, E. H. (1988). Flow limitation on expiration induces central particle deposition and disrupts effective flow of airway mucus. *Ann. Occup. Hyg.* **32**(S1), 101–111.

Heyder, J., Armbuster, L., Gebhart, J., Gren, E., and Stahlhofen, W. (1975). Total deposition of aerosol particles in the human respiratory tract for nose and mouth breathing. *J. Aerosol Sci.* **6,** 311–328.

Itoh, H., Ishii, Y., Maeda, H., Todo, G., and Smaldone, G. C. (1981). Clinical observations of aerosol deposition in patients with obstructive lung disease. *Chest* **80S,** 837–840.

Leith, D. E. (1985). The development of cough. *Am. Rev. Respir. Dis.* **131S,** S39–S42.

Lippmann, M. (1977). Regional deposition of particles in human respiratory tract. *In* "Handbook of Physiology Section 9, Reactions to Environmental Agents" (D. H. K. Lee, ed.), pp. 213–232. Williams & Wilkins, Baltimore.

Lippmann, M., Albert, R. E., and Peterson, H. J. (1971). The regional deposition of inhaled aerosol in man. *In* "Inhaled Particles III" (W. H. Walton, ed.), pp. 105–120. Unwin, London.

Macklem, P. T., and Wilson, N. J. (1965). Measurement of intrabronchial pressure in man. *J. Appl. Physiol.* **20,** 653–663.

Macklem, P. T., Fraser, R. G., and Brown, W. G. (1965). Bronchial pressure measurements in emphysema and bronchitis. *J. Clin. Invest.* **44,** 897–905.

Martonen, T. (1983). Deposition of inhaled particulate matter in upper respiratory tract, larynx and bronchial airways: Mathematical description. *J. Toxicol. Environ. Health* **12,** 787–800.

Moersch, H. J., and McDonald, J. R. (1953). The significance of cell types in bronchogenic carcinoma. *Dis. Chest* **23,** 621–633.

Smaldone, G. C., and Messina, M. S. (1985). Flow limitation, cough, and patterns of aerosol deposition in humans. *J. Appl. Physiol.* **59,** 515–520.

Smaldone, G. C., Itoh, H., Swift, D. K., and Wagner, H. N. (1979). Effect of flow-limiting segments and cough on particle deposition and mucociliary clearance in the lung. *Am. Rev. Respir. Dis.* **120,** 747–758.

Smaldone, G. C., Itoh, H., Swift, D. L., Kaplan, R., Florek, R., Wells, W., and Wagner, Jr., H. N. (1983). Production of pharmacologic monodisperse aerosols. *J. Appl. Physiol.* **54,** 393–399.

Chapter 12

The Physical Properties of Mainstream Cigarette Smoke and Their Relationship to Deposition in the Respiratory Tract

B. J. Ingebrethsen
Research and Development
Bowman Gray Technical Center
R. J. Reynolds Tobacco Co.
Winston-Salem, North Carolina 27102

I. INTRODUCTION

Mainstream cigarette smoke is a high-number-density aerosol containing liquid particles in the 1-μm diameter size range. Studies of the deposition efficiency of dilute nonhygroscopic model aerosols with particle size similar to cigarette smoke have developed a reasonably consistent description of the deposition process based on lung physiology, respiration patterns, and aerosol particle mechanics (Heyder *et al.,* 1986). These studies predict a mass deposition efficiency in the 20% range for particles the size of cigarette smoke. The predictions are not born out by studies of mainstream smoke mass deposition, which report efficiencies in the 20–90% range, Table 12-1. In the latter experiments much of the spread is attributed to the different methods employed to measure retention, but the results are generally higher than what one would expect on the basis of the model aerosol research.

While a partial explanation for the discrepancy between the smoke and model aerosol studies may lie in the differences between the controlled breathing patterns used in the model aerosol studies and the puffing maneuver a smoker employs, it has also been suggested (Phalen, 1984; Prodi

Extrapolation of Dosimetric Relationships for Inhaled Particles and Gases

Table 12-1 Some Reported Mainstream Smoke Particulate Mass Deposition Efficiencies

Inhaled %[a]		Mouth %[b]		
Range	Average	Range	Average	Reference
22–75	47	—	—	Hinds *et al.* (1983)
—	96	—	16	Dalhamn *et al.* (1968)
73–98	88	48–78	67	Baumberger (1923)
63–89	84	—	—	Polydorova (1961)
—	82	—	37	Beekmans (1965)
68–90	82	24–51	37	Mitchell (1962)

[a] Smoke inhaled normally and exhaled.
[b] Smoke held only in mouth and exhaled.

and Mularoni, 1984) that physical phenomena, unique to the dense volatile cigarette smoke aerosol, may be involved. The objective of this article is to review research on the physical properties of cigarette smoke related to the explanations for the discrepancy between reports of model aerosol and smoke deposition. The discussion will focus on the particle size and fluid dynamic behavior of smoke during inhalation and on deposition mechanisms other than those that are usually considered for dilute nonhygroscopic model aerosols.

II. PHYSICAL PROPERTIES OF CIGARETTE SMOKE

A. Particle Size Distribution

Fresh mainstream smoke is an extremely dynamic system due to the high-number concentration and volatile nature of the particulate matter. Consequently, many of the particle sizing techniques employed for more stable aerosols are not appropriate for studying fresh mainstream smoke. Nevertheless, a wide range of methodologies has been applied in attempts to determine the particle size distribution of mainstream smoke. Represented in these studies (Ingebrethsen, 1986a) are not only a variety of analytical techniques, but a broad range of aging times and dilution levels. As a result, reported average diameters vary from a number mean of 0.18 μm to a mass median aerodynamic diameter (MMAD) of 0.96 μm. Some of the more reliable measurements where attempts were made to avoid coagulation during aging and evaporation during dilution (Keith, 1982; Okada and Matsunuma, 1974; Ingebrethsen, 1986b) give a more consistent picture of a number mean diameter in the .20 to .30 μm range.

To obtain a more reliable measurement of the size distribution and number concentration of fresh undiluted mainstream smoke, a light scattering procedure was specially designed (Ingebrethsen, 1986b). By positioning a photometer at the mouth end of the cigarette and aerodynamically focusing a smoke sample to a fine stream, the need for dilution was eliminated and the aging time minimized. With this method, time-resolved (20-msec intervals) size distribution and number concentration measurements of 0.1-sec aged, undiluted smoke were obtained during standard puffs. A summary of the puff-averaged number mean diameters for a series of cigarettes is in Table 12-2. Over a range of puff numbers and cigarette ventilation levels, a number mean diameter of 0.21 to 0.29 μm was found, corresponding to a MMAD of 0.30 to 0.40 μm.

These results verify that fresh mainstream smoke particle size falls in the 0.1-μm range, with insignificant mass greater than 1.0 μm, and that discrepancies between smoke and model aerosol deposition studies cannot be traced to inaccurate estimates of fresh smoke particle size.

B. Coagulation

The number concentration of fresh mainstream smoke has been reported to be as high as 10^{10} particles/cm^3 (Okada and Matsunuma, 1974). At number concentrations of this order, coagulation is expected to proceed rapidly resulting in a decrease in number concentration and an increase in average particle diameter. During the smoking process, undiluted mainstream smoke is first drawn into the oral cavity without respiratory tract

Table 12-2 Number Mean Puff Average Diameter (μm)

		Filter Cigarette		
Puff	Nonfilter Cigarette	0% Ventilation	46% Ventilation	84% Ventilation
1	0.24	0.24	0.27	0.24
2	0.24	0.24	0.27	0.25
3	0.24	0.24	0.26	0.25
4	0.24	0.25	0.26	0.25
5	0.24	0.25	0.26	0.25
6	0.24	0.24	0.26	0.25
7		0.24	0.26	0.25
8		0.24	0.25	0.25
9		0.24	0.25	0.24
10				0.23
11				0.22

expansion, followed by inhalation and dilution of the smoke with the additional inhaled air (Hinds *et al.,* 1983). Coagulation is expected to proceed rapidly in the oral cavity and more slowly after dilution during inhalation. The question relevant to the present discussion is whether or not the increase in particle size due to coagulation of the undiluted smoke in the oral cavity is sufficient to influence the expected deposition compared to that for fresh smoke.

Estimates of the extent of coagulation in the oral cavity can be made with a knowledge of the particle size distribution and number concentration of fresh smoke by performing numerical simulations using coagulation theory. Calculations of this type have been made employing a theoretical treatment of coagulation by Gelbard and Seinfeld (1980). An example of these calculations is illustrated in Fig. 12-1 for an assumed initial number mean diameter of 0.20 μm and three mass concentrations corresponding to 0.5, 1.0, and 2.0 mg of particulate matter per 35-cm^3 puff of smoke. Figure 12-1 indicates that although the average particle diameter increases rapidly and at a higher rate for the more concentrated smoke, the predicted size does not exceed 0.5 μm after 2 sec. The increase in diameter

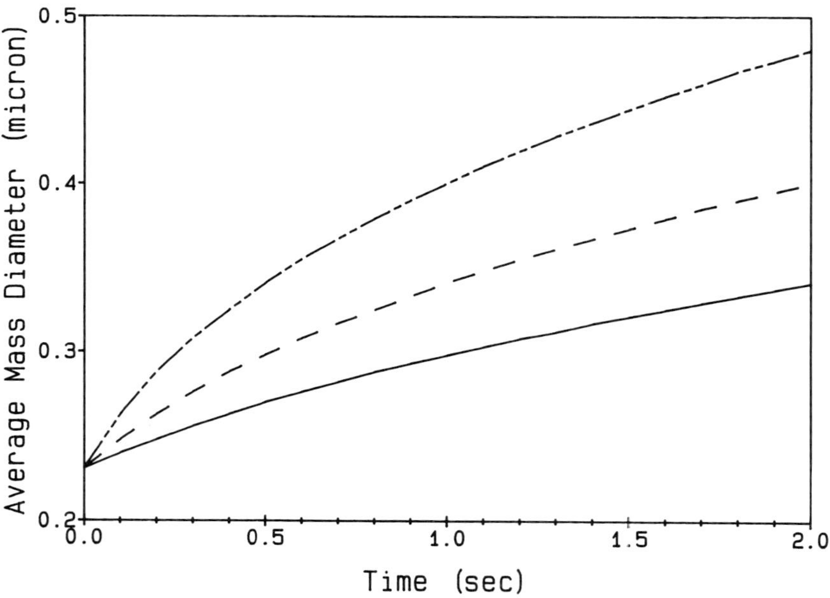

Fig. 12-1 Calculated average mass diameter versus time for undiluted mainstream smoke of mass concentration 2.0 -·-·-, 1.0 ----, and 0.5 — mg particulate matter per 35 cm^3 aerosol volume.

to 0.5 μm would correspond to a situation in which a very concentrated puff was held for a full 2 sec in the mouth before inhalation. The latter smoking procedure represents an extreme case relative to normal smoking behavior and one that would maximize the influence of coagulation. Even for the extreme case, the average particle size is expected to remain in the 0.1-μm size range.

Measurements of the average particle diameter for fresh, mouth puffed, and inhaled mainstream smoke have been made for full flavor [16.5 mg "Tar" Federal Trade Commission (FTC) standard] and low tar (5.5 mg "Tar" FTC) cigarettes. A reverse puff smoking machine was used to force a measured puff of smoke out the mouth end of the cigarette. The smoke exiting the mouth end of the cigarette was either directly captured with an air sampling device or delivered to a smoker who either mouth puffed or inhaled the smoke and exhaled into the collection device. Samples of the smoke were diluted with dry air and delivered as aerosols to an optical particle counter (LAS-X, Particle Measuring Systems) that measured particle size distributions. The high dilution level required for the optical counter introduces the possibility of particle evaporation. However, since all samples were subjected to the same dilution and measurement procedure, the results are informative on a comparative basis and were in good agreement on an absolute basis with other more reliable size measurements.

The experimental results for several subjects are shown in Table 12-3. In all cases, mouth puffing resulted in an increase in particle size compared to fresh smoke. The proportionate increase is smaller for the less concentrated smoke from the low tar cigarette, consistent with the hypothesis that

Table 12-3 Average Mass Diameter in μm from Mainstream Smoke

Cigarette[a]	Subject	Machine Smoking[b]	Mouth Puffing[b]	Inhaled[b]
M	1	.287 (.0079)	.412 (.0111)	.354 (.0094)
L	1	.284 (.0080)	.367 (.0092)	.322 (.0086)
M	2	.288 (.0087)	.383 (.0120)	.353 (.0156)
L	2	.285 (.0057)	.345 (.0108)	.331 (.0087)
M	3	.289 (.0040)	.405 (.0095)	.345 (.0108)
L	3	.284 (.0069)	.358 (.0088)	.322 (.0062)
M	4	.273 (.0065)	.403 (.0317)	.326 (.0104)
L	4	.284 (.0031)	.348 (.0099)	.302 (.0214)
M	5	.283 (.0052)	.389 (.0184)	.348 (.0116)
L	5	.286 (.0046)	.366 (.0094)	.328 (.0056)

[a] M, Moderate; L, low-delivery cigarette.
[b] Standard deviations in parentheses.

coagulation is the cause of the size increase. The relative increase in average particle diameter for both cigarettes and for all subjects is substantially less than that for the extreme case treated theoretically above. Again, one is led to the conclusion that coagulation during inhalation does not cause a shift in particle size of fresh mainstream smoke to a significantly more efficiently deposited size range.

C. Electrical Charge

It has been suggested that mainstream smoke may carry a high level of particle electrical charge and that enhanced deposition due to electrostatic effects may account for some of the discrepancy between model aerosol and smoke deposition studies described above (Phalen, 1984). While other authors have stated that electrostatic deposition is probably a minor contributor to aerosol deposition in humans in general (Fraser, 1966; Melandri *et al.,* 1977), an effect would only be expected for charge levels well above the low equilibrium charge level all aerosols will acquire by natural charging mechanisms (Chan and Yu, 1982). A condensation aerosol like cigarette smoke is not expected to carry a high degree of charge as would be expected for aerosols generated by powder dispersion or atomization methods, and a high level of charge, if present, would be expected to be dissipated by the rapid coagulation the dense smoke undergoes after formation and before inhalation.

Norman and Keith (1965) have measured the charge on mainstream smoke using an electrostatic classification apparatus. They found that mainstream smoke carries a charge level roughly equal to or below that expected for equilibrium charging. It would appear that enhanced deposition of mainstream smoke due to electrostatic effects is unlikely.

D. Growth by Water Condensation

The effect of hygroscopic particle growth on aerosol deposition in humans has been the subject of numerous theoretical and experimental investigations (Persons *et al.,* 1987). Direct measurements of model aerosol particle growth in tubes at high humidities have been reported by several authors (Scherer *et al.,* 1979; Martonen, 1982; Hicks *et al.,* 1986) as have *in vivo* measurements of the growth of model aerosols at the mouths of subjects (Anselm *et al.,* 1986). While these studies provide a basic understanding of the growth process of certain types of particles, particularly atmospheric salts, the response of cigarette smoke particles to high humidities has been less frequently studied and consequently is not well understood. Attempts have been made to estimate indirectly growth factors for cigarette smoke particles from deposition efficiency measurements, estimates of fresh smoke particle size, and theoretical size versus deposition efficiency relationships (Hicks *et al.,* 1986; Porstendorfer and Schraub, 1972). In addition to hygroscopic growth, the other mechanisms discussed in this report can contribute to deviations of mainstream smoke deposition efficiencies from those expected on the basis of fresh smoke particle size and single-particle mechanics, so that the contribution of hygroscopic growth cannot be unambiguously extracted from deposition efficiency measurements.

Predictions of the rate of growth and the equilibrium size of particles as a function of relative humidity can be made theoretically by simultaneous solution of the equations for particle heat balance, water vapor diffusion rate, and particle surface water vapor concentration (Cocks and Fernando, 1982). Meaningful analyses of this type require, in addition to some general physical constants, a knowledge of the aqueous solution properties of the water-soluble components of the particulate phase. The activity of water in the particle solution is required to calculate the partial pressure of water at the particle surface which determines the rate of diffusion of water to the particle and the equilibrium particle size. For particles of simple inorganic materials (e.g., H_2SO_4, $(NH_4)_2SO_4$, NH_4HSO_4), reliable values for aqueous solution properties are available. For cigarette smoke particles, the situation is considerably less well defined.

The particulate phase of mainstream smoke contains water-soluble compounds numbering in the thousands, making calculations of water activity from measurements on pure constituent components an intractable problem. Analyses of filter or impactor deposits have been used to estimate the solution properties of cigarette smoke particulate phase (Ishizu *et al.,* 1980; Cinkotai, 1968; Kousaka *et al.,* 1982). This latter approach involves determining the water-soluble fraction of the deposits through extractions and gravimetric measurement of the extent of water uptake at equilibrium as a function of relative humidity. While these experiments are informative, they do not directly address the physical system of interest, the fresh smoke aerosol in the lung. In the aerosol, an equilibrium between the vapor and particulate phases exists not only for water but for numerous other compounds. Disruption of these equilibria by particle collection is expected to alter the solution properties of the condensed phase. The collected samples were aged before analysis, creating the opportunity for chemical changes in the complex condensed phase mixture and further complicating the basis for predicting the properties of the fresh aerosol.

Ishizu *et al.* (1980) have made measurements of mainstream smoke particle growth at relative humidities up to 90% with an ensemble light-scattering procedure. These calculations were made on the aerosol, maintaining the properties of the dispersed system and avoiding the complicating effects introduced by collection of the condensed phase. The experimental results were compared to theoretical predictions based on information obtained from analysis of filter deposits in the manner outlined above. The measured particle growth agreed well with the theoretical analysis within the range of relative humidities studied; the diameter increased by a factor of 1.3 at 90% relative humidity (r.h.). Extrapolation of the theoretical curve to humidities higher than those studied experimentally suggested a sharp increase in size beginning at approximately 98% r.h. and reaching a factor of 1.7 diameter increase at 99.5 r.h.

It can be concluded that mainstream smoke particles are hygroscopic and are expected to increase their average diameter when exposed to the high humidities of the respiratory tract. This increase in size is expected to change the deposition efficiency and pattern from that which would be predicted on the basis of the fresh smoke particle size distribution. The extent of particle growth by water condensation and consequently the significance of this phenomenon's effect on smoke deposition has not been unambiguously defined either by experiment or theoretical analysis.

E. Evaporative Transfer

Comparisons between deposition efficiencies for model aerosols and cigarette smoke must be made using the same method of particulate quantitation, either particulate mass or number concentration. Most smoke studies report deposition on a particle mass basis, while the model aerosol studies most frequently report particle number deposition efficiencies. Volatile cigarette smoke constituents that are assigned to the particulate phase by filter or impactor collection can be deposited by evaporation and vapor adsorption during inhalation. In this case, the condensed particles are not deposited by contact with surfaces but particle mass is transferred through the vapor phase. Since mainstream smoke contains compounds covering a broad range of volatilities (Falk, 1977), evaporative transfer is a potentially important process involved in cigarette smoke deposition.

Reports of the speed and efficiency of nicotine deposition during smoking support the hypothesis that evaporative transfer contributes to smoke deposition. While reported particulate mass deposition efficiencies are in the range of 20–90+%, nicotine deposition efficiencies are consistently reported in the 90+% range (Greenberg *et al.,* 1952; Artho and Grob, 1964; Isaac and Rand, 1972; Armitage *et al.,* 1974). Nicotine deposition was found to exceed significantly particulate deposition in experiments where the two were measured simultaneously (Landahl and Tracewell, 1957). Although there are other possible explanations for these results, e.g., particle size-dependent composition with size-dependent deposition, evaporative transfer of even a relatively high boiling (b.p. = 245°C) compound like nicotine seems a more likely mechanistic cause.

The size measurements of exhaled smoke listed in Table 12-3 indicate a decrease in particle size as a result of inhalation. This could be explained by the selective deposition of large particles or by the shrinkage of all particles by evaporative transfer from the particulate phase. Taken together with the results reported, the latter explanation seems quite plausible and supports the hypothesis that evaporative transfer contributes to mainstream smoke deposition.

Dalhamn *et al.* (1968) made a comparative study of the deposition rates of several relatively volatile (boiling points 35–125°C) mainstream smoke components. They found a slight increase in deposition efficiency with increased volatility. In the author's laboratory, a study of the deposition efficiencies of eight less volatile smoke components (Table 12-4) has been performed. In these experiments, controlled amounts of fresh mainstream smoke were delivered to subjects who either mouth puffed or inhaled the smoke. The subjects then exhaled through filters for collection. The filters were extracted, and gas chromatographic analysis was performed for the compounds listed in Table 12-5. Deposition efficiencies were determined relative to the amounts of each component surviving the mouth puffing, i.e., deposition efficiency was calculated as 1 − (mass exhaled after inhalation/mass exhaled after mouth puffing). The results of these experiments are summarized in Table 12-5. While all the deposition efficiencies were relatively high and there was considerable inter-subject variability, one does observe a trend consistent among subjects toward lower deposition for the compounds in the order listed in Table 12-5. This approximately corresponds to ordering by decreasing volatility, but other factors, like water solubility, appear to be involved. These data demonstrate the proba-

Table 12-4 Physical Properties of Mainstream Smoke Constituents Studied in Retention Experiments

Compound	Boiling Point °C (Pressure in mm Hg)[a]	Solubility H_2O[a]
Phenol	182(760)	Sol
Nicotine	245(729)	Inf
Triacetin	260(760)	Slig
Propylene glycol	189(760)	Inf
3-Hydroxypyridine	Melts 125 sublimines	Sol
Neophytadiene Phytadiene	N/A 187(14)	I I
Hydroquinone	285(730)	Sol
Glycerol	290(760)	Inf

[a] Inf, soluble in all proportions; Sol, soluble; Slig, slightly soluble; I, insoluble; N/A, not available.

Table 12-5 Deposition Efficiency (% Inhalation Relative to Mouth Puffing) for Selected Mainstream Smoke Constituents for Six Subjects

	Subject Number[a]					
	1	2	3	4	5	6
Phenol	100.0(0.0)	100.0(0.0)	100.0(0.0)	98.6(3.9)	100.0(0.0)	100.0(0.0)
Nicotine	96.5(1.1)	99.7(0.1)	99.2(0.2)	97.6(0.7)	99.9(0.6)	98.4(0.5)
Triacetin	99.0(0.7)	99.6(0.2)	99.4(0.5)	99.4(0.2)	99.2(0.7)	99.3(0.3)
Propylene Glycol	98.8(0.5)	99.8(0.3)	99.5(0.4)	99.0(0.3)	99.0(0.7)	99.5(0.5)
3-Hydroxypyrodine	97.7(0.8)	99.8(0.5)	99.4(1.0)	98.0(0.4)	97.7(0.8)	97.3(1.7)
Neophytadiene	87.9(3.7)	99.5(0.3)	96.6(1.8)	90.0(1.8)	93.9(1.9)	97.5(1.8)
Hydroquinone	82.7(3.5)	99.5(0.9)	96.4(1.5)	86.9(2.8)	91.2(4.2)	92.4(2.7)
Glycerol	73.3(7.1)	99.3(0.7)	92.2(2.5)	78.2(4.8)	86.7(3.7)	91.9(2.9)

[a] Standard deviations in parentheses.

bility that evaporative transfer is a contributing process to mainstream smoke deposition, but they do not define quantitatively the overall contribution of this mechanism to smoke deposition.

F. Cloud Effect

Hydrodynamic interactions between particles in a dense aerosol cloud can entrain the phase surrounding the particles and effectively reduce the resistance of the gas medium to the motion of the particles (Fuchs, 1964). If the cloud is unbounded, i.e., surrounded by particle-free gas, and the magnitude of the hydrodynamic interactions is sufficient, the cloud can move as a single large low-density object rather than (as would be expected) on the basis of the size of the constituent particles and the laws of single particle mechanics. The cloud or mass subsidence effect could conceivably result in fronts of inhaled mainstream smoke moving as large objects which, under the right flow conditions, could be more efficiently deposited than the constituent particles moving independently. The latter mechanism will influence mainstream smoke deposition if during inhalation the smoke is of sufficient density and the flow patterns are such that a cloud of the aerosol can move through a surrounding gas of different density. It is possible that the cloud effect is an important contributing mechanism to mainstream smoke deposition.

Observations of cloud effect behavior for several aerosol systems have been reported (Slack, 1962; Essenhigh *et al.*, 1960; Mitchell *et al.*, 1966; Fuchs, 1964). Included among these are reports of fresh smoke settling at velocities between 1 and 10 cm/sec (Mitchell *et al.*, 1966) which correspond to settling velocities of single particles with a diameter of 18 to 60 μm. Another report determined a smoke particle size of 24 μm from the settling velocity of fresh mainstream smoke (Prosad and Sen, 1938). These diameter estimates are so unreasonably high that it seems certain that fresh smoke settling in still air does display cloud effect behavior. No experimental studies of the cloud effect under conditions relevant to smoke inhalation appear in the literature, but certain predictions of aerosol mechanics combined with descriptions of lung physiology are informative regarding the plausibility of cloud effects during mainstream smoke inhalation.

Fuchs (1964) has developed equations to estimate the importance of the cloud effect by considering a spherical cloud of aerosol in motion through particle-free gas. He treats two cases: (1) the cloud blows through, i.e., the medium flows through the cloud and around each particle; and (2) the cloud does not blow through, i.e., the medium flows around the exterior of the cloud boundary. The forces acting on the cloud for case (1) and (2), F_1 and F_2, are given by

$$F_1 = 6\pi r\eta U 4/3\ R^3 n \tag{1}$$

$$F_2 = \psi\pi R^2 \gamma_g U^2/2 \tag{2}$$

where r = particle radius,
η = viscosity of medium,
U = medium velocity,
R = cloud radius,
n = particle number concentration,
ψ = drag coefficient for cloud radius R, and
γ_g = density of medium.

If $F_1 \gg F_2$, then the cloud cannot be blown through, the medium between the particles is entrained, and the cloud moves as a single entity. When $F_1 \ll F_2$, the cloud blows through and the particles move independently relative to the medium as is predicted by particle mechanics. Fuchs also gives expressions for the settling velocities due to gravity for cases (1) and (2), $Vs1$ and $Vs2$, respectively.

$$Vs1 = (2r^2\gamma p^b)/(9\eta) \tag{3}$$

$$Vs2 = \sqrt{\frac{32Rn\pi r^3\gamma g}{(9\psi\gamma g)}} \tag{4}$$

where γp = density of particle, and
g = gravitational constant.

$Vs1$ is the settling velocity of the constituent particles, and $Vs2$ is the settling velocity of the aerosol cloud acting as a single entity. Equations (1), (3), and (4) are modified for application to mainstream smoke by inclusion of the slip correction factor (C) required in the 0.1-μm size range and by the addition of a term to account for the difference between the density of air and that of the gas phase of mainstream smoke ($\Delta\gamma$). These modifications yield

$$F_1 = 6\pi r\eta U4/3\pi R^3 n/C \tag{5}$$

$$Vs1 = (2r^2\gamma p^g C)/(9\eta) \tag{6}$$

$$Vs2 = \sqrt{\frac{8Rg(4/3n\pi r^3\gamma p + (\Delta\gamma))}{3\psi\gamma g}} \tag{7}$$

With reasonable estimates of respiratory tract dimensions, respiration pattern and mainstream smoke properties F_1, F_2, $Vs1$, and $Vs2$ can be calculated and the influence of the cloud effect during mainstream smoke inhalation estimated.

Cloud effect movement is manifested in systems where the dense aerosol moves through particle-free surrounding air. During mainstream smoke inhalation, appropriate conditions for cloud movement are created at the interface between the inhaled aerosol and the reserve air in the bronchi and bronchioles. As inhaled aerosol enters bronchi filled with reserve air, parabolic velocity fronts of aerosol develop at appropriate combinations of bronchi diameter, length, and flow rate. Whether or not these curved flow fronts act as large low-density objects or the constituent particles act independently will depend on the values of F_1/F_2 and $Vs1/Vs2$. It can be speculated that cloud movement of the curved front could result in impaction of the cloud at bifurcations or in deeper penetration and mixing of the aerosol with the reserve air. In either of the latter two cases, enhanced deposition would be expected compared to inhalation of the same size particles at sufficiently dilute concentrations to eliminate cloud movement. It also seems possible that cloud movement in the upper airways where parabolic flow fronts have not developed could alter deposition from that expected for dilute aerosols.

Calculations of F_1/F_2 and $Vs1/Vs2$ have been made using the Horsfield *et al.* (1971) respiratory tract model and representative values for respiration parameters, smoke particulate properties (drawn from the data cited above), and mainstream gas phase composition (Keith and Tesh, 1965). In order to test the case of maximum possible dilution, a 35-cm^3 puff of fresh smoke was assumed to be fully diluted into the inspired air volume and new dilute particle mass and number concentrations were calculated as were diluted vapor phase mass densities. Fractional development lengths given by

$$L/Le = (34.8L)/(DRe_f) \tag{8}$$

where L = generation length,
Le = development length,
D = generation diameter, and
Re_f = Reynolds number of the flow.

were calculated for each Horsfield generation. L/Le is an estimate of the fraction of a tube's length required for the development of a parabolic velocity profile under laminar flow conditions. Drag coefficients (Ψ) for the cloud fronts were estimated by approximating the front with a sphere of diameter (d) either one-half of or equal to the generation diameter (D) and using the correlation (Fuchs, 1964)

$$\Psi = \left(\frac{24}{Re_p}\right) + \left(\frac{4}{(Re_p)^{\frac{1}{3}}}\right) \tag{9}$$

where $Re_p = Vd_{\gamma C}/\eta$ = particle Reynolds number, and γ_p = density of cloud.

Figures 12-2 and 12-3 show the results of calculations of (L/Le) and

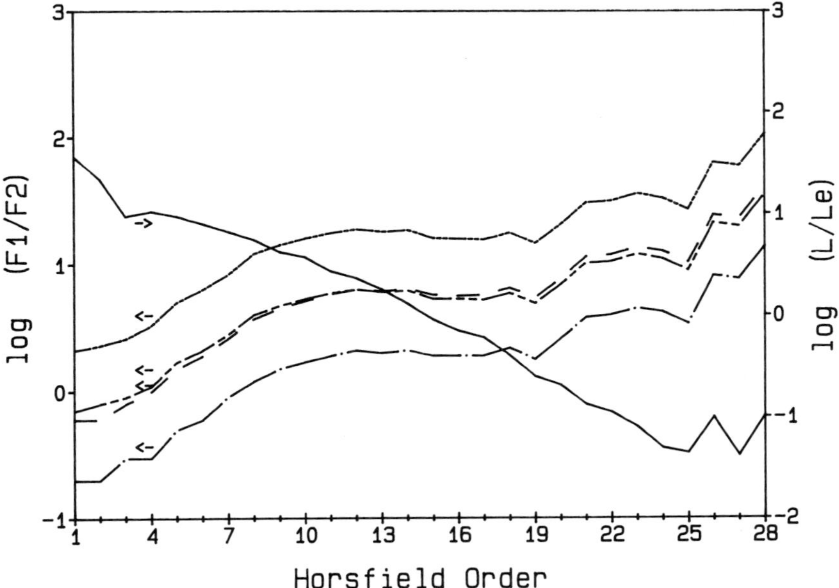

Fig. 12-2 Calculated log base 10 of fractional development length (L/Le) —, and log base 10 of ratio of cloud effect forces (F_1/F_2) for 800 cm³ total volume at 400 cm³/sec. (d/D) (d = cloud front diameter, D = bronchi diameter) and WTPM (particulate mass per 35 cm³ aerosol) values are, respectively, 0.5, 0.5 —·—, 0.5, 1.5 ----, 1.0, 0.5 ———, and 1.0, 1.5 -··-.

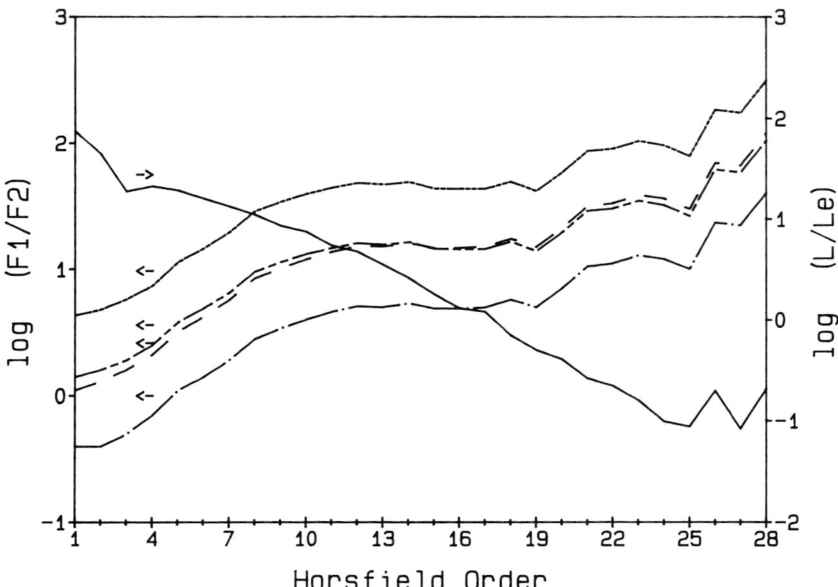

Fig. 12-3 Calculated log base 10 of fractional development length (L/Le) —, and log base 10 of ratio of cloud effect forces (F_1/F_2) for 400 cm^3 total volume at 200 cm^3/sec. (d/D) (d = cloud front diameter, D = bronchi diameter) and WTPM (particulate mass per 35 cm^3 aerosol) values are, respectively, 0.5, 0.5 —·—, 0.5, 1.5 ----, 1.0, 0.5 ———, and 1.0, 1.5 -··-.

(F_1/F_2) plotted on log scales for several combinations of inhalation pattern, cloud front diameter, and smoke particulate mass density. In the Horsfield model, generation 28 represents the trachea and generation 1 the terminal bronchiole. For the two respiration patterns considered, parabolic flows are expected below generations 14–17 at the average flow rates indicated. Actual respiration is, of course, transient and flow rates covering a wide range are experienced during each breathing cycle. The ratio F_1/F_2 decreases with declining generation number but is greater than 1 for most generations in all cases considered, indicating the possibility of cloud movement. In the upper airways, the large F_1/F_2 values suggest that cloud movement may be a factor in deposition at bifurcations even though parabolic flow profiles are not developed. There is an intermediate region in each case where the analysis suggests that curved flow fronts are present and F_1/F_2 values are large enough for cloud movement to be expected. In the lower airways, F_1/F_2 values decrease and cloud movement is predicted to be less likely.

For all cases considered at all generations, $Vs1/Vs2$ values were less than 0.02, indicating that the settling velocity of the aerosol cloud was predicted to be at least 50 times greater than that of the constituent particles. This calculation predicts that at any generation where settling of the aerosol cloud through reserve air is possible, the settling velocity will be substantially greater than that expected on the basis of constituent particle size.

While the calculations discussed above cannot be taken as quantitative predictions of the influence of cloud movement during mainstream smoke inhalation, they do strongly suggest that during mainstream inhalation certain parameters are well within the range necessary to produce cloud movement. Therefore cloud behavior must be included as a possible important factor influencing mainstream smoke deposition.

III. CONCLUSIONS

Six possible explanations, based on smoke properties rather than breathing patterns, have been discussed for the discrepancy between reported mainstream smoke deposition efficiencies and those for dilute nonhygroscopic aerosols of comparable particle size. Three items were shown not to be good explanations for the discrepancy. Careful, artifact-free measurements of the particle size distribution of fresh mainstream smoke indicate a MMAD in the 0.3- to 0.4-μm range. While coagulation is found to enlarge the particle size during inhalation, the increase is not sufficient to move the average size into a range where substantially different deposition efficiencies are expected. Electrical charge, both for intuitive reasons and on the basis of a single reported measurement of the charge on mainstream smoke, is not expected to be an important factor in mainstream smoke deposition.

The latter three items are probable contributors to smoke retention, but each is yet to be unambiguously and quantitatively investigated experimentally. Particle growth by water adsorption during smoke inhalation most certainly takes place, but theoretical analysis has been limited by the complex smoke chemistry and definitive experiments have not yet been performed. Evaporative transfer as a contributing mechanism to smoke deposition is supported by selective deposition experiments and particle shrinkage during inhalation, but its contribution to total particulate mass deposition has not been quantitatively determined. Finally, realistic estimates of smoke properties, respiratory tract dimensions, and inhalation patterns allow calculations indicating that during mainstream smoke inhalation proper conditions exist to suggest that cloud movement may be an important mechanism of mainstream smoke deposition.

REFERENCES

Anselm, A., Gebhart, J., Heyder, J., and Ferron, G. (1986). Human inhalation studies of growth of hygroscopic particles in the respiratory tract. *In* "Aerosols: Formation and Reactivity," pp. 252–255. Pergamon, Oxford.

Armitage, A. K., Dollery, C. T., George, C. F., Houseman, T. H., Lewis, P. J., and Turner, D. M. (1974). Absorption and metabolism of nicotine by man during cigarette smoking. *Br. J. Clin. Pharmacol.* **1,** 180–181.

Artho, A. J., and Grob, K. (1964). Nicotine absorption from cigarette smoke. *Z. Praeventionsmed.* **9,** 14–25.

Baumberger, J. P. (1923). The amount of smoke produced from tobacco and its absorption in smoking as determined by electrical precipitation. *J. Pharmacol. Exp. Ther.* **21,** 47–57.

Beekmans, J. M. (1965). The deposition of aerosols in the respiratory tract. II. Deposition in cigarette smoking. *Can. J. Physiol. Pharmacol.* **43,** 707–714.

Chan, T. L., and Yu, C. P. (1982). Charge effects on particle deposition in the human tracheobronchial tree. *Ann. Occup. Hyg.* **26,** 65–76.

Cinkotai, F. F. (1968). The growth of cigarette smoke particles suspended on fine platinum wire in moist air. *Beitr. Tabakforsch.* **4,** 189–195.

Cocks, A. T., and Fernando, R. P. (1982). The growth of sulfate aerosols in the human airways. *J. Aerosol Sci.* **13,** 9–19.

Dalhamn, T., Edfors, M. L., and Rylander, R. (1968). Retention of cigarette smoke components in human lungs. *Arch. Environ. Health* **17,** 746–748.

Essenhigh, R. H., Buckle, E. R., and Picknett, R. G. (1960). General Discussion. *Faraday Discuss. Chem. Soc.* **30,** 145–147.

Falk, H. L. (1977). Chemical agents in cigarette smoke. *In* "Handbook of Physiology, Sect. 9, Reaction to Environmental Agents," p. 199. *Am. Physiol. Soc.,* Bethesda, Maryland.

Fraser, D. A. (1966). The deposition of unipolar charged particles in the lungs of animals. *Arch. Environ. Health* **13,** 152.

Fuchs, N. A. (1964). "The Mechanics of Aerosols," pp. 46–49. Pergamon, New York.

Gelbard, F., and Seinfeld, J. H. (1980). Simulation of multicomponent aerosol dynamics. *J. Colloid Interface Sci.* **78,** 485–501.

Greenberg, L. A., Lester, D., and Haggard, H. W. (1952). The adsorption of nicotine in tobacco smoking. *J. Pharmacol. Exp. Ther.* **104,** 162–167.

Heyder, J., Gebhart, J., Rudolf, G., Schiller, C. F., and Stahlhofen, W. (1986). Deposition of particles in the human respiratory tract in the size range 0.005–15 μm. *J. Aerosol Sci.* **17,** 811–825.

Hicks, J. F., and Megaw, W. J. (1985). The growth of ambient aerosols in the conditions of the respiratory tract. *J. Aerosol Sci.* **8,** 521–534.

Hicks, J. F., Pritchard, J. N., Black, A., and Megaw, W. J. (1986). Experimental evaluation of aerosol growth in the human respiratory tract. *In* "Aerosols: Formation and Reactivity," pp. 244–247. Pergamon, Oxford.

Hinds, W., First, M. W., Huber, G. L., and Shea, J. W. (1983). A method for measuring the deposition of cigarette smoke during smoking. *Am. Ind. Hyg. Assoc. J.* **44,** 113–118.

Horsfield, K., Dart, G., Olson, D. E., Filley, G. F., and Cumming, G. (1971). Models of the human bronchial tree. *J. Appl. Physiol.* **31,** 207–217.

Ingebrethsen, B. J. (1986a). Aerosol studies of cigarette smoke. *Recent Adv. Tob. Sci.* **12,** 54–142.

Ingebrethsen, B. J. (1986b). Evolution of the particle size distribution of mainstream cigarette smoke during a puff. *Aerosol Sci. Technol.* **5,** 423–433.

Isaac, P. F., and Rand, M. J. (1972). Cigarette smoking and plasma levels of nicotine. *Nature (London)* **236,** 308–310.

Ishizu, Y., Ohta, K., and Okada, T. (1980). The effect of moisture on the growth of cigarette smoke particles. *Beitr. Tabakforsch.* **4,** 189–195.

Keith, C. H. (1982). Particle size studies on tobacco smoke. *Beitr. Tabakforsch.* **11,** 123–133.

Keith, C. H., and Tesh, P. G. (1965). Measurement of the total smoke issuing from a burning cigarette. *Tob. Sci.* **9,** 61–64.

Kousaka, Y., Okoyama, K., and Wang, C. (1982). Response of cigarette smoke particles to change in humidity. *J. Chem. Eng. Jpn.* **15,** 75–76.

Landahl, M. D., and Tracewell, T. N. (1957). An investigation of cigarette smoke as an aerosol with special reference to retention in the lungs in human subjects. *Trans. Ill. State Acad. Sci.* **50,** 213–220.

Martonen, T. B. (1982). Analytical model of hygroscopic particle behavior in human airways. *Bull. Math Biol.* **44,** 425–442.

Melandri, C., Prodi, V., Tarroni, G., Formignani, M., De Zaiacomo, T., Bompane, G. F., Maestri, G., and Maltoni, G. G. (1977). On the deposition of unipolarly charged particles in the human respiratory tract. *In* "Inhaled Particles IV" (W. H. Walton, ed.), pp. 193–201. Pergamon, Oxford.

Mitchell, R. I. (1962). Controlled measurement of smoke particle retention in the respiratory tract. *Am. Rev. Respir. Dis.* **85,** 526–533.

Mitchell, R. I., Gieseke, J. A., and Putnam, A. A. (1966). A study of the settling of concentrated aerosol clouds. Technical Report Air Force Armament Laboratory AFAT2-TR-66-105.

Norman, V., and Keith, C. H. (1965). Charged particles in cigarette smoke. *Tob. Sci.* **9,** 75–79.

Okada, T., and Matsunuma, K. (1974). Determination of particle size distribution and concentration of cigarette smoke by a light scattering method. *J. Colloid. Interface Sci.* **48,** 461–473.

Persons, D. D., Hess, G. D., Muller, W. J., and Scherer, P. W. (1987). Airway deposition of hygroscopic heterodispersed aerosols: Results of a computer calculation. *J. Appl. Physiol.* **63,** 1195–1204.

Phalen, R. F. (1984). "Inhalation Studies: Foundations and Techniques." CRC Press, Boca Raton, Florida.

Polydorova, M. (1961). An attempt to determine the retention of tobacco smoke by means of membrane filters. *In* "Inhaled Particles and Vapours" (C. N. Davies, ed.), pp. 142–144. Pergamon, Oxford.

Porstendorfer, J., and Schraub, A. (1972). Concentration and mean particle size of the main and side stream of cigarette smoke. *Staub-Reinhalt, Luft* **32,** 33–36.

Prodi, V., and Mularoni, A. (1984). Particulate deposition in smoking. *In* "Smoking and the Lung" (G. Cumming, ed.), pp. 249–285. Plenum, New York.

Prosad, K., and Sen, D. (1938). On the trajectory of ash particles in air and the determination of their size. *Philos. Mag.* **25,** 993.

Scherer, P. W., Haselton, R. F., Hanna, L. M., and Stone, D. R. (1979). Growth of hygroscopic aerosols in a model of bronchial airways. *J. Appl. Physiol.: Respir. Environ. Exercise Physiol.* **47,** 544–550.

Slack, G. W. (1962). Sedimentation of a large number of particles as a cluster in air. *Nature (London)* **200,** 1300.

Swift, D. L., Carpin, J. C., and Mitzner, W. (1982). Pulmonary penetration and deposition of aerosols in different gases: Fluid flow effects. *Ann. Occup. Hyg.* **26,** 109–117.

Chapter 13

Localization of 14-C Dotriacontane-Labeled Cigarette Smoke Particulate in the Dog Lung

J. E. McNamee
Department of Physiology
University of South Carolina School of Medicine
Columbia, South Carolina 29208

A. B. Boykin
Department of Medicine
University of South Carolina School of Medicine
Columbia, South Carolina 29208

I. INTRODUCTION

Some forms of pulmonary disease are closely linked to the chronic inhalation of mainstream cigarette smoke. Lung cancer attributable to cigarette smoking is estimated to claim 88,000 lives each year (Surgeon General, 1982). Another 55,000 smokers are believed to die yearly from pulmonary emphysema acquired as a consequence of cigarette smoking (Surgeon General, 1984). While the steps in the initiation of these two lung diseases are becoming better understood, their topography remains unexplained. Smoking-related lung cancer and emphysema are not found uniformly throughout the lung. Each tends to occur in a characteristic portion of lung tissue. In particular, lung cancer shows a predilection for the upper lobes (Lisa *et al.*, 1965; Daly *et al.*, 1984).

One explanation for the increased incidence of cancer observed in superior lung regions may be that relatively more tobacco smoke deposits in apical lung tissues (Pearson *et al.*, 1985). There is a strong correlation between the likelihood of lung cancer or emphysema and the duration and

Extrapolation of Dosimetric Relationships for Inhaled Particles and Gases

intensity of exposure to mainstream cigarette smoke (Doll and Peto, 1978; Higenbottam *et al.*, 1982; Lee and Garfinkle, 1981). Within the lung, it may be that localized deposition of "tar," the particulate phase of tobacco smoke, increases the likelihood of disease.

The hypothesis that apical lung regions receive a disproportionate amount of particulate matter during cigarette smoking was investigated by tracking the regional deposition of cigarette smoke aerosol in the intact lung. Acute human smoking was simulated in an animal model. The deposited mass of tracer–labeled cigarette smoke particulate and its resultant tissue concentration were determined regionally throughout the lung.

II. METHODS

A. Cigarette Preparation

A2 king-size filter cigarettes were selected for study because their yield of 0.9 mg of nicotine and 13 mg of tar is similar to the sales-weighted average yield of cigarettes marketed in the United States (Gori and Lynch, 1985). They were kindly furnished by the National Cancer Institute.

A radioactive tracer was used to track the deposition of the particulate phase of cigarette smoke inside the lung. Dotriacontane, a 32-carbon paraffin normally present in tobacco smoke, is frequently used as a marker for the particulate phase of mainstream cigarette smoke. Jenkins *et al.* (1970) have previously established that 14-C dotriacontane is primarily confined to the particulate phase of tobacco smoke with no more than 10% in the gas phase. Furthermore, 14-C dotriacontane-labeled tobacco produced a mainstream cigarette smoke whose specific activity (14-C activity per g of smoke particulate matter) was nearly constant among puffs. Jones *et al.* (1975) showed that 14-C dotriacontane's specific activity was equal in all cigarette smoke aerosol particles regardless of particle size.

Research Products International (Mount Prospect, Illinois) provided the 14-C dotriacontane. Cigarettes were made radioactive by adding 5–10 μCi of 14-C dotriacontane into their tobacco. Radiolabeled dotriacontane was dissolved in a 10- to 20-μl volume of cyclohexane and ethyl acetate (4 : 1) and injected axially along the cigarettes' tobacco rods with the aid of a Hamilton syringe. The small volume of solvent quickly evaporated leaving the tracer evenly dispersed along the length of the cigarette.

The behavior of 14-C dotriacontane-labeled A2 cigarette smoke was investigated *in vitro.* It was found that approximately two thirds of the radioisotope applied to a cigarette was incorporated into sidestream smoke and one third was incorporated into mainstream smoke. Half of mainstream smoke label was subsequently sequestered by the cigarette's own filter. Smoke particulate was sized using a Marple 6-stage cascade impactor (Andersen Samplers, Atlanta, Georgia) which could resolve particles as small as 0.25 μm in diameter. Within the limits of resolution of this device, approximately 95% of both smoke aerosol (by weight) and 14-C dotriacontane tracer (by activity) were found to be confined to impactor stages corresponding to particle sizes less than 0.5 μm in diameter.

The manner in which a cigarette is smoked directly influences the mass of smoke particulate it delivers (Wynder and Hoffman, 1968). Consequently, all A2 cigarettes were smoked in accordance with the Federal Trade Commission (FTC) method to ensure consistent testing conditions. Briefly, the FTC testing method required that cigarettes were conditioned in an atmosphere of room air maintained at $20 \pm 2°C$ and $60 \pm 5\%$ relative humidity for at least 24 hr before smoking. Cigarettes were smoked by drawing a 35-ml puff of 2-sec duration with a SEPCO syringe pump (Model 40A, Scientific Equipment Products, Baltimore, Maryland). This pump produced puffs having a sinusoidal flow profile. Puffs were drawn once each min until the tobacco rod burned to a distance of 3 mm from the cigarette's filter tipping.

The puff volume employed in the FTC testing method is substantially less than the puff volume smokers normally take from their cigarettes (Tobin and Sackner, 1982). To preserve the use of the FTC puff volume while reproducing a smoke–air mixture comparable to human smoking conditions it was necessary to reduce the volume of air in which cigarette smoke was entrained during inhalation by approximately 40%. A lung of proportionately smaller size than the human lung was also required.

B. *In Vivo* Testing

The anesthetized dog was selected as a model of the smoker. Lungs of a 20-kg dog are approximately half the size of human lungs. We observed that these lungs accommodated inspired volumes of 550 ml from functional residual capacity with respiratory pressures generally less than 15 cm H_2O. An erect smoking posture was adopted to reproduce gravitational effects which may act on the upright human lung during smoking. This posture was achieved in the anesthetized dog by vertically suspending the animal from its forelimbs.

Five animals were anesthetized with 30 mg/kg of pentobarbital sodium. Each animal received a venous catheter for subsequent infusion of anesthetic and sampling of venous blood. A tracheal cannula was inserted through a tracheostomy to permit inhaled smoke to bypass the naso- and oropharyngeal portions of the respiratory tract. Additional anesthetic was given to minimize spontaneous respiratory efforts. Ventilation was subsequently provided by a mechanical respirator (Harvard Apparatus, Cambridge, Massachusetts). Animals were provided with a maintenance level of minute ventilation using a tidal volume of 350 ml and a respiratory rate of 10 breaths/min which in all instances was sufficient to meet gas exchange needs and eliminate spontaneous respiratory movements.

Once a minute, maintenance breathing was interrupted so that a standardized puff of A2 cigarette smoke could be delivered to the tracheal cannula. The pulmonary deposition of an aerosol is known to be a function of both aerosol size and the manner of breathing (Valberg *et al.*, 1982). For this reason it was necessary to establish a breathing pattern for cigarette smoke that was identical throughout all experiments. Acute smoking was simulated by first drawing a 35-ml smoke puff into a 50-ml chamber. Within 0.40 ± 0.02 sec after drawing a puff a second mechanical respirator flushed a 550-ml tidal breath of air through the chamber to entrain the puff in inspired air flow. Inspiration lasted 2.89 ± 0.05 sec.

Cigarette puffing and puff inhalation were coordinated and measured by a microcomputer (APPLE II+, Apple Computer Products, Cupertino, California). The computer also controlled the actions of the maintenance respirator together with accessory valves to direct the flow of air between pumps, the cigarette, and the tracheal cannula. Figure 13-1 diagrams the experimental apparatus used to produce maintenance breathing with periodic inhalation of cigarette smoke.

A total of four cigarettes were smoked in each experiment and the first two contained no tracer. They were used to verify the proper operation of the apparatus and to initiate any acute physiologic reactions which might occur in response to cigarette smoke inhalation. The last two cigarettes were labeled with 14-C dotriacontane.

After smoking both labeled cigarettes, each animal was given an overdose of anesthetic followed in 2 min by an intravenous bolus of saturated potassium chloride solution to arrest the heart. Occlusion of the tracheal cannula held the lungs at resting volume. The chest was opened, the lungs carefully removed and passively drained of blood. The time from the last inhaled puff to the removal of the lungs was always less than 20 min. Nonpulmonary tissue was carefully trimmed from the isolated lungs. Lung weight ranged from 98–151 g. The lungs were placed on a flat support grid taking care to spread and separate adjacent lobes and were then frozen in this position.

In this position, lung regions distant from the tracheal bifurcation were also distant from the large conducting airways. Figure 13-2 illustrates the spatial relationship among lung lobes in a typical experiment. The frozen lungs were sectioned perpendicular to the surface of the support grid forming 2.5-cm squares of lung tissue. Individual lung tissue squares were

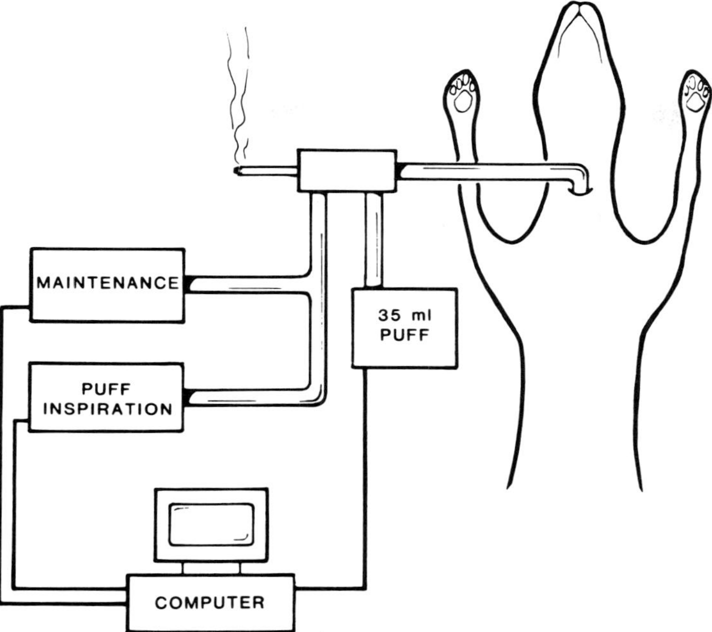

Fig. 13-1 Schematic of the *in vivo* experimental apparatus.

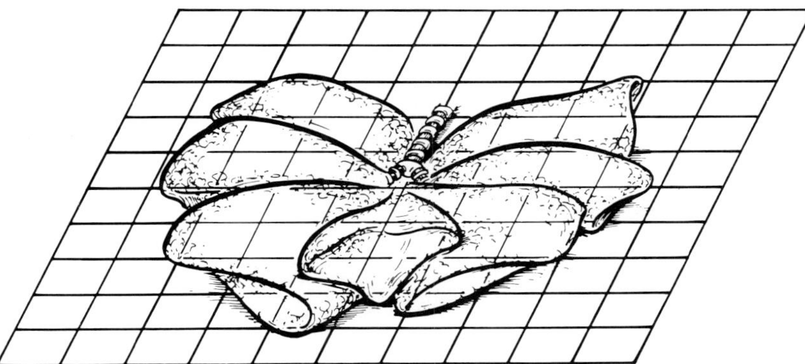

Fig. 13-2 Lungs from a dog as they appeared when spread on the sampling grid. The pleural surfaces visible in this ventral view normally wrap around the heart when the lungs are in the chest.

weighed before analysis. Representative 1-g samples of lung were cut from each square of tissue. Tissue samples were homogenized in 4 ml of cyclohexane to extract the tracer. After centrifugation, the supernatant was removed and assayed for 14-C activity using a liquid scintillation counter (Mark III, Searle Analytical, Des Plaines, Illinois). Preliminary experiments determined that this method recovered 76 ± 9% of 14-C dotriacontane from lung tissue samples. Occasionally, squares weighing more than 5 g were sampled several times to determine the extent of variation that might occur by randomly sampling tissue squares rather than analyzing the entire tissue block.

III. RESULTS

Successive randomly cut 1-g samples of lung tissue from the same lung square indicated there was a sample-to-sample variation of as much as 14%, much of which was presumably due to the process of tracer recovery during homogenization. Subject to this degree of variability, the activities measured in a 1-g sample were assumed to be representative of the tissue contained in the square.

The mass of deposited tracer was calculated for each square of lung by multiplying the mass of the lung in that square by the 14-C activity in the 1-g sample taken from that square. Figure 13-3 shows the spatial distribution of cigarette smoke mass from the lungs pictured in Fig. 13-2. The height of each column on the grid is proportional to the total quantity of radioactive tracer contained in lung tissue within that square. Tracer mass was located predominantly in the center of the lung and decreased toward

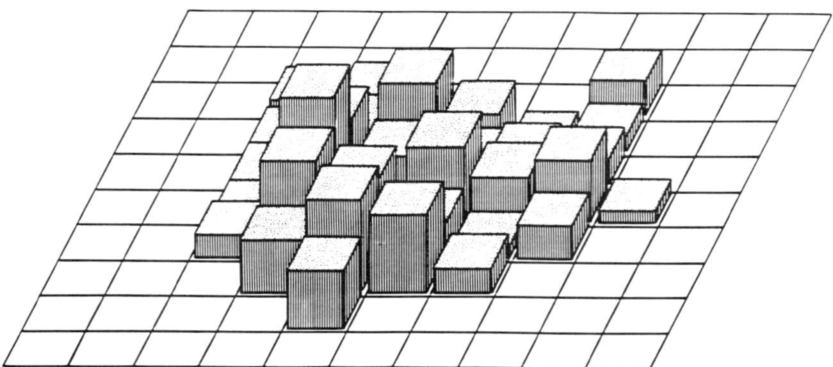

Fig. 13-3 Histogram representing the mass of 14-C dotriacontane present in each grid square for the same lung depicted in Fig. 13-2.

the periphery. Although there was measurable activity in lung squares which contained the trachea and segmental bronchi, this activity came almost entirely from overlying peripheral lung tissue. The larger conducting extrapulmonary airways in this lung and all others were nearly devoid of tracer.

Tissue tracer concentration was the activity measured in representative 1-g samples taken from each lung square. Figure 13-4 shows the spatial distribution of 14-C tissue tracer concentration in the same lungs shown in Fig. 13-2 and 13-3. The height of the tallest column was normalized to the height of the tallest column in Fig. 13-3 to facilitate comparison between concentration and mass data. Tissue tracer concentration increased from apex to base. In five lungs, middle lobes averaged a concentration 3.6-fold that of apical lobes, and lower lung lobes averaged a tracer concentration 5.0-fold that of apical lobes. Tracer concentration was evenly distributed in the lateral direction.

IV. DISCUSSION

These experiments endeavored to determine the fate of inhaled mainstream cigarette smoke within the lung. It has been shown in past experiments that under a wide variety of smoking and inhalation conditions 14-C dotriacontane is a reliable marker for the particulate phase of smoke (Jenkins *et al.,* 1970; Jones *et al.,* 1975).

Figure 13-3 depicts a typical deposition pattern for the mass of mainstream cigarette smoke in the erect, mechanically ventilated anesthetized

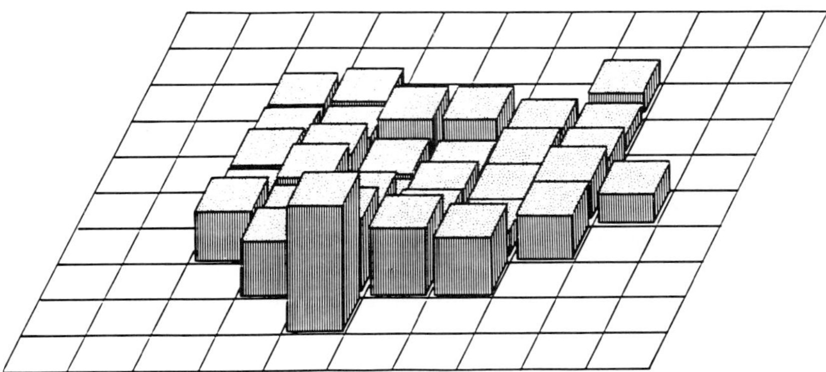

Fig. 13-4 Histogram representing the tissue concentration of 14-C dotriacontane in each grid square for the same lung depicted in Fig. 13-2 and 13-3.

dog. Mainstream cigarette smoke deposition was greatest in lung regions surrounding the bifurcation of the trachea. It declined gradually toward the periphery of the lung. This pattern of smoke deposition paralleled the distribution of lung tissue mass itself. While these data show that cigarette particulate matter deposited to a greater degree in the central portion of Fig. 13-3, this observation should not be interpreted as showing a greater deposition in central airways. It must be remembered that areas from central grid regions of Fig. 13-3 contain not only conducting airways but also a substantial amount of overlying alveolar tissue. Direct sampling of lobar and larger conducting airways revealed only trace amounts of 14-C dotriacontane.

In some respects, the distribution of tracer in these experiments resembles the deposition pattern observed in the study of Pearson *et al.* (1985). In both cases more cigarette smoke was deposited in dependent lung lobes than in superior ones and more was deposited in central than in peripheral lung regions. But, unlike Pearson's findings, intensive deposition of smoke tracer in central conducting airways was never observed. In this respect the present results more closely resemble those of Kendrick *et al.* (1976) and Phelps *et al.* (1984) who found the preponderance of inhaled 14-C dotriacontane was confined to lung parenchyma instead of extrapulmonary airways.

The reason for the lack of central airway deposition in this and previous inhalation experiments of others, compared to those of Pearson *et al.*, is unclear. The smoke particulate measured in the present experiments was different from Pearson's labeled aerosol–smoke mixture. Pearson estimated the size of their labeled particles to be nearly 0.8 μm in diameter whereas 95% of our tracer mass was detected in particles less than 0.5 μm in diameter. In theory, this difference in particle diameter should have minimal influence upon deposition (Task Group on Lung Dynamics, 1966).

Perhaps of greater significance than size is the stability of the labeled particles used in these two studies. Pearson mixed a very stable saline aerosol containing ^{99m}Tc sulfur colloid with fresh cigarette smoke. They assumed the two aerosols coalesced to yield a homogeneous radiolabeled smoke, but that assumption was not explicitly tested. In the present study a radiolabeled marker, 14-C dotriacontane, was incorporated into cigarette smoke during tobacco pyrolysis, thereby directly labeling smoke aerosol droplets (Jones *et al.,* 1975). It is possible that cigarette smoke labeled during tobacco pyrolysis behaves more like unlabeled smoke in its physical and chemical properties. Smoke particles can increase in size in the high humidity environment of the lower respiratory tract (Ingebrethsen, Ch. 12, this volume). During exhalation such swollen particles might be removed by peripheral lung units before they could significantly sediment or diffuse to conducting airway surfaces. The more stable aerosol used by Pearson may have experienced a different deposition fate.

Other explanations for the differences in deposition topography may be due to the various anatomic configurations of the canine and human lung or the active versus passive means of inhalation. Until direct comparisons are made between cigarette smoke aerosols labeled by each method, the disparity between these two outcomes will remain unresolved.

Since cigarette smoke constituents act on lung tissue, the measurement of tissue tracer concentration (activity per unit weight of wet lung) should be a more relevant index of pulmonary tissue exposure than deposited cigarette smoke mass. Figure 13-4 maps the pattern of tissue concentration for deposited mainstream cigarette smoke. In contrast to the topography of total smoke mass in Fig. 13-3, the tissue concentration of tracer shown in Fig. 13-4 is greatest in the lower lobes and least at the apices. Tissue concentration of smoke remained nearly constant from the midline to the lateral margins of the lung.

The preponderance of tracer mass in central lung regions suggests that these regions should experience the greatest injury from cigarette smoke inhalation. The tissue concentration of smoke tracer is actually less in the central and in apical regions than at the base of the lung. Under present experimental conditions the base of the lung receives the greatest dose per g of tissue, while middle lobes receive less and the lung apices experience the smallest dose. This concentration data would suggest the lung bases may be at greatest risk of smoking-related lung disease.

The erect posture in these experiments was employed to elicit gravitational effects on inspired mainstream smoke. However, the means of achieving the erect posture may have distorted the thorax and interfered with the distribution of inhaled air. The scarcity of tracer in upper lobes may have been exaggerated by lower than normal ventilation and hence lower delivery of inhaled smoke aerosol to apical lung regions. This means of achieving the erect posture may not accurately simulate the human smoking experience. Passive insufflation may not deliver smoke to the same lung regions as spontaneous or purposeful inhalations. There are data suggesting that each of these factors may affect the distribution of gas and aerosol in the lung (Engel, 1986; Brain and Valberg, 1979). Further work is needed to determine more precisely how posture, breathing pattern, the mode of ventilation, and anesthesia influence the deposition of inhaled mainstream cigarette smoke.

Data from these experiments indicate that neither deposited smoke mass nor lung tissue concentration of smoke is greater in apical lobes than elsewhere in the lung under these simulated smoking conditions. These results fail to explain the predisposition of apical lung regions to smoking-related respiratory diseases.

REFERENCES

Brain, J. D., and Valberg, P. A. (1979). Deposition of aerosol in the respiratory tract. *Am. Rev. Respir. Dis.* **120,** 1325–1373.

Daly, B. D. T., Faling, L. J., Pugatch, R. D., Jung-Legg, Y., Gale, M. E., Bite, G., and Snider, G. L. (1984). Computed tomography: An effective technique for Medistinal staging of lung cancer. *J. Thorac. Cardiovasc. Surg.* **88,** 486–494.

Doll, R., and Peto, R. (1978). Cigarette smoking and bronchial carcinoma: Dose and time relationships among regular smokers and non-smokers. *J. Epidemiol. Community Health* **32,** 303–313.

Engel, L. A. (1986). Dynamic distribution of gas flow. *In* "Handbook of Physiology, Section 3, The Respiratory System, Vol. III, Mechanics of Breathing, Part 2," pp. 575–594. American Physiological Society, Waverly Press, Baltimore.

Gori, G. B., and Lynch, C. J. (1985). Analytical cigarette yields as predictors of smoke bioavailability. *Regul. Toxicol. Pharmacol.* **5,** 314–326.

Higenbottam, T., Shipley, M. J., and Rose, G. (1982). Cigarettes, lung cancer, and coronary heart disease: The effects of inhalation and tar yield. *J. Epidemiol. Community Health* **36,** 113–117.

Jenkins, R. W., Newman, R. H., Carpenter, R. D., and Osdene, T. S. (1970). Cigarette smoke formation studies. I. Distribution and mainstream products from added 14-C-dotriacontane-16,17. *Beitr. Tobakforsch.* **5,** 295–298.

Jones, R. T., Lugton, W. G. D., Massey, S. R., and Richardson, R. B. (1975). The distribution with respect to smoke particle size of dotriacontane, hexadecane and decachlorobiphenyl added to cigarettes. *Beitr. Tobakforsch.* **8,** 89–92.

Kendrick, J., Nettlesheim, P., Guerin, M., Caton, J., Dalbey, W., Griesemer, R., Rubin, I., and Maddox, W. (1976). Tobacco smoke inhalation studies in rats. *Toxicol. Appl. Pharmacol.* **37,** 557–569.

Lee, P. N., and Garfinkle, L. (1981). Mortality and type of cigarette smoked. *J. Epidemiol. Community Health* **35,** 16–22.

Lisa, J. R., Trinidad, S., and Rosenblat, M. B. (1965). Site of origin, histogenesis and cytostructure of bronchogenic carcinoma. *Am. J. Clin. Pathol.* **44,** 375–384.

Pearson, M. G., Chamberlain, M. J., Morgan, W. K. C., and Vinitski, S. (1985). Regional deposition of particles in the lung during cigarette smoking in humans. *J. Appl. Physiol.* **59,** 1828–1833.

Phelps, D. W., Veal, J. T., Filipy, R. E., Wehner, A. P., and Buschborn, R. L. (1984). Tobacco smoke inhalation studies: A dosimetric comparison of different cigarette types. *Arch. Environ. Health* **39,** 359–363.

Surgeon General (1982). "The Health Consequences of Smoking. Cancer: A Report of the Surgeon General." U.S. Dept. of Health and Human Services. DHHS (PHS) 82-50179.

Surgeon General (1984). "The Health Consequences of Smoking: A Report of the Surgeon General. Chronic Obstructive Lung Disease." U.S. Dept. of Health and Human Services, Washington, D.C., U.S. Govt. Printing Office, Washington, D.C.

Task Group on Lung Dynamics (1966). Deposition and retention models for internal dosimetry of the human respiratory tract. *Health Phys.* **12,** 173–207.

Tobin, M. J., and Sackner, M. A. (1982). Monitoring smoking patterns of low and high tar cigarettes with inductive plethysmography. *Am. Rev. Respir. Dis.* **126,** 258–264.

Valberg, P. A., Brain, J. D., Sneddon, S. L., and LeMott, S. R. (1982). Breathing patterns influence aerosol deposition sites in excised dog lungs. *J. Appl. Physiol.: Respir. Environ. Exercise Physiol.* **53,** 824–837.

Wynder, E. L., and Hoffman, D. (1968). "Tobacco and Tobacco Smoke." Academic Press, New York.

Chapter 14

Human Lung Clearance Following Bolus Inhalation of Radioaerosols

W. Stahlhofen
Gesellschaft für Strahlen- und Umweltforschung mbH
Institut f. Biophysikalische Strahlenforschung
D-6000 Frankfurt am Main, Paul-Ehrlich-Str.20

I. INTRODUCTION

Health effects resulting from particle inhalation are dependent upon the site of deposition in the respiratory tract and the mechanism of clearance from each site. When insoluble particles are deposited in the respiratory tract, two main phases of clearance from the thorax are usually observed. The fast phase is completed in about a day and is assumed to represent mucociliary transport of particles deposited in the tracheobronchial tree, while the slow phase is assumed to represent clearance of the alveolar deposit.

These assumptions form the basis of measurements of regional deposition using labeled aerosols (Stahlhofen *et al.,* 1980). To investigate the assumptions, lung clearance experiments were carried out using 3 μm Fe_2O_3 particles labeled with gamma emitting radionuclides which were administered as a 50-cm^3 pulse or "bolus" of aerosol. The aerosol was injected at a predetermined point of the breathing cycle to restrict particle deposition to a relatively small region of the lung (Stahlhofen *et al.,* 1986a, b, 1987a,b). The fraction of the thoracic deposit retained in the slow clearance phase, *A,* as determined by the zero time intercept of the slowly

This work is dedicated in honor of Prof. Dr. Wolfgang Pohlit on the occasion of his 60th Birthday.

cleared portion of the deposit, was found to be surprisingly large when the pulse was injected at or near the end of inspiration. A was still about 50% when the bolus front depth, V_F, was only 45 cm^3, i.e., about half the volume of the anatomical dead space.

A delayed clearance from a portion of the conducting airways could be responsible for these experimental findings. Alternatively, several mechanisms which could cause part of the bolus to reach alveoli under these conditions have been proposed (Stahlhofen *et al.,* 1987a): the pathways to some alveoli may be relatively short (asymmetric airway branching); the main flow of inspired air in an airway may be confined to the axis (core flow); or some lobes may be filled before others (asynchronous or sequential ventilation).

In the size range of about 3 μm almost all particles reaching respiratory bronchioles and alveoli are deposited, whereas those contained in the larger conducting airways are partially expired again. In this study, aerosol boluses were inspired while the subject was in a supine position and were followed by a period of breath-holding to increase particle deposition in these larger airways. In addition, a variety of particle materials were employed to negate any effect of a select particle material on this method of clearance measurement.

II. MATERIALS AND METHODS

Monodisperse iron oxide particles with 1 or 3 μm aerodynamic diameter were produced by atomizing an aqueous Fe_2O_3 colloid with a spinning top generator. For radioactive labeling, the Fe_2O_3 colloid was mixed with colloidal ^{198}Au. No dissolution of ^{198}Au from the labeled Fe_2O_3 particles suspended in simulated body liquids could be found. The leakage of ^{198}Au in such a solution is less than 1% within a period of 2 weeks (Stahlhofen *et al.,* 1979). Monodisperse Teflon particles labeled with ^{111}In were produced from a colloidal solution of Teflon by means of a spinning disk technique (Philipson, 1977; Camner, 1971). The aerodynamic diameter of these particles was 3.4 μm. Studies have shown that the leakage of ^{111}In from the Teflon particles was less than 1% within 3 days (Stahlhofen *et al.,* 1981). The technique for preparing fused aluminosilicate particles (FAP) labeled with ^{198}Au was based on the method of Bailey and Strong (1980). Colloidal gold labeled with ^{198}Au was mixed with a suspension of montmorillonite clay and diluted to give 0.01 mg/ml gold and 2 mg/ml clay in a 2:1 mixture of water and ethanol. Uniform droplets produced with a spinning top were dried and heated to 1150°C.

The inhalation apparatus used for these studies allows continuous monitoring of particle concentration, c(t), and flow rate, V(t), close to the mouth and has been described elsewhere by Heyder *et al.* (1980). Figure 14–1 shows the breathing maneuver on a time scale.

The subject initiated each experiment by breathing clean air until particles were no longer observed in the expired air. Then, the subject started from the normal functional residual capacity (FRC) and inhaled filtered air at a constant flow rate of 250 $cm^3 s^{-1}$ with a tidal volume of 1000 cm^3. At a preset volume, an aerosol bolus of about 50 cm^3 half-width was introduced into the inspired air and drawn into the airways to a known volumetric lung depth. With additional periods of breath-holding at end-inspiration, the number of particles deposited was sufficiently large to obtain a detectable radioactive signal. After the inhalation procedure, the subject expired to a residual volume with the same flow rate as during inspiration.

Inspired and expired volumes and the corresponding particle concentration profiles were recorded with an ultraviolet recorder. Alternatively, volume (V(t)) and concentration (c(t)) could be stored in an oscilloscope and plotted versus each other. Figure 14–2 illustrates the definitions of the bolus position during inhalation called lung depth (penetration) and front depth.

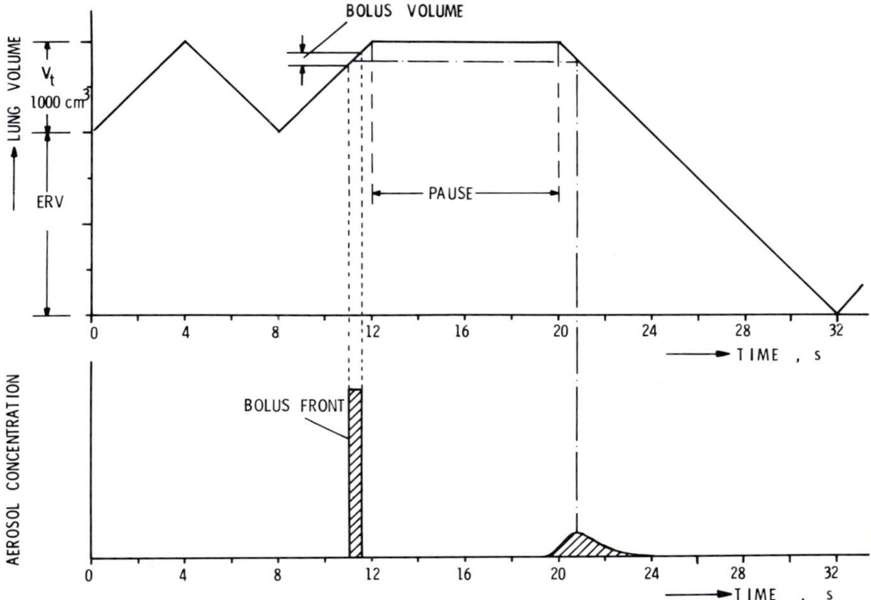

Fig. 14–1 Time-scale of a breathing maneuver.

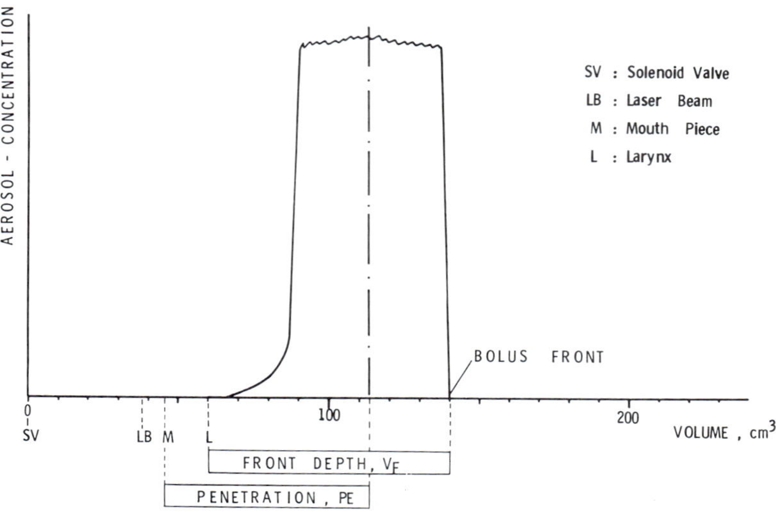

Fig. 14–2 Characteristic parameters of an aerosol bolus: front depth, V_F; penetration, PE; lung depth, VL; PE = VL.

The front depth (V_F) of a bolus is the volumetric distance between larynx (L) and the front of the bolus at end-inspiration. The penetration (PE) or the lung depth (V_L) of a bolus is the volumetric distance between the mouth piece (M) and the maximum particle concentration within the bolus. Since the oral cavity represents a large dead space which would cause convective mixing during inhalation, it was bridged by a tube. The cavity outside the tube was filled with a preformed material, so the dead space was reduced to the combined volume of the tube, the pharynx, and the larynx. In Fig. 14–2, the front depth is 80 cm^3. This value is based on an assumption that the dead space between the mouth piece (M) and the larynx (L) was 15 cm^3. This is assumed to be a minimum value, so for larger dead spaces the front depth becomes even smaller.

After administration of a bolus of radioactively labeled particles using the procedure described above, the thoracic retention was measured with a body gammacounter. From these retention measurements the deposition of particles in the bronchial and alveolar regions was determined according to the definition by Stahlhofen *et al.* (1980).

III. RESULTS

The dependence of retention values on time (clearance) is shown in Fig. 14–3. Even for front depths between 45 and 150 cm^3, the fraction of

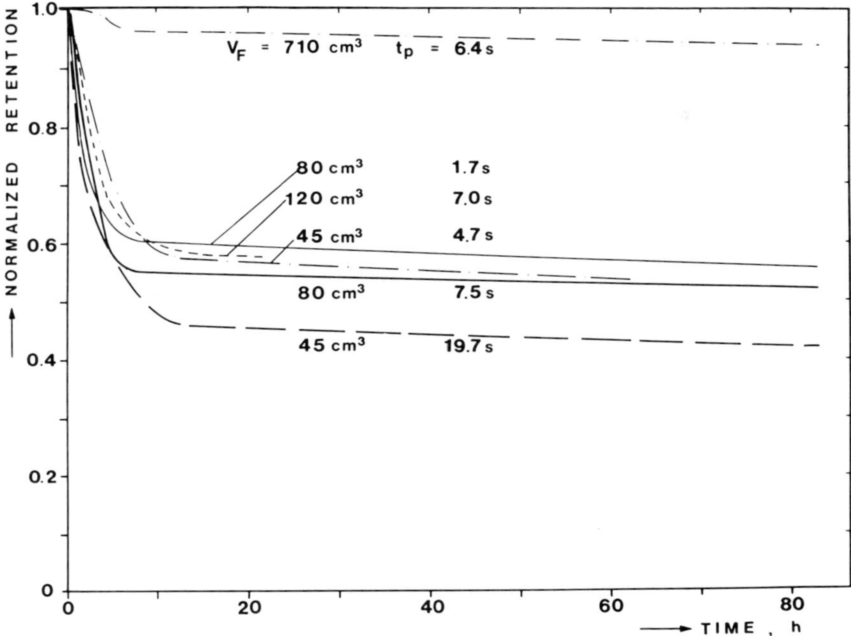

Fig. 14–3 Thoracic retention for different front depths of the inhaled aerosol measured with iron oxide particles (aerodynamic particle diameter, 3 μm).

deposited particles not rapidly cleared within the subsequent six days was around 40%.

If the assumed dead space of 15 cm^3 is actually larger, the front depth can be even smaller. The biological half-life of this slow clearance is on average 20 $\pm$ 10 days. The breath-holding time has only a minor effect on these clearance curves. This is demonstrated in Table 14–1 for a front depth ranging between 30 and about 150 cm^3. Despite different periods of breath-holding resulting in different deposition, the A values are within the experimental scatter. This gives evidence that the fraction of particles deposited from the bolus does not affect clearance behavior.

An experiment with 20 successive breaths containing boluses of 1.3-μm particles at a front depth of 50 cm^3 without breath-holding yielded no measurable deposition at all. As in the previous experiments, the flow rate was chosen to be 250 cm^3 s^{-1} and the tidal volume to be 1000 cm^3. This indicates that the deposition of these particles during inhalation and exhalation can be neglected and that the solenoid and poppet valves of the apparatus are tight.

A speculation has been made that after each breath some particles may remain airborne in the respiratory tract, be pushed into the deeper lung

Table 14-1 Fraction *A* of Thoracic Deposit Retained in Slow Clearance Phase[a]

SUBJECT	V_F, cm^3	t_p, s	A
6	27	65	0.62
4	35	31	0.47
2	35	9.5	0.57
1	40	5	0.40
25	45	4	0.57
25	45	5	0.59
25	45	20	0.47
25	50	47	0.62
1	50	17	0.42
6	50	13	0.57
6	50	14	0.56
1	70	9	0.58
1	70	9	0.54
1	70	9	0.45
4	70	8.5	0.60
25	80	1.5	0.61
25	80	8	0.55
25	80	8	0.59
2	85	20	0.70
25	120	7	0.60
4	135	7	0.56
6	145	16	0.57
1	200	12	0.79
4	275	14	0.88
25	710	6.5	0.97

[a] Correlated with front depth, V_F, and breath-holding time, t_p (zero time intercept).

during the following breaths, and be deposited. However, the experiment with the 1.3-μm particles demonstrates that this kind of particle transport to the lung periphery is not feasible.

To investigate the influence of the particle material, retention curves have been measured under equal breathing conditions with highly insoluble particles consisting of iron oxide, Teflon, and fused aluminosilicate. Figure 14–4 shows the thoracic retention of iron oxide and Teflon parti-

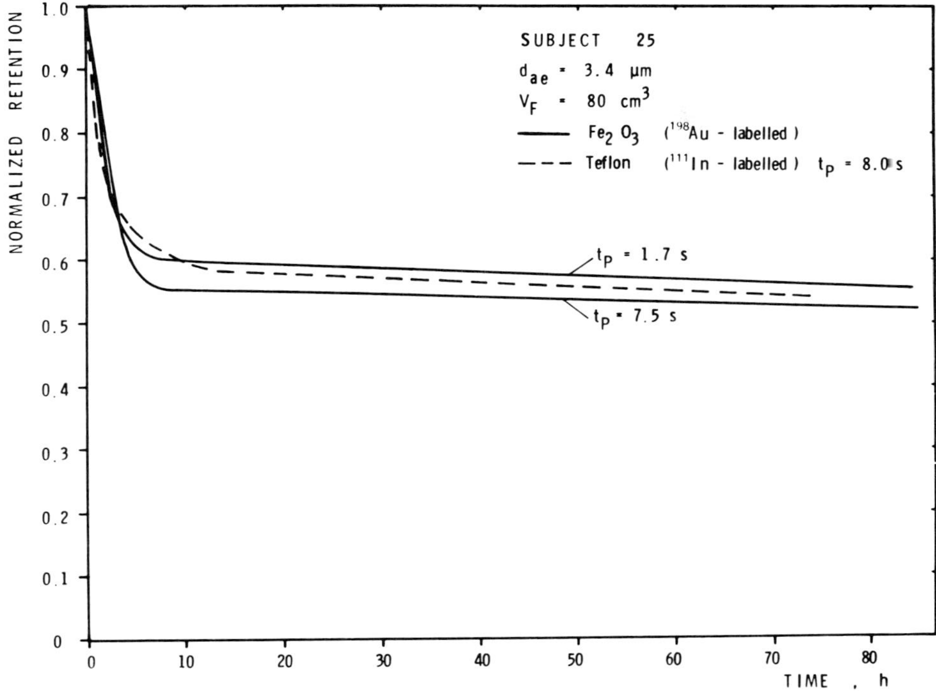

Fig. 14–4 Thoracic retention of FeO_3 and Teflon particles after a bolus inhalation of these aerosols in the same lung depth (d_{ae}, aerodynamic particle diameter).

cles after a bolus inhalation of these aerosols into the same lung depth. Approximately the same chest retention function is obtained for both kinds of particles, indicating that retention following bolus inhalation of the aerosol used in the study is not influenced by solubility or leakage of the particle material, which also has been suggested by experiments with body liquids (Stahlhofen *et al.,* 1986b). The same results are observed if the thoracic retention of Fe_2O_3 and FA particles are compared (Fig. 14–5).

The experiments which are described in Fig. 14–3, 14–4, and 14–5 are replicated in different subjects. In all these experiments, the subject inhaled the bolus in a sitting position. In subsequent experiments, the aerosol was inhaled while the subject was supine, thus reducing the distance the particles must travel for deposition by sedimentation in the major airways. Five subjects inhaled a single bolus with a V_F between 30 and 130 cm^3, followed by a breath held as long as possible (Table 14–2).

In the supine subject, complete deposition of the inhaled bolus was achieved in four out of five cases, and even in the fifth one, deposition was

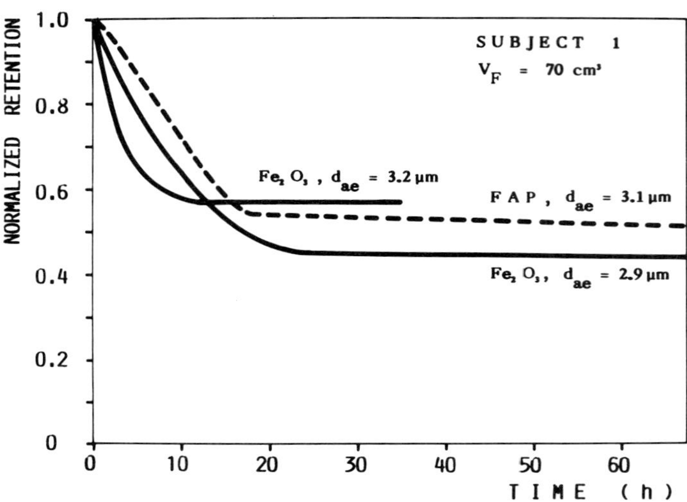

Fig. 14–5 Thoracic retention of FeO_3 and FA particles after a bolus inhalation of these aerosols and a breath-holding time of tp = 9 sec (d_{ae}, aerodynamic particle diameter).

higher than in the sitting subject. A larger deposition of the particles in proximal airways and a lower *A* value were expected in supine subjects. However, in all five subjects (most of them had participated in the earlier experiments), *A* was higher, ranging from 0.72 to 0.83, even though V_F varied (Table 14–2). A typical example illustrating the effect of the subject's position on thoracic retention is given in Fig. 14–6. While *A* was 0.55 in sitting subjects, it increased to 0.72 in the supine subjects.

Table 14-2 Experimental Parameters and Results from Bolus Inhalations in Supine Horizontal Position

SUBJECT	POSITION	d_{ae} µm	V_F cm^3	t_p s	DE	A
25	lying lateral	3.1	45	50	1.00	0.80
6	lying supine	2.9	35	52	1.00	0.73
30	lying supine	3.0	70	6	0.70	0.72
2	lying supine	3.0	70	49	1.00	0.83
4	lying supine	3.0	125	38	1.00	0.79

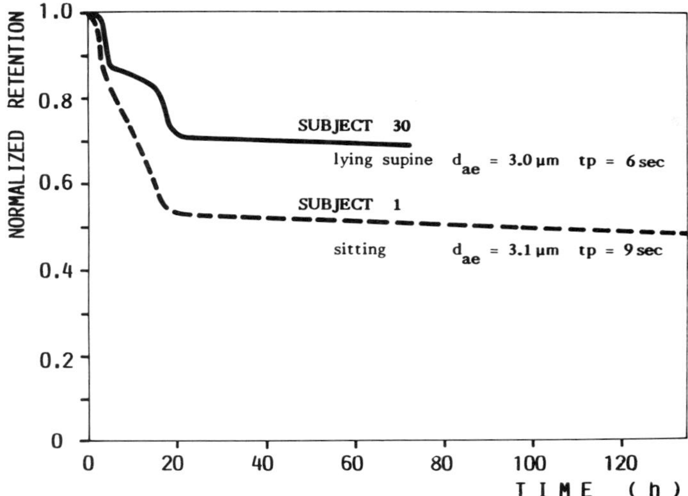

Fig. 14–6 Thoracic retention of FA particles after a bolus inhalation in sitting and supine positions. Front depth of the bolus, $V_F = 70\ cm^3$ (d_{ae}, aerodynamic particle diameter).

IV. DISCUSSION

Two kinds of explanations are possible for the experimental results of a high *A* value for shallow boluses. First, particles may be deposited in bronchial airways but a substantial fraction may be cleared much slower than with the conventionally assumed half-time of a few hours. Second, particles may penetrate into the deeper lungs even for shallow boluses due to the following reasons:

1. Asymmetry in the branching of bronchial airways may enable aerosol particles to reach respiratory structures along shorter pathways (Fig. 14–7b).
2. Convective mixing of tidal and reserve air in the bronchi (axial dispersion) and axial core flow in the airways may lead to a penetration into the alveolar region (Fig. 14–7c).
3. Asynchronous filling of different lobes (nonuniform ventilation) may reduce the dead space and again result in alveolar penetration (Fig. 14–7d).

Briscoe *et al.* (1954) hypothesized that axial streaming in the airways might result in penetration of inspired air deeper into the lungs. Their experiments were similar to those described in this paper, but they used

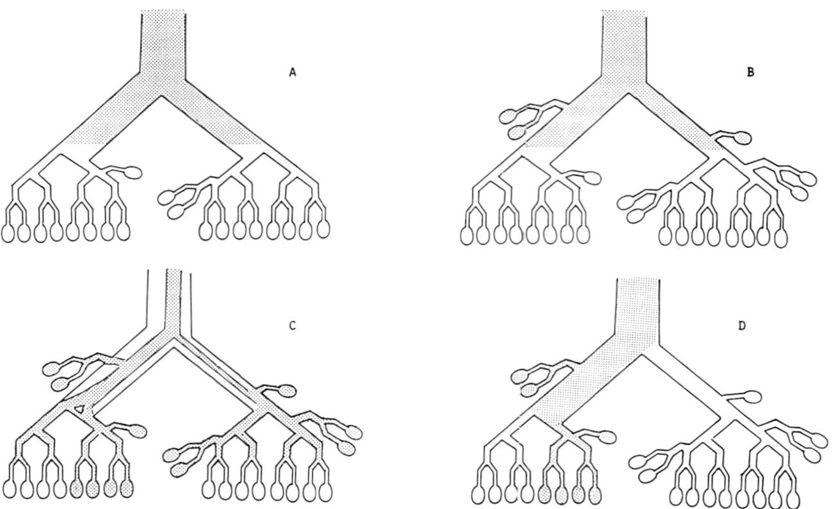

Fig. 14–7 Different modes of lung filling during breathing: A, symmetric lung model and uniform filling; B, asymmetric lung model and uniform filling; C, asymmetric lung model and axial core; D, asymmetric lung model and nonuniform filling (Chang, 1984).

helium gas instead of an aerosol. They found helium deep in the respiratory volume when it was inhaled only to a depth of 60 cm^3, however, ventilation mechanisms alone do not likely account for these results. If *A* corresponds to the alveolar deposit for a front depth of 40 cm^3, then approximately 30 cm^3 of the bolus reached alveoli while 10 cm^3 remained in the conducting airways. Even if totally asynchronous ventilation is assumed, the aerosol would have to pass through the trachea (volume about 20 cm^3), one main bronchus (volume 5 cm^3), and a number of smaller airways to reach the alveoli. For 10 cm^3 of the bolus to remain in the conducting airways, an extremely well defined core flow or an unusual asynchronous filling mechanism of the lungs must exist. The importance of these possible mechanisms can only be determined by a combination of further experimental work and improvement of theoretical models of airflow dispersion and its effect on particle transport and deposition within the lungs.

To estimate the bolus site at end-inspiration, another technique derived by Heyder *et al.* (1980) was applied by Scheuch and Stahlhofen (1987). This technique uses the deposition rate of monodisperse aerosol particles during breath-holding to evaluate average airway diameters. In these experiments the particles had an aerodynamic diameter of 3 μm.

Figure 14–8 and Table 14–3 show the influence of the subject's body position on the aerosol-derived effective airway diameters for a definite volumetric front depth of the bolus.

The different curves for different body positions can be explained by two effects:

1. The bolus was distributed over several airway generations, i.e., part of it remained in the trachea and part in the main bronchi. Particles in the trachea of subjects in the sitting position have a much longer sedimentation distance than those in subjects in the supine position, so the

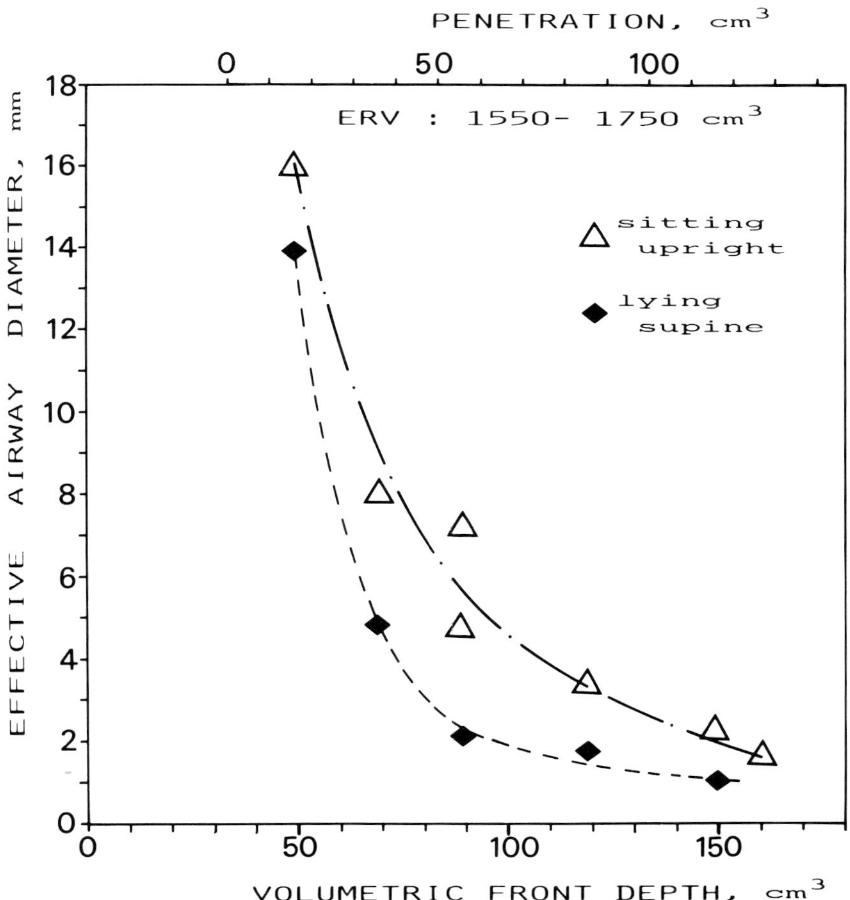

Fig. 14–8 Dependence of the effective airway diameter on the volumetric front depth for different breathing positions of the subject (Scheuch and Stahlhofen, 1987).

Table 14-3 Influence of Body Position, Aerosol Recovery, and Effective Airway Diameter

	SITTING UPRIGHT	LYING LATERAL	LYING SUPINE
ERV	1.60	1.55	1.65
$RE_{(5)}$	0.69	0.65	0.53
$RE_{(10)}$	0.50	0.42	0.26
D_A	4.7	3.3	2.0

VOL. FRONT DEPTH: 90 cm^3

ERV = EXPIRED RESERVE VOLUME

$RE_{(5)}$ = RECOVERY AFTER 5 S BREATH-HOLDING

$RE_{(10)}$ = RECOVERY AFTER 10 S BREATH-HOLDING

D_A = EFFECTIVE AIRWAY DIAMETER

calculated diameter is influenced by the spatial orientation of the airways. In other words, if a fraction of aerosol is still in the trachea and the main bronchi, then the deposition is higher, and the calculated airway diameters are smaller in the supine position.

2. The bolus penetrated deeper into the lungs by a different ventilation pattern observed in subjects in the supine position. However, when using a low flow rate of 250 cm^3 s^{-1}, no difference between apex and base filling of the lungs in either the sitting or supine position was found (Sybrecht *et al.* 1976). The normal expiratory reserve volume (ERV) in a sitting position is higher than in the supine case, but the breathing

experiments discussed here were all done with the same ERV (Table 14–3).

As the data in Fig. 14–8 and Table 14–3 indicate, the 3-μm particles were deposited primarily in the tracheobronchial region because the effective airway diameters ranged from 2 to 15 mm. Furthermore, the experiments with shallow boluses of 1.3-μm particles without breath-holding demonstrate that mixing between tidal and reserve air cannot transport particles into the deeper lungs. The conclusion is that the slowly cleared material contains particles deposited in the tracheobronchial region.

In addition to questions of the distribution of aerosols and ventilation pattern in the lungs, the influence of heart beat during breath-holding is uncertain (Palmes *et al.,* 1973). The possibility that particles may penetrate deeper into the lungs because of the influence of heart beat cannot be excluded.

REFERENCES

Bailey, M. R., and Strong, J. C. (1980). Production of monodisperse radioactively labelled aluminosilicate glass particles using a spinning top. *J. Aerosol Sci.* **11,** 557.

Briscoe, W. A., Forster, R. E., and Comroe, J. H. (1954). Alveolar ventilation at very low tidal volumes. *J. Appl. Physiol.* **7,** 27–30.

Camner, P. (1971). The production and use of test aerosols for studies of human tracheobronchial clearance. *Environ. Phys.* **1,** 137–154.

Chang, H. K. (1984). Mechanisms of gas transport during ventilation by high-frequency oscillation. *J. Appl. Physiol.: Respir. Environ. Exercise Physiol.* **56,** 553–563.

Heyder, J., Gebhart, J., and Stahlhofen, W. (1980). Inhalation of aerosol particles: Deposition and retention. *In* "Generation of Aerosols" (K. Willeke, ed.), pp. 65–103. Ann Arbor Science Publ., Ann Arbor, Michigan.

Palmes, E. D., Wang, C., Goldring, R. M., and Altshuler, B. (1973). Effects of depth of inhalation on aerosol persistence during breath-holding. *J. Appl. Physiol.* **34,** 356–360.

Philipson, K. (1977). Monodisperse, labelled aerosol for studies of lung clearance. Dissertation, University of Uppsala, Sweden.

Scheuch, G., and Stahlhofen, G. (1987). Particle deposition of inhaled aerosol boluses in the upper human airways. *J. Aerosol. Sci.* **18,** 725–728.

Stahlhofen, W., Gebhart, J., Heyder, J., and Stuck, B. (1979). Herstellung von monodispersen Fe_2O_3-Testaerosolen mit Hilfe der Zentrifugal-Zerstäubung. *Staub-Reinhalt. Luft* **39,** 73–108.

Stahlhofen, W., Gebhart, J., and Heyder, J. (1980). Experimental determination of the regional deposition of aerosol particles in the human respiratory tract. *Am. Ind. Hyg. Assoc. J.* **41,** 385–398a.

Stahlhofen, W., Gebhart, J., Heyder, J., Philipson, K., and Camner, P. (1981). Intercomparison of regional deposition of aerosol particles in the human respiratory tract and their long-term elimination. *Exp. Lung Res.* **2,** 131–139.

Stahlhofen, W., Gebhart, J., Rudolf, G., and Scheuch, G. (1986a). Measurements of lung clearance with pulses of radioactively labelled particles. *J. Aerosol Sci.* **17,** 333–336.

Stahlhofen, W., Gebhart, J., Rudolf, G., Scheuch, G., and Philipson, K. (1986b). Clearance from the human airways of particles of different sizes deposited from inhaled aerosol boli. *In* Aerosol: Formation and Radioactivity. *Proc.* 2nd *Intl. Aerosol Conf.*, pp. 192–196. Pergamon Press, Berlin.

Stahlhofen, W., Gebhart, J., Rudolf, G., and Scheuch, G. (1987a). Retention of radioactively labelled Fe_2O_3 particles in the human lungs. *Proc.* 2nd *Int. Symp. Deposit. Clearance Aerosols Hum. Respir. Tract, Salzburg, Sept. 18–20, 1986* p. 123.

Stahlhofen, W., Gebhart, J., Rudolf, G., Scheuch, G., and Bailey, M. R. (1987b). Human lung clearance of inhaled radioactively labelled particles in horizontal and vertical position of the inhaling person. *J. Aerosol Sci.* **18,** 741–744.

Sybrecht, G., Landau, L., Murphy, B. G., Engel, L. A., Martin, R. R., and Macklem, P. T. (1976). Influence of posture on flow dependence of distribution of inhaled ^{133}Xe boli. *J. Appl. Physiol.* **41,** 489–496.

Chapter 15

Effects of Ventilatory Patterns and Pre-existing Disease on Deposition of Inhaled Particles in Animals

J. D. Brain
T. D. Sweeney
Respiratory Biology Program
Department of Environmental Science and Physiology
Harvard School of Public Health
Boston, Massachusetts 02115

I. INTRODUCTION

A major goal of the Lung Dosimetry Symposium was to understand and predict species differences in response to similar concentrations of inhaled gases or particles. However, different species (e.g., mice vs dogs) have varying exposure–dose relationships. Thus, even if the inspired concentration of a particle or gas is constant, the dose deposited in the lungs may be quite variable due to metabolic rate, breathing pattern, and lung structure. The impact of changes in breathing pattern and the effects of chronic lung disease on particle retention are the themes of this paper.

The deposition and retention of aerosols in mammalian lungs have been extensively investigated. Published symposia serve as excellent reviews in this area, e.g., the First (Oxford), Second (Cambridge), Third (London), Fourth (Edinburgh), Fifth (Cardiff), and Sixth (Cambridge) International

Extrapolation of Dosimetric Relationships for Inhaled Particles and Gases

Symposia on Inhaled Particles (Davies, 1961, 1967; Walton, 1971, 1977, 1982; Dodgson *et al.,* 1988). Another review is that of Lippmann (1977). We have prepared several papers and books that review deposition and clearance processes (Brain and Valberg, 1974, 1979, 1985; Brain, 1977).

The distinction between retention (the amount of an aerosol present in the lungs at any time) and deposition (the initial attachment of suspended particles to a surface) should be kept in mind. Retention, but not deposition, is influenced by clearance and translocation. Even during a brief exposure to an aerosol (30–60 min), there may be loss and redistribution of deposited particles, especially in the large ciliated airways. To acknowledge the possibility of particle redistribution during a measurable exposure period, we use the term retention to refer to the amount and distribution of particles in animal lungs measured at any time after an exposure to an aerosol.

Various parameters which influence aerosol deposition have been studied. Total deposition in the lungs depends on particle size (Morrow, 1966), tidal volume (Taulbee *et al.,* 1978), breathing frequency (Altshuler, 1961; Muir and Davies, 1967; Valberg *et al.,* 1982), and lung volume (Davies *et al.,* 1972). To date, little is known about the actual anatomic distribution of deposited particles within the lungs, especially at the level of small airways and parenchyma. Since the rates and pathways of clearance are determined by the sites of aerosol deposition, it is necessary to study factors such as exercise and disease that strongly influence the distribution of retained aerosol in the lungs.

II. EFFECTS OF EXERCISE AND CHANGES IN BREATHING PATTERN ON PARTICLE DEPOSITION

There are compelling reasons why exercise and the resulting changes in breathing pattern should influence particle deposition in the respiratory tract. First, the level of ventilation is a major determinant of the mechanism of deposition. Inertial impaction depends on the velocity of the air stream, while the importance of diffusion and settling depends on residence time in the lungs and the distance to reach lung surfaces. Inertial impaction is more important in central airways where linear velocities are high; diffusion and settling are more important in peripheral airways and alveoli where residence time is longer and distances are smaller. Second, minute volume (respiratory frequency × tidal volume) determines the total particle mass that enters the respiratory tract. Hence, even if the deposition fraction or collection efficiency is constant, increasing the minute volume will increase particle deposition in the lungs. Dennis (1971) has reported that during exercise, the deposition fraction increases with minute volume, particularly for larger particles (1.0 to 3.0 μm) in some human subjects; thus, the amount of aerosol deposited in the lung per unit time increases in two ways. Both the amount entering the respiratory tract as well as the percentage deposited were elevated.

Landahl and associates (1952) demonstrated that the percentage of airborne particles deposited in the respiratory tract rises with increasing minute volume. In exercising hamsters (Harbison and Brain, 1983), the deposition of a 0.4 μm ^{99m}Tc-labeled aerosol increased as their oxygen consumption rose. As shown in Fig. 15–1, exercising animals consumed twice as much oxygen as sedentary animals, but retained 2.5 to 3 times as many particles in their lungs. Enhanced retention probably reflects both an increase in ventilation and an increase in collection efficiency. It is likely

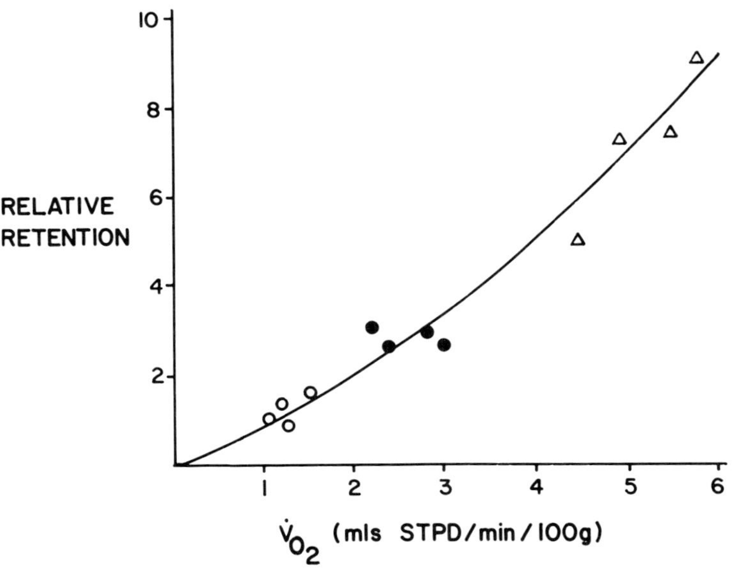

Fig. 15–1 Relative retention of a 0.4-μm radioactive aerosol is plotted against oxygen consumption ($\dot{V}_{O_2}$) in anesthetized Syrian golden hamsters (○). Also shown are resting hamsters (●) or running hamsters (△). $\dot{V}_{O_2}$ on the horizontal axis is given as the volume of oxygen consumed per min per 100 g body weight. Gas volume is corrected to standard temperature and pressure. The relative retention on the vertical axis is the total amount of aerosol retained in the lungs. It is corrected from experiment to experiment to correct for variability in the specific activity of the particles. (Adapted from Harbison and Brain, 1983.)

that different relationships may pertain with larger particles. Then, increased flow rates may be more effective at depositing particles in the nose, pharynx, larynx, and large airways, leading to diminished particle deposition deep in the lungs. Zeltner *et al.* (1988) found the same absolute number of 1.0 μm particles deposited in the parenchyma of exercising hamsters, compared to resting animals, while deposition in the upper airways increased. It is possible that for still larger particles, exercise could actually lower particle dose to the parenchyma.

In humans who are vigorously exercising, minute volumes can exceed 120 liters/min, greatly increasing the amount of aerosol inspired. Even at rest, inspiratory flows are elevated during speech (Bouhuys *et al.,* 1966; Bunn and Mead, 1971). The high velocities (2 m/sec) achieved in the main bronchi enhance inertial impaction and turbulence. Experimenting with 0.5 μm monodisperse particles, Muir and Davies (1967) found that the percentage of particles deposited increased linearly with tidal volume and decreased with the square root of breathing frequency. Expiratory reserve volume also influenced this percentage (Davies *et al.,* 1972).

Valberg and co-workers (1982) studied excised dog lungs that were ventilated with a 0.4-μm radioactive aerosol using various breathing patterns. After a half-hour exposure, the lungs were fully expanded, dried, and sliced transversely at 1-cm intervals; each slice was then weighed and

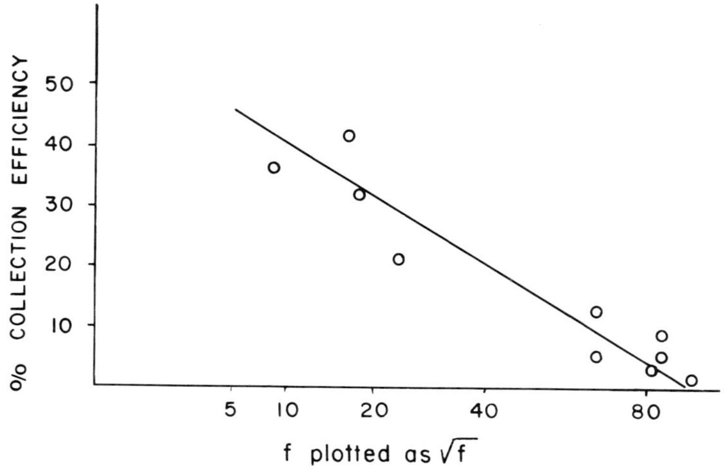

Fig. 15–2 Effect of breathing patterns on particle deposition in excised dog lungs. The collection efficiency (vertical axis) for all the lungs studied is plotted as a function of the square root of breathing efficiency (horizontal axis). The percentage collection efficiency (deposition fraction) is defined as total aerosol activity in the lungs divided by the total aerosol activity inspired during the exposure. The data clearly show that collection efficiency decreases as breathing frequency increases. (Adapted from Valberg *et al.,* 1982.)

photographed. The distribution of retained radioactive aerosol was determined in three ways, gamma camera, autoradiography, and further dissection of each slice into 1-cm^3 pieces that were then scored for airway content, assayed for radioactivity, and weighed. Alterations in breathing pattern profoundly changed the collection efficiency (C.E.) for the inhaled particles. As shown in Fig. 15–2, the C.E. (also known as the deposition fraction) varied between 1 and 42% as the breathing pattern slowed from 93 breaths/min to 17. Variations in tidal volume and frequency also altered the distribution of retained particles. During slow deep breathing the parenchyma received an increasing share of deposited particles and large airways received a decreasing share. In contrast, during panting (rapid shallow breathing), large airways received more deposited activity per unit volume than parenchyma. Also rapid-shallow patterns produced more heterogeneity in distribution than did slow-deep patterns. Although some breathing patterns tested may not be experienced by humans, the results clearly indicate that different tidal volumes and frequencies markedly affect both collection efficiency and the anatomic distribution of retained particles.

III. EFFECTS OF EXERCISE ON RESPONSE TO INHALED TOXINS

Many published articles suggest that activity level influences the response to inhaled particles or gases. Punte *et al.* (1958) noted that exercise increased the toxicity of the inhaled organophosphorus insecticides paraoxon (diethyl 4-nitrophenyl phosphate) and sarin (isopropyl methylphosphonofluoridate) in rats. These authors postulated that an increase in respiratory retention was responsible for the increase in toxic effects.

In our laboratory, we found altered alveolar macrophage function in exercising hamsters exposed for 40 min to 50 ppm sulfur dioxide (Skornik and Brain, 1989). Phagocytosis of gold colloid particles, measured 1 hr after the sulfur dioxide exposure, was 38% lower in exercising hamsters when compared to normal unexposed animals. In contrast, macrophage phagocytosis in resting hamsters exposed to sulfur dioxide was normal. Among exercising animals, depression of phagocytosis due to sulfur dioxide was linearly related to oxygen consumption ($\dot{V}_{O_2}$); as $\dot{V}_{O_2}$ increased, phagocytosis decreased.

Similarly, we found that exercise altered the response to ozone inhalation in mice. The concentration that produced 50% mortality (LC_{50}) shifted from 9.0 ppm to 2.5 ppm with exercise during a 6-hr exposure (J.D. Brain, unpublished data). These results are consistent with those of Stokinger and associates (1956) who observed an LC_{50} of 7.5 ppm for nonexercising rats and 1.0 ppm for those animals that were intermittently exercised. More importantly, equivalent amounts of ^{131}I-labeled albumin (a measure of increased permeability) accumulated in respiratory tract surfaces at far lower ozone concentrations in animals that exercised than in resting animals.

IV. MOUTH VERSUS NOSE BREATHING

The ability of the upper airways to remove water soluble gases or larger particles changes significantly when there is a switch between nose and mouth breathing, or when the nose is bypassed by a tracheostomy or an endotracheal tube. The surface area of the human nose is small relative to the 140 m^2 of the total respiratory tract, but it has a rich vasculature that is well designed for heating or cooling, and for rapid exchange of dissolved substances between the mucus, tissues, and blood vessels. Most capillaries have fenestrated endothelial linings enclosed by porous basement membranes.

Proctor (1973) estimates that the nasal surface area between the bony opening into the nasal cavity and the posterior end of the turbinates is 160 cm^2, or 0.016 m^2. The volume of air within the human nasal passages is about 20 ml. At an inspiratory flow rate of 0.4 liters/sec, the residence time of the inspired gas in the nose is only 0.05 sec. In spite of this brief contact time with the nasal mucosa, special anatomic features help condition the inspired gas. At modest flows, inspired gas is warmed and humidified before it reaches the subglottal area. However, the upper airways are less effective at higher flows, especially when the inspired air is cold and dry. The nose also has a major role as a collector of inhaled aerosol particles. A small cross section for airflow and the resulting high linear velocities, sharp curves, and interior nasal hairs all help to promote particle impaction.

In animals and humans, particle deposition and clearance rates are markedly affected by the choice of pathway to the trachea. Shifts from the nasal to the oral route are of major significance. An excellent review of particle deposition in the nose is that of Hounam and Morgan (1977). A discussion of gas uptake by the nose is also available (Brain, 1970). Deposition and clearance of particles in the head during nose breathing have been extensively studied by Pattle (1961), Hounam and co-workers (1969), Lippmann (1970), and Martens and Jocobi (1974). Collectively, these reports indicate that substantial quantities of particles larger than 5.0 μm are deposited in the nasopharyngeal region. There are large individual variations, depending on the structure and aerodynamic properties of the nose. Hounam *et al.* (1969) showed an increase in the percentage of inspired particles deposited in the nose as the resistance to airflow increased. Forsyth and associates (1983) have demonstrated that nasal resistance decreases dramatically with exercise, probably because of mucosal shrinkage mediated by the effects of the sympathetic nervous system on the nasal microcirculation. Lower nasal resistance would decrease particle collection efficiency in the nose and increase particle penetration to the pulmonary airways and parenchyma. The nose would become a less efficient filter.

The nasal route is more efficient at removing large particles from inspired air than is the oral route. In bypassing the nose, mouth breathers deposit more particles in their lungs and may concomitantly increase the risk of pulmonary damage. Mouth breathing is common under conditions of heavy exercise, during speech, or singing (Bouhuys *et al.,* 1966; Bunn and Mead, 1971), and in persons who have decreased nasal conductance caused by infection or allergy. This may result in markedly increased exposure of the airways and alveoli to large particles.

Some inhaled toxins may induce physiologic changes that in turn affect the degree to which they penetrate. The uptake of other toxins may also be influenced. For example, sulfur dioxide and sulfate particles alter the dimensions of the upper and lower airways in both humans and experimental animals. The principal effect in acute experiments is to increase pulmonary flow resistance because this reflects the bronchoconstrictive action of the pollutant. Chronic exposure may lead to chronic bronchitis and airways may be narrowed. In either case, the reduction in airway cross section increases linear flow velocities and turbulence, with the result that more particles deposit by impaction. Thus, any factor that increases the penetration of sulfur dioxide to airways, such as exercise or mouth breathing, is likely to be associated with additional secondary changes in particle and gas uptake.

The loss of the filtering capacity of the nose may contribute to the phenomenon of exercise-induced asthma. With rising levels of exercise, enhanced metabolism, and increasing ventilation, the high-resistance nasal pathway is progressively abandoned in favor of the low-resistance oral pathway. When the nose is bypassed, more antigenic particles will deposit in the airways and lungs, and this rise in exposure increases the risk of allergic responses, particularly for large particles such as pollen. Work of McFadden and Ingram (1982) shows that airway cooling is usually a component of exercise-induced asthma. Nevertheless, if allergens are present in the inspired air, penetration and deposition may also increase because of mouth breathing. The nose is an efficient filter for large particles, but the mouth is not. It would not be surprising if tracheal and airway doses of 5 to 15μm diameter pollen particles increased by 10-fold or more as subjects switch from nose to mouth breathing.

V. EFFECTS OF PRE-EXISTING DISEASE ON PARTICLE DEPOSITION

Pre-existing pulmonary disease may modify particle deposition in the respiratory tract. One factor that determines where small particles are deposited in the lungs is the distribution of ventilation. Particles are carried into the lungs with the inhaled air and therefore reach only areas that are ventilated. We know that nonventilated regions of the lungs exist in a variety of pathological conditions (Milic-Emili, 1974). Because diseased lungs are nonuniformly affected, they are more likely to have large alterations in the distribution of ventilation than normal lungs. Disease may impair ventilation in some regions because of airway closure, atelectasis, or obstructing mucous plugs (Macklem, 1971), and fewer particles will reach these regions. Conversely, the remaining ventilated areas of the lungs may be exposed to increased toxic loads when a significant amount of the lung is not ventilated. Maintaining patent airways and open alveoli requires a balance between surface tension and tissue forces (Macklem, 1971; Weibel and Gil, 1977; Hoppin *et al.*, 1986). Alterations in these forces can cause alveoli to collapse, especially when breathing at low lung volumes (Bachofen *et al.*, 1979; Gil *et al.*, 1979). Thus, airway and alveolar patency during breathing depends upon surface and tissue forces, and any disruption in lung structure can lead to the creation of nonventilated areas of the lung.

Chronic pulmonary disease alters the architecture of the lungs. Emphysema involves alveolar wall destruction, which leads to enlarged air spaces. When radial traction forces of the alveolar septa are reduced and surface tension is diminished because of loss of alveolar surface area, intrapulmonary bronchi are poorly supported and tend to collapse during expiration (American Lung Association, 1981). Similarly, the increase of secretions in chronic bronchitis favors airway obstruction by mucous plugs, and the regional thickening of alveolar walls in interstitial fibrosis makes alveoli less compliant and thus favors local hypoventilation. In each of these cases, the changes in airway and alveolar properties cause changes in the distribution of ventilation. Although airway closure has been noted in chronic lung disease (Macklem, 1971), little is known about the anatomic distribution of poorly ventilated regions of the lungs and how this distribution changes during the disease process.

Radioactive aerosols have been used in humans to detect airway obstruction in Chronic Obstructive Pulmonary Disease (COPD) (Lourenco *et al.*, 1972; Goldberg and Lourenco, 1973; Ramanna *et al.*, 1975; Dolovich *et al.* 1976; Pavia *et al.*, 1977; Taplan *et al.*, 1977; Itoh *et al.*, 1981; Kim *et al.*, 1983a,b). Scintigraphic analysis of particle deposition shows that deposition is greater in the central airways of patients with airway obstruction than in airways of normal patients (Lourenco *et al.*, 1972; Ramanna *et al.*,

1975; Taplan *et al.*, 1977; Itoh *et al.*, 1981). Investigators have quantified this phenomenon by calculating a penetration index to assess how deeply the aerosol enters the lungs (Thomson and Pavia, 1973; Dolovich *et al.*, 1976; Agnew *et al.*, 1981). Penetration of aerosol has been found to correlate inversely with airway obstruction. Other investigators have used aerosol rebreathing techniques to measure total deposition in normal persons versus patients with obstructed airways (Kim *et al.*, 1983a). Theoretical analyses of these data suggest that aerosol deposition may be a more sensitive indicator of airway abnormalities than is measurement of overall airway resistance (Kim *et al.*, 1983b). Total deposition in the lungs of humans with airway obstruction is enhanced when compared to normal patients (Kim *et al.*, 1983a).

Since human disease is progressive, experiments such as those described above are complicated and difficult to interpret. Nevertheless, some conclusions can be made from these studies. Two types of airway obstruction can occur in COPD; physical obstruction of the airways with mucous plugs, as in chronic bronchitis, and obstruction due to small bronchioles that collapse during expiration. Airway collapse or closure is not present in normal individuals breathing at rest, but can occur during normal breathing in humans with emphysema (Macklem, 1971).

Obstruction of small airways has three effects on aerosol deposition. First, particles will deposit more centrally because they fail to reach the terminal air units. Second, deposition will be enhanced at the site of obstruction in both obstructed airways and in those that collapse during expiration. Smaldone and associates (1979) examined the latter type of obstruction and showed that deposition is enhanced around the site of closure because of changes in airflow patterns. Turbulent eddies will enhance deposition at the constriction site; collapse will bring particles in close contact with airway walls.

Third, Macklem and associates (1973) have postulated that obstructed airways will cause inspired gas to be diverted to the healthier regions of the lungs, thus, particles will preferentially deposit in healthy regions. Because of increased linear velocity of the gas in these healthy regions and enhanced penetration to alveoli, deposition would be enhanced (Macklem *et al.*, 1973).

VI. ANIMAL MODELS OF PULMONARY DISEASE

Animal models have been used to explore how aerosol deposition is influenced by lung pathology. In these studies variable factors in humans, like age, environmental exposure, and extent of disease, can be controlled. Gross and associates (1965, 1969, 1971) studied aerosol deposition in rats, hamsters, and guinea pigs with pulmonary emphysema induced by inhalation or intratracheal instillation of papain, a proteolytic enzyme that attacks connective tissue in the lungs. After papain treatment, which produced emphysema accompanied by mild inflammation, animals were exposed to an aerosol of quartz or bituminous coal. Lungs from each group of animals were examined immediately after the aerosol exposure. The lungs of animals with papain-induced emphysema retained less dust than did lungs of the nonemphysematous animals. The investigators noted that most of the dust appeared to be collected in nonemphysematous portions of the lungs. Diseased regions of the lung had little or no dust accumulation. However, dust content was not quantified, and the aerosol used in the study was not well characterized.

In 1979, Hahn and Hobbs studied both papain-induced emphysema and elastase-induced emphysema in hamsters. Intratracheal instillation of elastase produced focal destruction and enlargement of alveoli, accompanied by loss of lung elastic recoil. These effects of elastase on the lung have also been well documented by others (Hayes *et al.,* 1975; Snider *et al.,* 1977, 1986). Papain aerosols caused focal destruction and enlargement of alveoli primarily around terminal bronchioles without loss of lung elastic recoil. However, some investigators have reported that elastic recoil is lost in the lungs of animals treated with either papain or elastase (Johanson and Pierce, 1972).

Hahn and Hobbs (1979) exposed hamsters to an aerosol of ^{137}Cs-labeled aluminum silicate particles 21 days after elastase or papain administration. The emphysematous hamsters, examined 3 hr after the aerosol exposure, retained 45 to 65% fewer particles in their lungs than did the control hamsters. This decrease in initial dose of aerosol in emphysematous lungs was also observed by Damon and associates (1983) in rats exposed to iron oxide particles and by Lundgren and associates (1981) in rats exposed to ^{169}Yb-labeled plutonium oxide particles.

How do the emphysematous changes influence the distribution of the deposited particles throughout the lungs? To find out, we exposed hamsters with elastase-induced emphysema to an insoluble ^{99m}Tc-labeled aerosol (0.45-μm activity median aerodynamic diameter) either 30, 60, or 90 days post-elastase (Sweeney *et al.,* 1987). Immediately after exposure, the animals were sacrificed. The lungs were excised, dried at total lung capacity, and sliced into 1-mm thick sections. Each slice was cut into pieces, which were counted for radioactivity and weighed. Then a measure of the uniformity of deposition, the evenness index (EI), was calculated. With perfect uniformity of particle retention, all EIs would be 1.0. We found fewer particles in the lungs of emphysematous hamsters, compared to controls at 60 or 90 days after elastase instillation. As shown in Fig. 15–3A, the

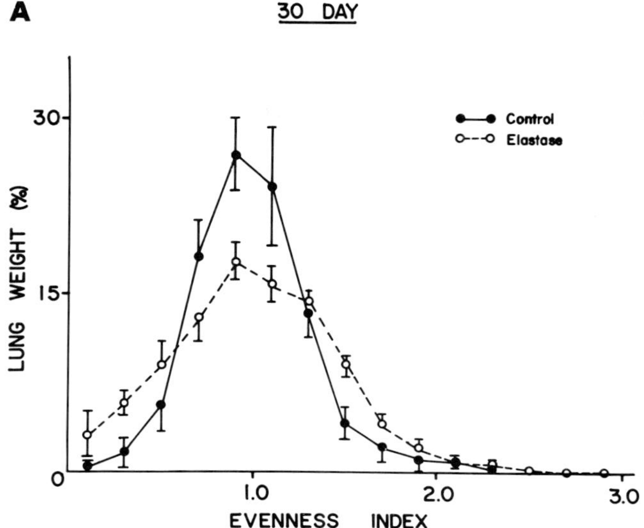

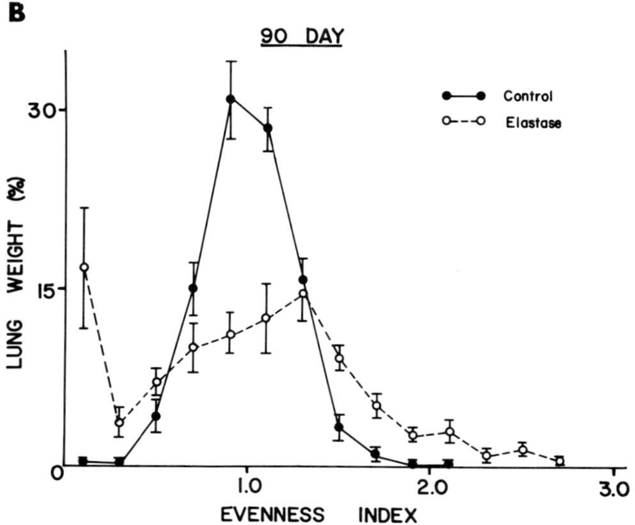

Fig. 15–3 Using an elastase model of emphysema in the hamster, diseased animals are compared to matched controls. The profile (analogous to a histogram) of the evenness index (X axis) is shown as a function of the percentage of total lung weight (Y axis) for both groups of animals. The data shown were measured 30 days (A) and 90 days (B) after the exposure to elastase. The evenness index is equal to the cpm/g piece divided by cpm/g lung. If the aerosol was distributed with perfect uniformity, all pieces would have an evenness index of 1.0. (Adapted from Sweeney *et al.,* 1987.)

deposited particles were distributed less uniformly throughout the emphysematous lungs than in controls. This was particularly true at 90 days after elastase as shown in Fig. 15–3B. To correlate deposition with changes in lung structure, dried lung pieces were embedded and examined histologically. We found no correlation between mean linear intercept (MLI), a measure of air space size, and EI, a measure of deposition. Frequently, particle deposition was prominent in airways. We conclude that the emphysematous lesions markedly alter the deposition of inhaled particles due to effects on airways as well as parenchyma.

Our studies also showed that emphysema decreased the total number of particles deposited in the lungs of these hamsters. There are two possible explanations for this result. First, the animals with emphysema may have inhaled fewer particles than the normal animals, that is, their minute ventilation (tidal volume per breath multiplied by number of breaths per min) may be decreased. Secondly, the emphysematous lungs may have collected a smaller fraction of particles from the inspired air than the normal lungs.

We exposed emphysematous and control hamsters to the same 0.45-μm aerosol while continuously monitoring their breathing patterns with a head-out pressure plethysmograph (Sweeney *et al.,* 1986). We found 41% fewer particles in the lungs of emphysematous hamsters than in lungs of controls, however, the minute ventilations of the two groups of animals were similar. The emphysematous hamsters breathed with a larger tidal volume and at a slower rate than the control hamsters. Since this type of slow, deep breathing would tend to favor higher parenchymal deposition, we believe that the changes in lung structure caused by elastase (increased air space size and diminished surface area) are the major factors causing lower particle deposition.

We also studied the deposition pattern of inhaled particles in another chronic lung disease, pulmonary fibrosis. Diffuse interstitial pulmonary fibrosis was produced in hamsters by a single intratracheal injection of bleomycin followed by exposure to 70% oxygen for 72 hr (Sweeney *et al.,* 1983a). This combination of agents resulted in diffuse alveolar damage that in 30 days resolved into interstitial pneumonitis involving 10% of the lungs, primarily in the peribronchial region (Tryka *et al.,* 1982, 1983). After 60 days, the amount of interstitial pneumonitis increased to 30%, and involved the subpleural parenchyma in addition to the peribronchial regions. Ninety days after treatment, the intervening parenchyma became involved and 38% of the lung was diseased. During this time, not only did the amount of fibrosis steadily increase, but pulmonary mechanics also deteriorated; dynamic compliance decreased and total pulmonary resistance increased.

Aerosol deposition was studied in control and bleomycin-oxygen-treated hamsters at 30, 60, and 90 days post-bleomycin (Sweeney *et al.,* 1983a, b). All groups were exposed while awake to an insoluble aerosol with a 0.45-μm activity median aerodynamic diameter for 25 min, and then sacrificed within 5 min after exposure. Their lungs were removed and the distribution of aerosol retention was examined. Retention was lower and particles were less uniformly distributed throughout the lungs of animals given bleomycin and oxygen than in the lungs of controls. This was particularly true at 30 days as shown in Fig. 15–4A. By 90 days (Fig. 15–4B), both the distribution of the disease and the particles became more uniform. At all times when examined histologically, individual lung pieces that had retained few particles were more diseased than those that had retained many particles.

Is decreased deposition a result of diminished ventilation (fewer particles enter the lungs) or of diminished collection efficiency (fewer inspired particles deposit in the lungs)? To find out, we exposed fibrotic and control hamsters to the 0.45-μm aerosol while monitoring breathing patterns with a head-out pressure plethysmograph (Sweeney *et al.,* 1983b). After the exposure, the animals were sacrificed, their lungs removed, and the number of particles in the lungs determined. The average volume of air inhaled per min ($\dot{V}_E$) was 54 ± 5.2 ml/min/100 g BW (SE) for control animals and 56.7 ± 2.0 (SE) for animals treated with bleomycin and oxygen. We found no differences in tidal volume or frequency between the two groups of animals. However, the collection efficiency decreased from 9.9 ± 1.5% (SE) in controls to 4.3 ± 1.0% (SE) in bleomycin-oxygen-treated animals. These data suggest that as with emphysema, aerosol retention is altered by anatomical changes due to fibrosis in the lungs rather than by the animal's breathing pattern.

Although there are similarities between our findings in fibrotic and emphysematous hamsters (lower total deposition and increased heterogeneity in deposition), there are also important differences. First, the heterogeneity observed in fibrotic lungs could be explained primarily by the distribution of disease in the lungs. This was not true in emphysematous animals. Second, heterogeneity was greatest at 30 days after bleomycin-O_2 and then became more uniform as the fibrosis progressed. In emphysema, heterogeneity in aerosol deposition was greater at 90 days after elastase instillation. Third, there was an increase in the amount of airway-associated deposition in emphysematous animals, but not in fibrotic animals. These variations reflect the differences in the two types of disease. Bleomycin-O_2-induced fibrosis is a progressive lesion primarily involving parenchyma and not airways. Although elastase-induced emphysema also produced a parenchymal lesion (increased air space size), the decrease in total surface area reduces the support system of the small airways in these lungs. This favors airway closure, enhanced airway deposition, and maldistribu-

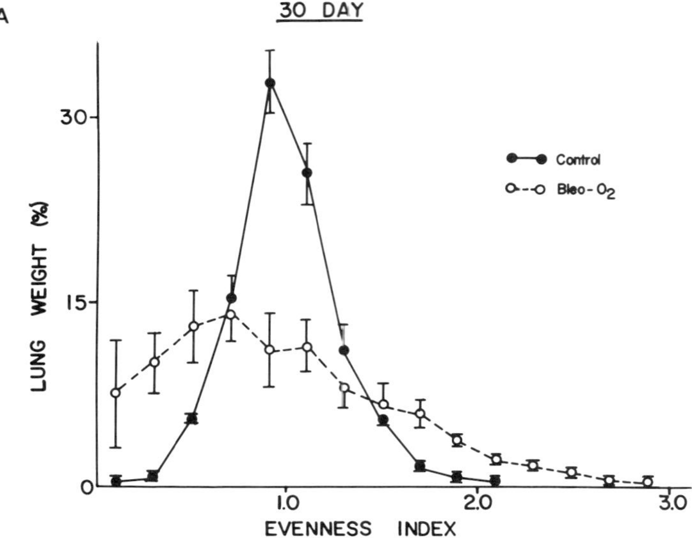

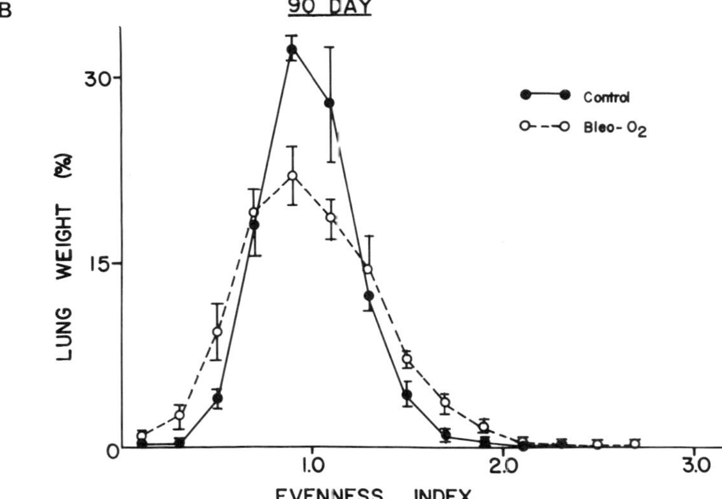

Fig. 15–4 Evenness index profiles (analogous to histograms) in lungs from animals with fibrosis produced by bleomycin and hyperoxia. Animals were exposed to particles at 30 days (A) or 90 days (B) after the exposure to intratracheal bleomycin and 3 days of 70% oxygen. (Adapted from Sweeney *et al.*, 1983a.)

tion of ventilation. Our data suggest that these effects are more prominent at the later times following elastase as the lesion becomes more severe.

Added evidence for the effect of airway changes on local deposition can be seen in animals with chronic bronchitis. We exposed rats 5 hrs/day, 5 days/week for 7 weeks to 236 ± 14 ppm SO_2 to produce chronic bronchitis (Sweeney *et al.,* 1985). SO_2 exposure produces prominent airway mucus, inflammation, and squamous metaplasia in central and peripheral airways. When these rats which had developed bronchitis were exposed to our 0.45-μm aerosol, we found lower deposition in the lung parenchyma, enhanced airway-associated deposition, and marked heterogeneity of deposition throughout the lungs, as shown in Fig. 15–5. The heterogeneity of deposition is primarily due to the airway changes caused by SO_2 exposure. Note the striking similarity of the EI distribution in the chronic bronchitic

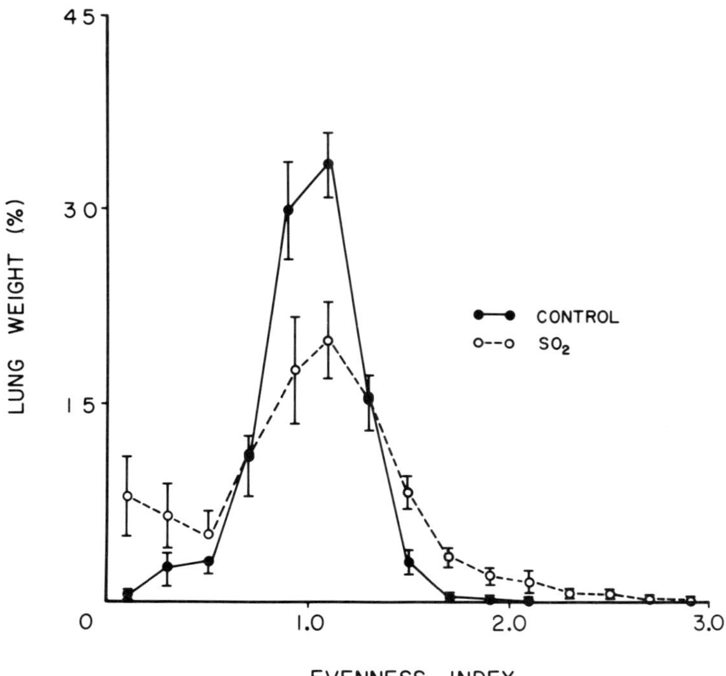

Fig. 15–5 Effect of chronic bronchitis on deposition evenness index profiles (analogous to histograms) in lungs from normal rats and those exposed to sulfur dioxide (236 ppm) for 5 hr/day, 5 days/week, for 7 weeks to produce chronic bronchitis. After SO_2, rats were exposed to radioactive particles and deposition patterns in their lungs were measured.

rats, a true airway lesion, to the EI distribution at 90 days post-elastase (Fig. 15–3B), a parenchymal disease that influences airway properties.

The anatomical changes in lung structure due to chronic lung disease dramatically alter the deposition of particles. In emphysema, fibrosis, and chronic bronchitis, disease is not distributed uniformly throughout the lungs. Normal areas of the lungs retain more particles than fibrotic areas (Sweeney *et al.,* 1983a). Fibrotic regions of the lungs do not expand as much as normal regions with the same change of pleural pressure. This would favor ventilation of the normal rather than the fibrotic lung regions. In emphysema, although the parenchyma is the primary site of the lesion, airways are also influenced in the later states of the disease. Thus, both parenchyma as well as airways will influence the site of deposition. Finally, in chronic bronchitis, the obstruction due to mucus plugging favors enhanced airway deposition.

These studies have many implications for the progression of chronic lung disease in humans. They suggest that individuals with fibrosis, emphysema, or chronic bronchitis will collect a smaller total mass of inhaled dust in their parenchyma than normal individuals. However, if proportionally more dust is deposited in areas of the lung free of disease, as in fibrosis, the disease process could be accelerated. Risk of damage to normal areas is increased. During airway diseases such as chronic bronchitis and later stages of emphysema, enhanced deposition is likely at the site of obstruction due to mucus plugging or airway closure. Dust associated risk in these types of diseases would be elevated at the site of the airway impairment and be more prominent in airways than in parenchyma. Nevertheless, extrapolation of these findings in animals to humans with chronic obstructive pulmonary disease or idiopathic pneumonitis with fibrosis, must be done cautiously. Our data in animals can be used to identify some factors influencing deposition, and thus risk, but are limited in predicting causal relationships in human disease.

REFERENCES

Agnew, J. E., Pavia D., and Clark, S. W. (1981). Airways penetration of inhaled radioaerosol: An index to small airways function? *Eur. J. Respir. Dis.* **62,** 239–255.

Altshuler, B. (1961). The role of the mixing of intrapulmonary gas flow in the deposition of aerosol. *In* "Inhaled Particles and Vapours" (C. N. Davies, ed.), pp. 47–54. Pergamon, Oxford.

American Lung Association (1981). "Chronic Obstructive Lung Disease," 5th Ed. American Lung Association, New York.

Bachofen, H., Gehr, P., and Weibel, E. R. (1979). Alterations of mechanical properties and morphology in excised rabbit lungs rinsed with a detergent. *J. Appl. Physiol.* **47,** 1002–1010.

Bouhuys, A., Proctor, D. F., and Mead, J. (1966). Kinetic aspects of singing. *J. Appl. Physiol.* **21,** 2–10.

Brain, J. D. (1970). The uptake of inhaled gases by the nose. *Ann. Otol. Rhinol. Laryngol.* **79,** 529–539.

Brain, J. D. (1977). The respiratory tract and the environment. *Environ. Health Perspect.* **20,** 113–126.

Brain, J. D., and Valberg, P. A. (1974). Models of lung retention based on the report of the ICRP Task Group. *Arch. Environ. Health* **28,** 1–11.

Brain, J. D., and Valberg, P. A. (1979). Deposition of aerosols in the respiratory tract. *Am. Rev. Respir. Dis.* **120,** 1325–1373.

Brain, J. D., and Valberg, P. A. (1985). Aerosols: Basics and clinical considerations. *In* "Bronchial Asthma: Mechanisms and Therapeutics" (E. B. Weiss, M. S. Segal, and M. Stein, eds.), 2nd Ed., pp. 594–603. Little Brown, Boston.

Bunn, J. D., and Mead, J. (1971). Control of ventilation during speech. *J. Appl. Physiol.* **31,** 870–872.

Damon, E. G., Mokler, B. V., and Jones, R. K. (1983). Influence of elastase-induced emphysema and the inhalation of an irritant aerosol on deposition and retention of an inhaled insoluble aerosol in Fischer-344 rats. *Toxicol. Appl. Pharmacol.* **67,** 322–330.

Davies, C. N., ed. (1961). "Inhaled Particles and Vapours." Vol. I. Pergamon, Oxford.

Davies, C. N., ed. (1967). "Inhaled Particles and Vapours," Vol. II. Pergamon, Oxford.

Davies, C. N., Heyder, J., and Ramu, M. C. S. (1972). Breathing of half-micron aerosols: I. Experimental. *J. Appl. Physiol.* **32,** 591–600.

Dennis, W. L. (1971). The effect of breathing rate on the deposition of particles in the human respiratory system. *In* "Inhaled Particles III" (W. H. Walton, ed.), pp. 91–102. Unwin, Surrey.

Dodgson, J., McCallum, R. I., Bailey, M. R. and Fisher, D. R. (eds.) (1988). "Inhaled particles VI." *Ann. Occ. Hyg.* **32.**

Dolovich, M. B., Sanchis, J., Rossman, C., and Newhouse, M. T. (1976). Aerosol penetrance: A sensitive index of peripheral airways obstruction. *J. Appl. Physiol.* **40,** 468–471.

Forsyth, R. D., Cole, P., and Shephard, R. J. (1983). Exercise and nasal patency. *J. Appl. Physiol.* **55,** 860–865.

Gil, J., Bachofen, H., Gehr, P., and Weibel, E. R. (1979). Alveolar volume–surface area relation in air- and saline-filled lungs fixed by vascular perfusion. *J. Appl. Physiol.* **47,** 990–1001.

Goldberg, I. S., and Lourenco, R. V. (1973). Deposition of aerosols in pulmonary disease. *Arch. Intern. Med.* **131,** 88–91.

Gross, P., and deTreville, R. T. P. (1969). Emphysema and pneumoconiosis: An experimental study on their interrelationship. *Arch. Environ. Health* **18,** 340–349.

Gross, P., Pfitzer, E. A., and Tolker, E. (1965). Experimental emphysema: Its production with papain in normal and silicotic rats. *Arch. Environ. Health* **11,** 50–58.

Gross, P., Tuma, J., and deTreville, R. T. P. (1971). Emphysema and pneumoconiosis. *Arch. Environ. Health* **22,** 194–199.

Hahn, F. F., and Hobbs, C. H. (1979). The effect of enzyme-induced pulmonary emphysema in Syrian hamsters on the deposition and long-term retention of inhaled particles. *Arch. Environ. Health* **34,** 203–211.

Harbison, J. P. R., Charles, T. J., Monie, R. D. H., Seaton, A., Taylor, W. H., Westwood, A., and Williams, J. D. (1981). Arterial plasma histamines after exercise in normal individuals and in patients with exercise-induced asthma. *Clin. Sci.* **61,** 151–157.

Harbison, M. L., and Brain, J. D. (1983). Effects of exercise on particle deposition in Syrian golden hamsters. *Am. Rev. Respir. Dis.* **128,** 904–908.

Hayes, J. A., Korthy, A., and Snider, G. L. (1975). The pathology of elastase-induced panacinar emphysema in hamsters. *J. Pathol.* **117,** 1–14.

Hoppin, F. G., Stothert, J. C., Greaves, I. A., Lai, Y.-L., and Hildebrandt, J. (1986). Lung recoil: Elastic and rheological properties. *In* "Handbook of Physiology, Section 3, The Respiratory System, Volume III, The Mechanics of Breathing, Part 1" (P. T. Macklem and J. Mead, eds.), pp. 195–215. American Physiology Society, Bethesda, Maryland.

Hounam, R. F., and Morgan, A. (1977). Particle deposition. *In* "Respiratory Defense Mechanisms, Part I" (J. D. Brain, D. F. Proctor, and L. M. Reid, eds.), pp. 125–156. Dekker, New York.

Hounam, R. F., Black, A., and Walsh, M. (1969). Deposition of aerosol particles in the nasopharyngeal region of the human respiratory tract. *Nature (London)* **221,** 1254–1255.

Itoh, H., Ishii, Y., Maeda, H., Todo, G., Torizuka, K., and Smaldone, G. C. (1981). Clinical observations of aerosol deposition in patients with airways obstruction. *Chest (Suppl.)* **80,** 837–839.

Johanson, W. G., and Pierce, A. K. (1972). Effects of elastase, collagenase, and papain on structure and function of rat lungs *in vitro*. *J. Clin. Invest.* **51,** 288–293.

Kim, C. S., Brown, L. K., Lewars, G. G., and Sackner, M. A. (1983a). Aerosol rebreathing method for assessment of airway abnormalities: Theoretical analysis and validation. *Am. Ind. Hyg. Assoc. J.* **44,** 349–357.

Kim, C. S., Brown, L. K., Lewars, G. G., and Sackner, M. A. (1983b). Deposition of aerosol particles and flow resistance in mathematical and experimental airway models. *J. Appl. Physiol.* **55,** 154–163.

Landahl, H. D., Tracewell, T. N., and Lassen, W. H. (1952). Retention of airborne particulates in the human lung: III. *Arch. Ind. Hyg. Occup. Med.* **6,** 508–511.

Lippmann, M. (1970). Deposition and clearance of inhaled particles in the human nose. *Ann. Otol. Rhinol. Laryngol.* **79,** 519–528.

Lippmann, M. (1977). Regional deposition of particles in the human respiratory tract. *In* "Reactions to Environmental Agents, Handbook of Physiology, Section 9" (D. H. K. Lee, ed.), pp. 213–232. American Physiological Society, Bethesda, Maryland.

Lourenco, R. V., Loddenkemper, R., and Carton, R. W. (1972). Patterns of distribution and clearance of aerosols in patients with bronchiectasis. *Am. Rev. Respir. Dis.* **106,** 857–866.

Lundgren, D. L., Damon, E. G., Diel, J. H., and Hahn, F. F. (1981). The deposition, distribution, and retention of inhaled $^{239}PuO_2$ in the lungs of rats with pulmonary emphysema. *Health Phys.* **40,** 231–235.

McFadden, Jr., E. R., and Ingram, Jr., R. H. (1982). Large and small airway effects with exercise and other bronchoconstrictor stimuli. *Eur. J. Respir. Dis. (Suppl.)* **63,** 99–104.

Macklem, P. T. (1971). Airway obstruction and collateral ventilation. *Physiol. Rev.* **51,** 368–436.

Macklem, P. T., Hogg, W. E., and Brunton, J. (1973). Peripheral airway obstruction and particulate deposition in the lung. *Arch. Intern. Med.* **131,** 93–97.

Märtens, A., and Jocobi, W. (1974). Die *in vivo* Bestimmung der Aerosolteilchendeposition im Atemtrakt bei Mund-bzw. Nasenatmung. *In* "Aerosole in Physik, Medizin und Technik" (W. Stahlhofen, ed.), pp. 117–121. Gesellschaft für Aerosolforschung, Bad Soden, West Germany.

Milic-Emili, J. (1974). Small airway closure and its physiological significance. *Scand. J. Respir. Dis. (Suppl.)* **85,** 181–189.

Morrow, P. F., Chairman, Task Group on Lung Dynamics (1956). Deposition and retention models for internal dosimetry of the human respiratory tract. *Health Phys.* **12,** 173–188.

Muir, D. C. F., and Davies, C. N. (1967). The deposition of 0.5 μm diameter aerosols in the lungs of man. *Ann. Occup. Hyg.* **10,** 161–174.

Pattle, R. E. (1961). The retention of gases and particles in the human nose. *In* "Inhaled Particles and Vapours" (C. N. Davies, ed.), Vol. 1, pp. 302–309. Pergamon, Oxford.

Pavia, D., Thomson, M., and Shannon, H. S. (1977). Aerosol inhalation and depth of deposition in the human lung. *Arch. Environ. Health* **32,** 131–137.

Proctor, D. F. (1973). The upper respiratory tract and the ambient air. *Clin. Notes Respir. Dis.* **12,** 2–10.

Punte, C. L., Owens, E. J., Krackow, E. H., and Cooper, P. L. (1958). Influence of physical activity on the toxicity of aerosols and vapors: Inhalation toxicity of paraoxon and sarin in rats. *AMA Arch. Ind. Health* **17,** 34–37.

Ramanna, L., Tashkin, D. P., Taplan, G. V., Elam, D., Detels, R., Coulson, A., and Rokaw, S. N. (1975). Radioaerosol lung imaging in chronic obstructive pulmonary disease. *Chest* **68,** 634–640.

Skornik, W. A., and Brain, J. D. (1989). Effect of sulfur dioxide on pulmonary macrophage endocytosis at rest and during exercise. *Am. Rev. Resp. Dis.* (Submitted).

Smaldone, G. C., Itoh, H., Swift, D. L., and Wagner, H. N. (1979). Effect of flow-limiting segments and cough on particle deposition and mucociliary clearance in the lung. *Am. Rev. Respir. Dis.* **120,** 747–758.

Snider, G. L., Sherter, C. B., Koo, K. W., Karlinsky, J. B., Hayes, J. A., and Franzblau, C. (1977). Respiratory mechanics in hamsters following treatment with endotracheal elastase or collagenase. *J. Appl. Physiol.* **42,** 206–215.

Snider, G. L., Lucey, E. C., and Stone, P. J. (1986). Animal models of emphysema. *Am. Rev. Respir. Dis.* **133,** 149–169.

Stokinger, H. E., Wagner, W. D., and Wright, P. G. (1956). Studies of ozone toxicity: I. Potentiating effects of exercise and tolerance development. *AMA Arch. Ind. Health* **14,** 158–162.

Sweeney, T. D., Brain, J. D., Tryka, A. F., and Godleski, J. J. (1983a). Retention of inhaled particles in hamsters with pulmonary fibrosis. *Am. Rev. Respir. Dis.* **128,** 138–143.

Sweeney, T. D., Skornik, W. A., Godleski, J. J., and Brain, J. D. (1983b). Collection efficiency is decreased in hamsters with pulmonary fibrosis. *Fed. Proc., Fed. Am. Soc. Exp. Biol.* **42,** 1349.

Sweeney, T. D., Brain, J. D., Skornik, W. A., and Godleski, J. J. (1985). Chronic bronchitis produced by SO_2 alters the deposition pattern of inhaled aerosol in rats. *Am. Rev. Respir. Dis.* **131,** A195.

Sweeney, T. D., Skornik, W. A. Godleski, J. J., and Brain, J. D. (1986). Collection efficiency of inhaled particles is decreased in hamsters with pulmonary emphysema. *Am. Rev. Respir. Dis.* **133,** A47.

Sweeney, T. D., Brain, J. D., Leavitt, S. A., and Godleski, J. J. (1987). Emphysema alters the deposition pattern of inhaled particles in hamsters. *Am. J. Pathol.* **128,** 19–28.

Taplan, G. V., Tashkin, D. P., Chopra, S. K., Anselmi, D. E., Elam, D., Calvarese, B., Coulson, A., Detels, R., and Rokaw, S. N. (1977). Early detection of chronic obstructive pulmonary disease using radionuclide lung-imaging procedure. *Chest* **71,** 567–575.

Taulbee, D. P., Yu, C. P., and Heyder, J. (1978). Aerosol transport in the human lung from analysis of single breaths. *J. Appl. Physiol.* **44,** 803–812.

Thomson, M. L., and Pavia, D. (1973). Particle penetration and clearance in the human lung. *Arch. Environ. Health* **29,** 214–219.

Tryka, A. F., Skornik, W. A., Godleski, J. J., and Brain, J. D. (1982). Potentiation of bleomycin-induced lung injury by exposure to 70% oxygen: Morphologic assessment. *Am. Rev. Respir. Dis.* **126,** 1074–1079.

Tryka, A. F., Godleski, J. J., Skornik, W. A., and Brain, J. D. (1983). Progressive pulmonary fibrosis in hamsters. *Exp. Lung Res.* **5,** 155–171.

Valberg, P. A., Brain, J. D., Sneddon, S. L., and LeMott, S. R. (1982). Breathing patterns influence aerosol deposition sites in excised dog lungs. *J. Appl. Physiol.* **53,** 824–837.

Walton, W. H., ed. (1971). "Inhaled Particles III." Unwin, Surrey.

Walton, W. H., ed. (1977). "Inhaled Particles IV." Pergamon, Oxford.

Walton, W. H., ed. (1982). "Inhaled Particles V." Pergamon, Oxford.

Weibel, E. R., and Gil, J. (1977). Structure-function relationship at the alveolar level. *In* "Lung Biology in Health and Disease, Vol. 3, Bioengineering Aspects of the Lung" (J. B. West, ed.), pp. 1–81. Dekker, New York.

Zeltner, T. B., Sweeney, T. D., Skornik, W. A., and Brain, J. D. (1988). Effects of exercise on the retention of inhaled 0.9 μm particles in hamsters. *Am. Rev. Respir. Dis.* **137,** 315.

Chapter 16

Nasopharyngeal Uptake of Ozone in Humans and Animals

T. R. Gerrity
U.S. Environmental Protection Agency
Health Effects Research Laboratory
Research Triangle Park, North Carolina 27711

I. INTRODUCTION

There have been numerous studies documenting the pulmonary effects of acute ozone (O_3) exposure in humans and of chronic O_3 exposure in laboratory animals (EPA, 1986). Despite these experiments there is no direct evidence linking long-term O_3 exposures of humans to chronic health effects. Because of obvious ethical problems, chronic exposures of human volunteers to O_3 in controlled environmental chambers are not possible. Thus, inferences of chronic health effects in humans from long-term exposures to environmental O_3 must be made by utilizing data from chronic animal exposure studies.

A requirement for extrapolation from chronic animal toxicity data to humans is a knowledge of the relative O_3 sensitivity in laboratory animals compared to humans. Relative sensitivity data can only be useful if there is information relating the O_3 dose delivered to sensitive tissue in both animals and humans. There have been several studies of respiratory uptake of

Disclaimer: The research described in this article has been reviewed by the Health Effects Research Laboratory, U.S. Environmental Protection Agency, and approved for publication. Approval does not signify that the contents necessarily reflect the views and policies of the Agency nor does mention of trade names or commercial products constitute endorsement or recommendation for use.

O_3 in animals (Miller *et al.,* 1979; Wiester *et al.,* 1988b; Yokoyama and Frank, 1972) and one in humans (Gerrity *et al.,* 1988). Figure 16–1 shows the results of these studies for uptake by the nasopharynx (NP); the lower respiratory tract (LRT), tracheobronchial and gas exchange regions of the respiratory tract; and the total respiratory tract.

As shown in Fig. 16–1, the amount of respiratory tract O_3 uptake measured in these studies varies widely, with no apparent patterns. This variability may be due to a number of factors:

1. The different techniques employed in these studies may have introduced systematic errors for which there has been no accounting.
2. There are likely to be species differences in local O_3 uptake coefficients that, coupled with interspecies anatomical differences, account for some of the variability.
3. The respiratory flows *(Q)* and volumes *(V)* which are important factors governing O_3 uptake used in some of these studies tend to be non-physiologic, thus potentially affecting O_3 uptake dynamics and contributing to variability. If the influences of *Q* and *V* can be isolated, estimations of experimental error and interspecies differences in O_3 uptake may be possible.

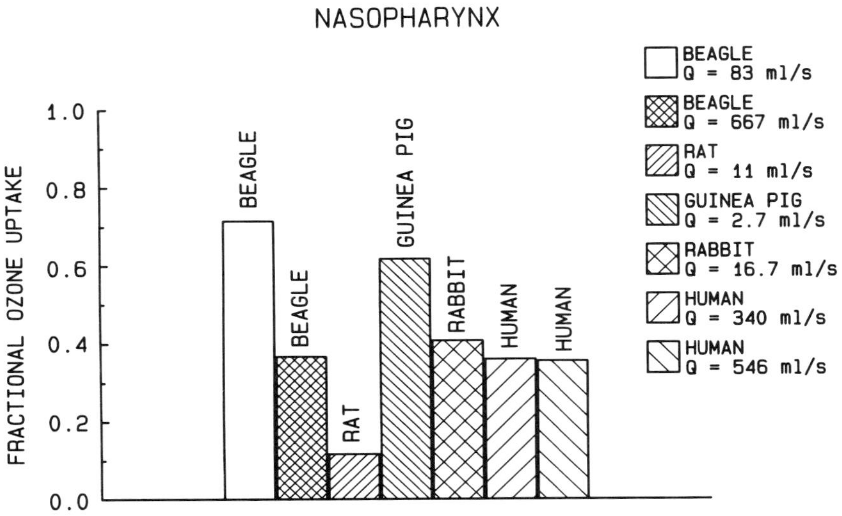

Fig. 16–1 O_3 uptake expressed as a fraction of the O_3 presented to the nasopharynx, lower respiratory tract, and total respiratory tract. *(Figure continues)*

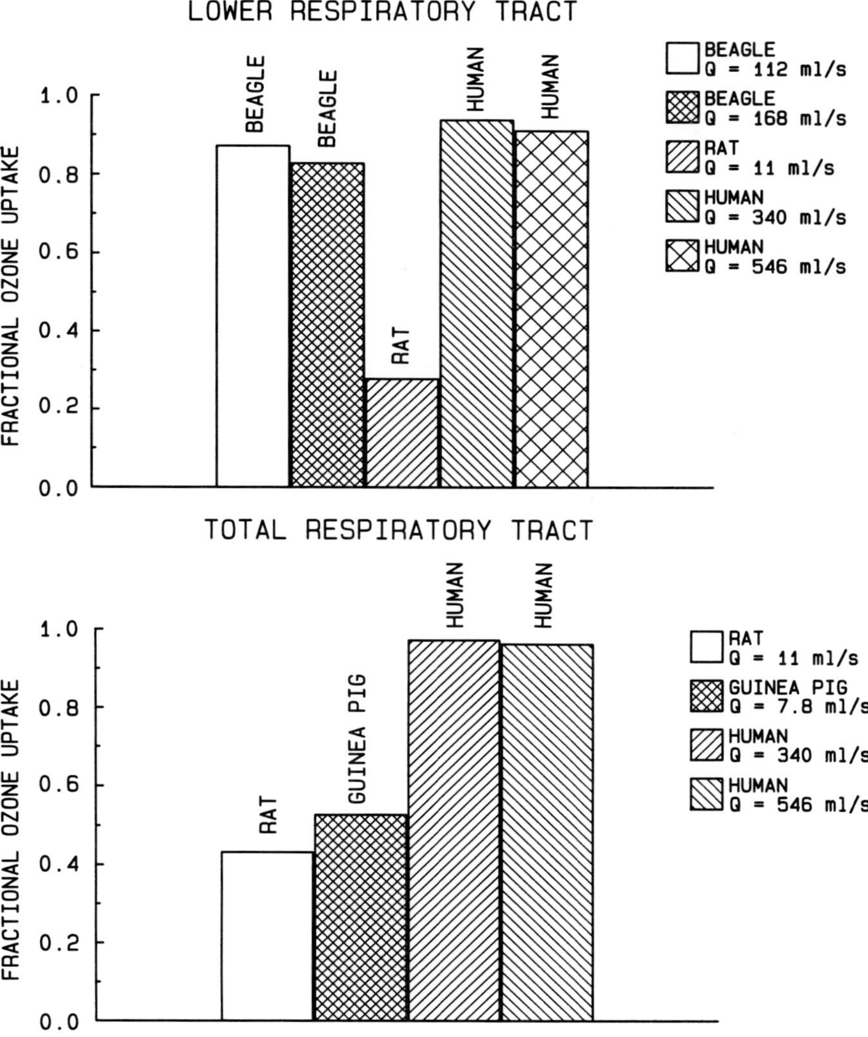

Fig. 16–1 *(continued)*

Models of O_3 uptake in the LRT, such as that of Miller *et al.* (1985), incorporate *Q* and *V* and are thus capable of examining their influences on O_3 uptake in this region. A complete picture of respiratory tract O_3 uptake, however, must include the upper respiratory tract (URT-nasopharynx, oropharynx, and larynx), since uptake by these structures will determine the amount of O_3 delivered to the LRT and thus greatly influence the O_3 dose to the LRT. Aharonson *et al.* (1974) developed a model of NP O_3 uptake that depends on Q, NP contact surface area, and a local O_3 uptake coefficient. The model, however, was only applied to the case of O_3 uptake in the canine NP.

In this paper the experimental data on NP uptake of O_3 in several animal species will be reviewed in the context of the model of Aharonson *et al.* (1974). Allometric equations relating breathing patterns among animal species will be incorporated into this model to help interpret the NP data and assess the consistency of data across all species studied.

II. NASOPHARYNGEAL O_3 UPTAKE DATA

Data on NP O_3 uptake, measured in a variety of animal species and humans are reviewed first. So the subsequent analysis will not be complicated by questions of the dependence of uptake on O_3 concentration, only data collected for concentrations at 0.3 and 0.4 ppm are examined. Data of Wiester *et al.* (1987) on regional O_3 uptake at 1 ppm are included since they are the only data of that type available in rats. Table 16–1 summarizes the ventilation and NP O_3 uptake data from these studies. The NP uptake data are presented as the fraction of O_3 delivered unidirectionally to the NP that is removed by the NP (F_{np}).

There are two broad types of studies represented here; unidirectional flow studies in anesthetized, tracheotomized, ventilated animals; and noninvasive studies conducted during unanesthetized, spontaneous breathing.

Table 16–1 Measurements of Ventilation and PNP O_3 Uptake in Animals and Humans[a]

	Beagle	Beagle	Rat[b]	Guinea Pig	Rabbit	Human	Human
$[O_3]$ (ppm)	0.3	0.3	0.3/1.0	0.4	0.4	0.4	0.4
W(grams)	16,000	16,000	299	430	2780	74,500	74,500
V_t(ml)	NA[c]	NA	2.6	NA	NA	782	666
f(bpm)	NA	NA	113	NA	NA	12	24
V_E(ml/min)	NA	NA	293	NA	NA	9700	15,800
Q_m(ml/sec)[3]	83.3	667	11.3	2.7	16.7	340	546
F_{np}	0.72	0.37	0.13	0.62	0.41	0.36	0.36

[a] Ventilation and NP uptake data in beagle dogs (Yokoyama and Frank, 1972), rats (Wiester *et al.*, 1987a, b), guinea pigs (Miller *et al.*, 1979; Wiester *et al.*, 1997a), rabbits (Miller *et al.*, 1979), and humans (Gerrity *et al.*, 1988).

[b] F_{np} was derived for rats from the total uptake data of Wiester *et al.* (1987a) at 0.3 ppm and the regional distribution data of Wiester *et al.* (1987b) at 1 ppm.

[c] NA, not applicable. These studies were unidirectional flow studies through the NP.

[d] O_m is either the unidirectional flow or average inspiratory flow.

Yokoyama and Frank (1972) measured O_3 uptake in the NP of anesthetized, ventilated beagle dogs at two different flow rates. Ozonated air was passed through the NP by a cannula affixed to the exterior nares. Air was sampled from a tracheal cannula to measure O_3 concentration and derive NP uptake. Miller *et al.* (1979) conducted similar experiments in guinea pigs and rabbits. Both of these experiments measured uptake by the larynx which for the purpose of this paper is assumed to have been small. Wiester *et al.* (1987a) measured total respiratory tract uptake in awake, unanesthetized, nose-breathing rats and guinea pigs. In these experiments animals were exposed nose-only to O_3 while sealed in a plethysmograph. Total respiratory tract uptake of O_3 was determined by measuring the difference between upstream and downstream O_3 mass. Wiester *et al.* (1987) measured the relative distribution of O_3 taken up within the respiratory tract of unanesthetized, nose-breathing rats that inhaled ^{18}O (a stable isotope of oxygen) enriched O_3 (Hatch and Aissa, 1987). After a nose-only exposure to the ^{18}O enriched O_3, animals were sacrificed. URT and LRT tissues were analyzed by mass spectrometry for relative increases in the amount of ^{18}O present. Since the work of Wiester *et al.* (1988) used the same experimental system and the same rat strain as Wiester *et al.* (1987), the O_3 uptake data from both rat studies are combined in Table 16–1 to yield F_{np} for rats. It is assumed that the relative distribution of O_3 between URT and LRT measured by Wiester *et al.* (1988) also occurred in the total uptake study of Wiester *et al.* (1988). Thus only the weight and ventilation data from the study of Wiester *et al.* (1988) are used. The one study in humans (Gerrity *et al.*, 1988) measured O_3 uptake by the URT and LRT during spontaneous breathing as a function of O_3 concentration, breathing frequency, and mode of breathing (nasal, oral, and oronasal). Subjects were exposed to O_3 in a controlled environmental chamber. A small sampling tube was inserted through the nose with the tip positioned in the posterior pharynx. Ozonated air was sampled continuously during spontaneous breathing into a rapidly responding O_3 analyzer. We will use only the data from this experiment for nose breathing.

III. MODEL PREDICTIONS OF DEPENDENCE OF NASOPHARYNGEAL O_3 UPTAKE ON SPECIES

A. Uptake Dynamics

The model of NP O_3 uptake of Aharonson *et al.* (1974) assumed quasi-steady, unidirectional flow. It was further assumed that the partial pressure of O_3 in the blood stream of nasal tissue can be neglected. With these assumptions, an expression for the partial pressure of O_3 exiting the NP was derived. This expression has been rearranged in the following form to yield F_{np}, the fractional uptake of O_3 in the NP:

$$F_{np} = 1 - e^{-RP_BA/Q_m} \tag{1}$$

where Q_m is the measured unidirectional flow, A is the surface area of the NP in contact with the airstream, R is the local O_3 uptake coefficient, and P_B is the barometric pressure.

The parameters R, P_B, and A are likely to be species dependent. Thus, this equation can be used for an interspecies comparison on NP uptake provided R, P_B, and A are known for each species of interest. The next section presents an approach to extending Equation (1) to accommodate interspecies comparisons.

B. Allometric Equations of Ventilation

In order to apply Equation (1) to the NP uptake data it must be reexpressed so that it scales with animal size. To accomplish this, allometric equations derived by Guyton (1947), relating resting ventilation to body weight (W), are used. Accordingly, the tidal volume (V_t) of any mammalian species is approximated by:

$$V_t = 0.0074W \tag{2}$$

where W is in g and V_t is in ml; and the breathing frequency (f) is approximated by:

$$f = 295W^{-0.25} \tag{3}$$

From these equations, predictions of minute volume (V_E) and average inspiratory flow (Q), under the assumption that $Q = 2V_E$, can be derived. Table 16–2 shows the predicted resting ventilation for each of the species of interest.

If it is assumed that the surface area (A) of the NP scales as $W^{2/3}$

Table 16–2 Resting Ventilation Predictions[a]

	Beagle	Rat	Guinea Pig	Rabbit	Human
W(g)	16,000	299	430	2780	74,500
V_t(ml)	118	2.2	3.2	20.6	551
f(bpm)	26	71	65	41	18
V_E(ml/min)	3068	156	208	845	9918
Q(ml/sec)	102	5.2	6.9	28.1	331

[a] Weights used are actual measured values. From Guyton (1947).

(Gunther, 1975) then Equation (1) can be rewritten as:

$$F_{np} = 1 - e^{-R'P_B r} \tag{4}$$

where r is the ratio of the predicted resting flow to the measured flow (i.e., $r = Q/Q_m$), and R' is the product of the local O_3 uptake coefficient R and a slowly varying function of W which behaves as $W^{-1/12}$.

Figure 16–2 is a plot of the natural log of $1 - F_{np}$ as a function of r. All of the data lie roughly along the same line, though a correlation analysis is not significant (r = −0.592). If the beagle data are dropped out, however, the correlation becomes significant (r = 0.898, $p < 0.05$).

Equation (4) predicts that all of the NP uptake data should fall roughly on the same curve when plotted as a function of r provided the local O_3 uptake coefficient (R) is both species and flow independent. Deviation of O_3 uptake, among species, from a single line is predicted to be induced by the small W dependence of Equation (4). To examine the effect of the W dependence predicted by Equation (4), a reference slope is arbitrarily established between the value of 1 at r = 0 and the single guinea pig point. The lower dashed line represents the predicted r behavior of rat data, and the upper dashed line represents the predicted r behavior of human data. If it is assumed that the guinea pig data are accurate, the region bounded by the dashed lines should encompass all of the other data, provided R is species independent. This, however, is not the case.

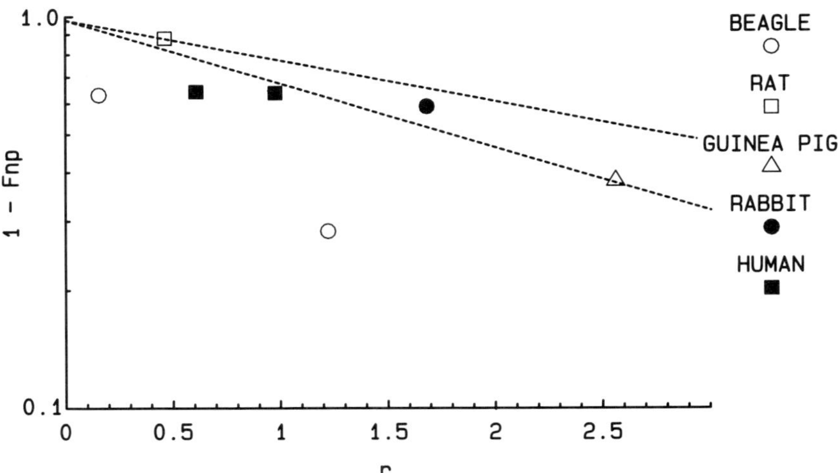

Fig. 16–2 The fraction of O_3 that passes through the nasopharynx ($1 - F_{np}$) as a function of flow relative to the predicted resting flows (r) of various species. The dashed lines represent the boundaries of $1 - F_{np}$ predicted by equation (4), referenced to the guinea pig data.

IV. DISCUSSION AND CONCLUSIONS

Figure 16–2 shows that there is some consistency of measured NP uptake among the various species, i.e., uptake falls with increasing flow. Based on the limited data available, it appears that there is species dependence of local uptake coefficients (R). If NP uptake, as predicted by Equation (4), only depends upon relative flow (r) and weight (W), all of the data should fall within the bounds, defined by the guinea pig data point, of the two dashed lines of Fig. 16–2. The rat and rabbit data are within these bounds. The human and beagle dog data, however, fall well outside. This suggests that there may be a significant dependence of R, especially when the species are widely separated phylogenetically, as is the case when comparing rodents, humans, and dogs. Variance of the data from Equation (4) may also be due to significant interspecies differences in tissue surface area. It was assumed that the weight dependence of NP surface area approximated that of the lower respiratory tract by scaling NP surface area as $W^{2/3}$. This dependence may not be strictly valid across species.

Aside from the issue of species dependence, another question that arises as a result of this analysis concerns the flow dependence of the local O_3 uptake coefficient (R). The data of Yokoyama and Frank (1972) and Gerrity *et al.* (1988) suggest that R for beagle dogs and humans is flow dependent, increasing with increasing flow. Aharonson *et al.* (1974) offer several explanations why this may occur. One is that increasing flow at a fixed concentration would increase the amount of O_3 available passing along the linear axis of the NP, thereby increasing the partial pressure difference between the airstream and the tissue and thus increasing the uptake coefficient. This explanation, however, is contradicted by data in rodents showing that when the partial pressure difference is raised by increasing O_3 concentration, no concentration dependence of O_3 uptake was observed (Miller *et al.,* 1979; Wiester *et al.,* 1988). Furthermore, an inverse dependence on concentration was seen in beagle dogs (Yokoyama and Frank, 1972). Another possible explanation is that the amount of NP surface area exposed to O_3 may increase with flow. Flow patterns around nasal turbinates may vary considerably with flow. Increasing flow may also decrease the boundary layer thickness, thus facilitating more rapid diffusion of O_3 to the tissue. Different flows may affect the hydration of nasal mucosa, potentially altering O_3 uptake. With the limited data available, no definitive conclusions can be drawn. Moreover, the existing data base suggests that a flow-dependent R may not occur in all species, further complicating the issue.

In summary, by normalizing the flows used in NP O_3 uptake studies to the predicted resting flows for each animal species, we have found that there is some consistency among various sets of data in spite of the variety of techniques employed to measure NP O_3 uptake. The present analysis further suggests that there may be a significant dependence of local O_3 uptake coefficients on species. The extent to which O_3 uptake differs between two animal species seems to depend upon their phylogenic proximity.

In all of this analysis it must be kept in mind that the extrapolation of the NP O_3 uptake model using allometric equations is subject to many sources of error, especially since the range of flows considered extends over two orders of magnitude between rats and humans. A more complete picture of interspecies differences of O_3 uptake will require a larger number of studies that cover a broad range of flow. The present work, however, points in the right direction toward methods of data comparison.

REFERENCES

Aharonson, E. F., Menkes, H., Gurtner, G., Swift, D. L., and Proctor, D. F. (1974). Effect of respiratory airflow rate on removal of soluble vapors by the nose. *J. Appl. Physiol.* **37,** 654–657.

Gerrity, T. R., Weaver, R. A., Berntsen, J. H., and O'Neil, J. J. (1988). Nasopharyngeal and lung removal of ozone during tidal breathing in man. *J. Appl. Physiol.* **65,** 393–400.

Gunther, B. (1975). Dimensional analysis and theory of biological similarity. *Physiol. Rev.* **55,** 659–699.

Guyton, A. C. (1947). Analysis of respiratory patterns in laboratory animals. *Am. J. Physiol.* **150,** 78–83.

Hatch, G. E., and Aissa, M. (1987). Determination of absorbed dose of ozone (O_3) in animals and humans using stable isotope (oxygen-18) tracing. *Proc. Annu. Meet. Air Pollut. Control Assoc.*

Miller, F. J., McNeal, C. A., Wirtz, J. M., Gardner, D. E., Coffin, D. L., and Menzel, D. B. (1979). Nasopharyngeal removal of ozone in rabbits and guinea pigs. *Toxicology* **14,** 273–281.

Miller, F. J., Overton, J. H., Jaskot, R. H., and Menzel, D. B. (1985). A model of the regional uptake of gaseous pollutants in the lung. I. The sensitivity of the uptake of ozone in the human lung to lower respiratory tract secretions and exercise. *Toxicol. Appl. Pharmacol.* **79,** 11–27.

U. S. Environmental Protection Agency (1988). "Air Quality Criteria for Ozone and Other Photochemical Oxidants." EPA-600/8-84-020eF.

Wiester, M. J., Tepper, J. S., King, M. E., Mènache, M. G., and Costa, D. L. (1988). Comparative study of ozone (O_3) uptake in three strains of rat and in guinea pig. *Toxicol. Appl. Pharmacol.* **96,** 140–146.

Wiester, M. J., Hatch, G. E., Aissa, M., and Norwood, J. (1987). Respiratory dosimetry of ozone following chronic exposure in rats. Abstract presented at Symposium on Extrapolation Modeling of Inhaled Particles and Gases: Lung Dosimetry, Duke University, October, 1987.

Yokoyama, E., and Frank, R. (1972). Respiratory uptake of ozone in dogs. *Arch. Environ. Health* **25,** 132–138.

Part Four

New Methods for Determining Dosimetry of Inhaled Particles and Gases

Chapter 17

New Methods: Introduction

J. J. Ospital
Southern California Edison
Rosemead, California 91770

The major goal of extrapolation modeling is to use dose-response relationships derived in laboratory experiments to predict quantitatively health effects in humans exposed to environmentally relevant particles and gases. To do this for a particular agent, we need to know the total dose, the differences in dose within different regions of the lung, and variations in dose and regional deposition that occur between animal species for exposure to a given concentration of particles or gas.

The three papers in this section discuss new methods that are being used to characterize inhaled particles and to measure deposition of particles and gases in the lung. This information is essential for constructing defensible models to extrapolate health effects from laboratory animals to humans.

Positron Emission Tomography (PET) methods are presented by Coleman and associates. Techniques employing PET are being used to assess a variety of physiological functions, including lung function. For inhalation dosimetry, PET uses short-lived isotopes that emit positrons, and produces computer-enhanced images of these isotopes after they deposit in the lungs. This allows researchers to determine sites of particle deposition. PET can be used with morphometric analyses of the lung to identify specific sites and structures associated with particle or gas impacts, and regions of the lung which are dosed can be determined for specific aerosols.

An advantage of this method is that it is noninvasive. It can be used repeatedly with different agents in the same animal or humans, can quantitate aerosol or gas deposition, and can be used to verify dosimetry models.

The next two chapters discuss ways to characterize inhaled aerosols. There are various methods to measure aerosol characteristics and concentration, but one limitation is that they do not provide real-time data.

The chapter by Mazumder and colleagues addresses these issues and describes a method that can provide information on the size and concentration of particles during the breathing cycle. The system uses a laser doppler technique called single particle aerodynamic relaxation time (SPART). This measures total deposition of aerosols on a breath-by-breath basis. Two laser beams at different wavelengths are used. Information can be obtained for polydisperse aerosols, and on particles of varying morphology.

In the last paper of this section, Gebhart presents a method using light scattering to measure the amount of aerosol in inspired and expired air. This method uses a single beam of light and measures the amount scattered to calculate the number of particles. It provides real-time data, but a disadvantage is that particles of spherical shape and known refractive index are needed. These requirements can be met for many laboratory-generated aerosols, such as sodium chloride, but not for many particles of interest that are present in the environment.

In summary, the methods described in this section can answer two important questions regarding dosimetry of inhaled particles and gases. Measuring aerosols by means of light scattering and single-particle aerodynamic relaxation time analysis can give us the total dose of inhaled particles that is retained in the lung. Use of positron emission tomography can give the amounts deposited at specific regions within the lung. These are powerful new tools that can be used in extrapolating dose-effect relationships from laboratory animals to humans.

Chapter 18

Regional Dosimetry of Inhaled Particles Using SPECT

R. E. Coleman
R. J. Jaszczak
K. L. Greer
Department of Radiology
Duke University Medical Center
Durham, North Carolina

R. R. Mercer
J. D. Crapo
Department of Medicine
Duke University Medical Center
Durham, North Carolina

I. INTRODUCTION

The reactions of inhaled particles and gases in the lung are related to the site of deposition (Bates, 1972). Radiolabeled aerosols (Brain *et al.*, 1976; Foord *et al.*, 1978; MacIntyre *et al.*, 1983, 1985; Susskind *et al.*, 1986; Kay *et al.*, 1986) have been used for many years to study the effects of particle size on lung deposition and retention. Gases labeled with positron-emitting isotopes have been used in studies of lung function (Ahluwalia *et al.*, 1981; Hughes *et al.*, 1985; Schuster *et al.*, 1985; Rhodes *et al.*, 1981). These studies have demonstrated the ability for quantitative study of regional lung function. Schuster *et al.* (1985) demonstrated that the regional hematocrit and distribution of extravascular lung water can be obtained in normal dogs using positron emission tomography (PET) and ^{15}O labeled water, carbon monoxide, and ^{68}Ga transferrin. Positron tomography has

Disclaimer: Although the research described in this article has been funded wholly or in part by the U.S. Environmental Protection Agency, it has not been subjected to the Agency's required peer and policy review and, therefore, does not necessarily reflect the views of the Agency, and no official endorsement should be inferred.

also been used to measure lung density in patients with fibrosing lung disease (Peacock *et al.*, 1984).

Positron-emitting isotopes available for studies with PET include ^{15}O ($T_{1/2} = 2$ min), ^{13}N ($T_{1/2} = 10$ min), ^{11}C ($T_{1/2} = 20$ min), and ^{18}F ($T_{1/2} =$ 109 min). These isotopes are cyclotron-produced and require the cyclotron to be on site since the half-lives of the isotopes are so short. Positron-emitting gases such as CO, CO_2, O_2, and NO are readily available for imaging in centers with cyclotrons since these have been used in a variety of studies (Madsen *et al.*, 1981; Wolf, 1981).

Both PET and single photon emission computed tomography (SPECT) are quantitative image modalities which have been used for studying lung function (Coleman *et al.*, 1986; Frankel and Coleman, 1983; Jaszczak *et al.*, 1985; Osborne *et al.*, 1981, 1982, 1983a, b, 1985). Although limitations related to correcting the emitted radiation for scatter and attenuation exist, quantitation of the distribution of activity in the lung is feasible (Jaszczak *et al.*, 1984; Osborne *et al.*, 1982).

The purpose of this study was to determine the site of deposition of inhaled particles using SPECT combined with a subsequent morphometric analysis of the lung tissue. This was done by correlating the distribution of the inhaled particles from the imaging study with morphometric analysis of matched anatomical sites in the lung. Selected regions of interest were chosen for analysis on the imaging study, and the same sites were compared with morphometric data carefully obtained from the same site to detect distribution of the aerosol deposition in the airways and alveoli. The imaging studies provide a determination of the relative magnitude of deposition of the radiolabeled agent within lung sample volumes of less than 1 cm^3. The morphometric studies were used to learn which anatomic structures were associated with the sites of high and low deposition of radioactivity. Since the localization data are obtained noninvasively, these studies can be repeated several times before the morphometric studies are done. This means that the distribution of different particles can be compared in the same animal.

II. METHODS

A. Imaging Studies

The preliminary studies were performed in mongrel dogs (12–15 kg) using ^{99m}Tc-labeled aerosols and imaging with single photon emission computed tomography. The animal was anesthetized with pentobarbital and intubated with a 7-mm i.d. endotracheal tube. After the animal was positioned in the SPECT scanner, the endotracheal tube was connected to a jet nebulizer (Ultravent Mallinckrodt Inc., St. Louis, Missouri) containing

100 mCi of ^{99m}Tc sulfur colloid. The radiolabeled aerosol was prepared by passing oxygen at 10 liters per min through the nebulizer. ^{99m}Tc sulfur colloid was used since it does not traverse the epithelial membrane (Osborne *et al.,* 1985). The nebulizer has been shown to generate an aerosol with an aerodynamic mass median diameter of 0.8 μm and a geometric standard deviation of 1.85 (Kay *et al.,* 1986). The aerosol was administered for a 10-min duration and then an 8-min SPECT scan was performed (Fig. 18–1).

The second phase of the study involved preparing the animal for a morphometric analysis of lung sites that could be closely correlated with the initial SPECT scan. To accomplish this, the animal was removed from the SPECT scanner and sacrificed with an overdose of anesthesia. The lungs were then vascular-perfusion-fixed for 30 min using buffered 2% glutaraldehyde plus 1% paraformaldehyde and then the anterior chest wall was removed. The animal was repositioned into the SPECT scanner and a 20-min SPECT scan was performed. Without moving the animal, 10 μl of buffer containing 1% osmium tetroxide and 70 μCi of ^{99m}Tc macroaggregates of albumin (MAA) were injected into each of nine sites. A 20-min SPECT scan was repeated (Fig. 18–2). This injection of markers gave a focal region of tissue that could be specifically localized on the SPECT scan and which was also fixed and darkly stained, making it easy to dissect for morphometric studies. The use of injections of combined radioactive and visual markers was required because previous attempts at correlating imaging and anatomic data from the lungs had been suboptimal. These attempts involved the use of external markers on the skin or lung surface of the dog which did not prove satisfactory for correlating the anatomic and imaging data. A pilot study demonstrated that the osmium tetroxide and either ^{99m}Tc sulfur colloid or ^{99m}Tc macroaggregates of albumin could be mixed and injected through a syringe.

The SPECT scans were reconstructed in a routine manner (Jaszczak *et al.,* 1985). The regions of interest were 2 cm $\times$ 2 cm $\times$ 0.3 cm (slice thickness) for a total volume of 1.2 cm^3. On the scans with the injected sources, 3-mm slices containing the point sources were identified. Regions of 2 cm $\times$ 2 cm were then placed around the point sources and the identical region was identified on the slice without the injected sources. The counts in these regions from the scan obtained prior to the injection of point sources were determined.

B. Morphometric Analysis

Morphometric analysis was performed after the fixed lung volumes were obtained by volume displacement. The nine osmium tetroxide stained sites were identified from the needle tracks. One cubic centimeter of tissue was obtained with its center containing the osmium tetroxide-labeled site. The

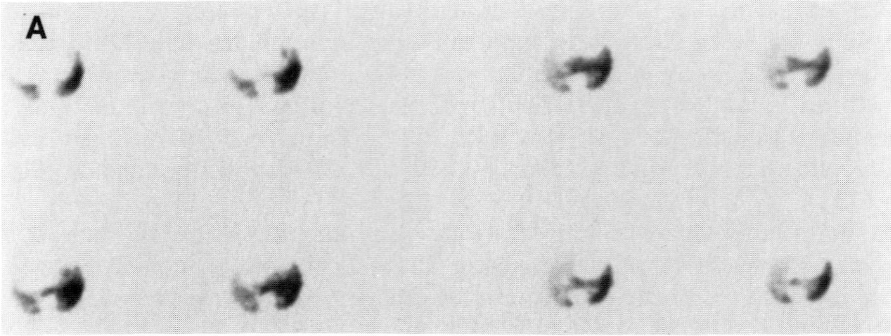

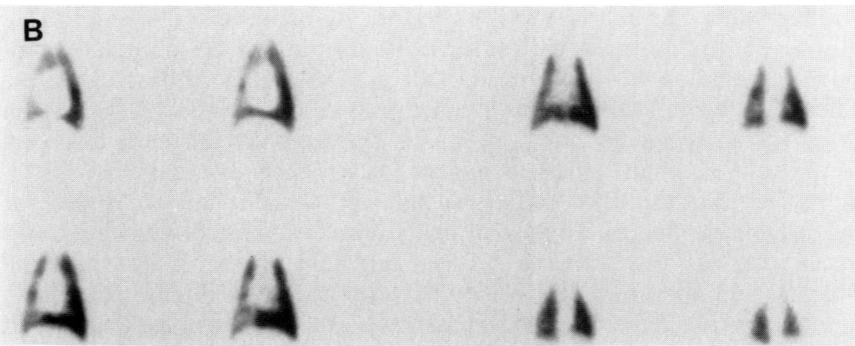

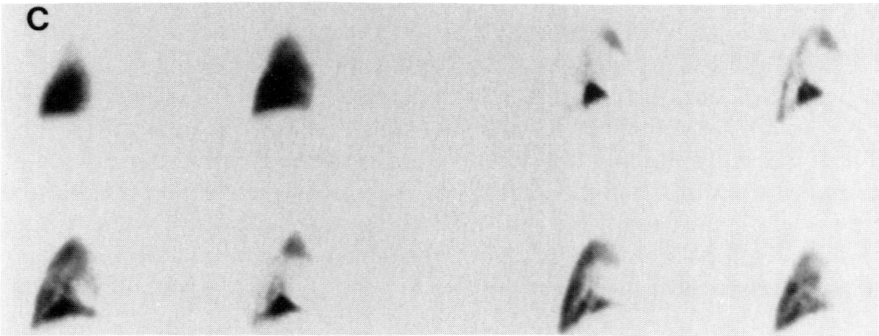

Fig. 18–1 **SPECT images of ^{99m}Tc aerosol in the transverse (A), coronal (B), and sagittal (C) views.** The transverse sections are 6 mm thick and begin near the lung bases in the upper left and extend into the cardiac region in the lower right image. The transverse images were formed by reformatting for the coronal images. They start anteriorly and move posteriorly in 12-mm thick images. The sagittal images start in the left lung and move through the right lung in 12-mm thick images.

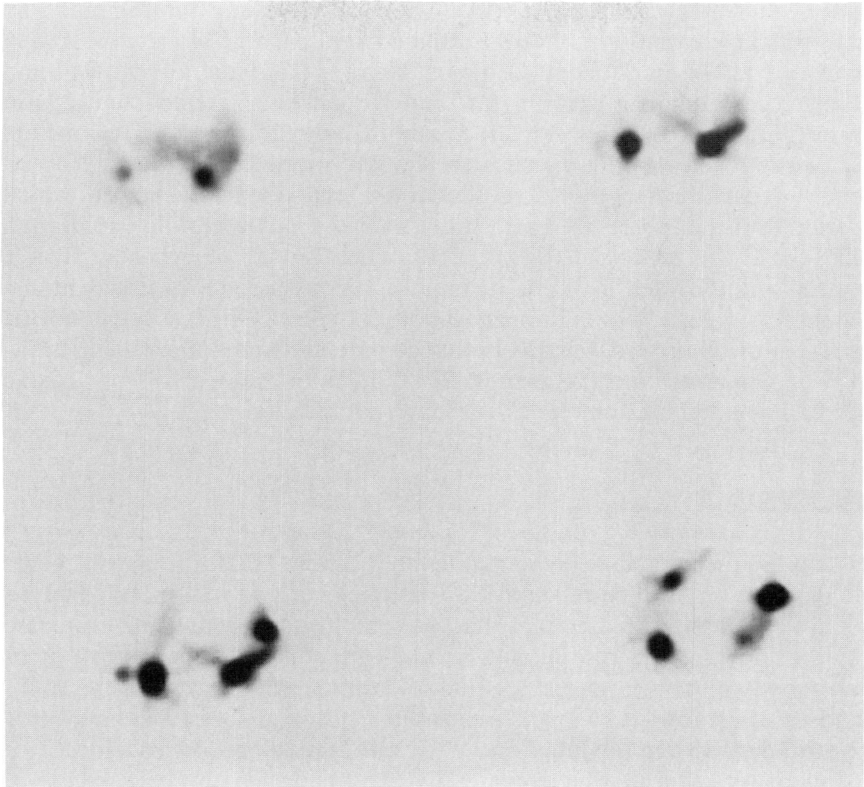

Fig. 18–2 Four transverse sections demonstrating labeling with point sources (dense black circles). These scans were used to localize image data that could subsequently be used to correlate with the morphometric analysis.

tissue was then embedded in glycol methacrylate, serially sectioned, and stained with azure B and methylene blue. Analysis of the sections was done using an automatic sampling stage light microscope. The volume fraction of airways of different size, proximal and distal gas exchange parenchyma, and blood vessels of various sizes were determined by point counting uniformly across all sections in the series.

III. RESULTS

The SPECT images were obtained as a series of continuous slices and displayed in three dimensions (Fig. 18–1). The marker sources were readily identified in the slices (Fig. 18–2). For the analysis reported in this initial study, airways of all sizes were considered as one compartment, all gas exchange parenchyma were combined, and all large blood vessels were combined. The fraction of the tissue samples that were airway wall and lumen, gas exchange parenchyma, and blood vessels greater than 25 μm (noncapillaries) and the SPECT counts in the corresponding regions are shown in Table 18–1. The sites chosen had more than a five-fold difference in particle deposition, as shown by relative SPECT counts which varied from 1.91 to 10.51. These sites also had about a ten-fold variation in the density of airways (0.02 to 0.19). The correlation between SPECT counts and the fraction of the site that was airway tissue plus airway lumen is shown in Fig. 18–3. The correlation between SPECT counts and the fraction of tissue that was gas exchange parenchyma is illustrated in Fig. 18–4. The areas with the greatest SPECT counts had less airway volume and the greatest amount of gas-exchange parenchyma.

IV. DISCUSSION

The preliminary studies demonstrated that careful attention must be given to detail in order to correlate SPECT imaging with morphometric analysis of tissue samples. To ascertain that the same regions were being evaluated by the imaging and morphometric analyses, point sources which can be identified on the histological sections and on the SPECT scans were mixed and injected into the lung. The osmium tetroxide was identified as a 3-mm^3 dark spot in the lung. The ^{99m}Tc radioactivity produced a focal area of increased activity which was easily identified over the background of ^{99m}Tc activity from the aerosolized particles.

Table 18–1 Results of Morphometric Analysis and SPECT Study

	Fraction			
Sample	Airway Wall plus Airway Lumen	Gas Exchange Parenchyma	Blood Vessels (25 μm)	SPECT Counts[a]
1	0.10	0.84	0.063	4.93
2	0.12	0.85	0.030	2.81
3	0.13	0.85	0.009	1.91
4	0.03	0.94	0.030	5.58
5	0.03	0.94	0.023	10.51
6	0.05	0.90	0.050	9.67
7	0.02	0.95	0.030	10.26
8	0.07	0.90	0.030	2.45
9	0.19	0.69	0.120	4.18

[a] Counts/sec in 1.2-cm^3 region of interest.

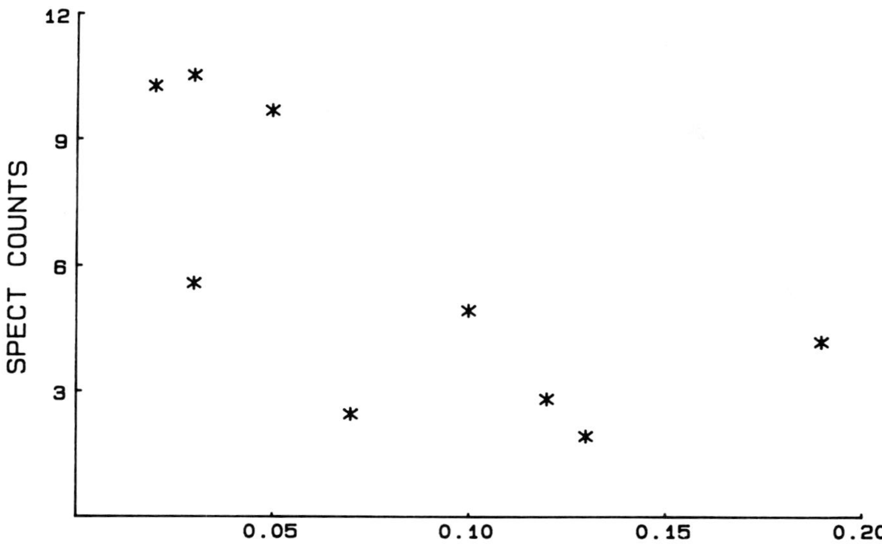

Fig. 18–3 Plot of SPECT counts and fraction of sample which is airway wall plus airway lumen.

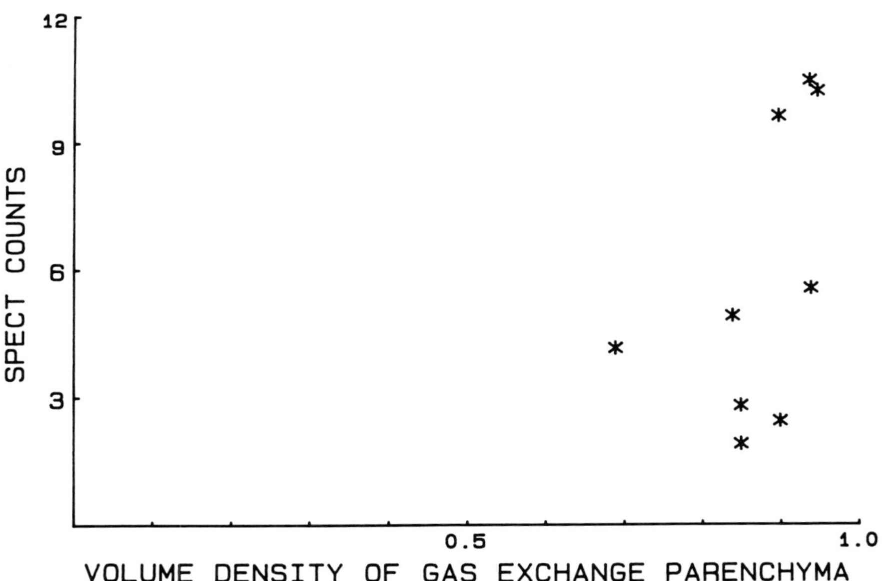

Fig. 18–4 Plot of SPECT counts and fraction of sample which is gas exchange parenchyma.

The results demonstrate that SPECT imaging can quantitate aerosol deposition. This can be done in specific lung regions by correlating SPECT counts with ratios of aerosol mass or number with the aerosol radioactivity as established by appropriate controls and phantoms. In addition, the distribution on the SPECT image can be correlated with a subsequent morphometric analysis of specific tissue sites. For the aerosol used in this study, an inverse relationship was found between the SPECT counts and the fraction of airway tissue plus airway lumen. A positive correlation was found between the SPECT counts and the fraction of gas exchange parenchyma. These results suggest that the aerosol was localized primarily in the distal alveolar region. These findings are consistent with previous studies in newborn piglets, which demonstrated that approximately three-fourths of a similar aerosol deposited distal to ciliated airways (Kay *et al.,* 1986).

This study demonstrates the feasibility of using high-resolution imaging techniques and site-specific morphometric analysis to assess dosimetry of inhaled particles in focused regions and to correlate the relative dosimetry with anatomical structures in the specific anatomic regions. The next step is to carry out more detailed correlations in which airways of different sizes and locations are separately identified. The two-compartment analysis used (airways vs parenchyma) can be extended to a large number of compartments, since anatomic recognition of airways with various sizes and definitions of different portions of the gas exchange parenchyma can be readily identified on the serial histologic sections. To get tight data on a greater number of variables, it will be essential to keep the size of the sites being studied as small as possible and to have extremely close correlations between the sites of particle deposition measured on the SPECT scan and the subsequent anatomic analysis of the site.

Methods for improving the correlation between the imaging studies and morphometric analysis are being sought. By putting three point sources in one lung, a plane can be identified. Along this plane, the intra-acinar distribution of particle deposition may be obtained by using a profile of counts down a bronchiole into the alveolar region. By doing a morphometric analysis of the distribution of anatomical structures along the same plane, the distribution of particle deposition could be directly related to the anatomy.

V. SUMMARY

SPECT imaging of radiolabeled aerosols was performed quantitatively and the imaging distribution was correlated with the related anatomic structures using morphometric analysis. The aerosol used in this study primarily localizes within the alveolar ducts and alveoli. Improved methods for accurate correlations between the SPECT images and anatomy are being researched. Successful completion of these studies will allow a site-specific determination of dosimetry for a wide variety of inhaled particles. The availability of PET scanning means that a similar analysis of selected inhaled gases (such as $^{13}NO_2$) can also be done. It is also particularly significant that this approach to analysis of regional dosimetry can be done repetitively on the same animal before the morphometric studies are completed. The analysis of deposition distribution is independent of dose. This means that the deposition distribution of different magnitudes of dose for the same gas or particle can be directly compared in the same animal. In addition, variations in the deposition patterns occurring in different species can be determined. The use of these high-resolution techniques open the possibility for doing experimental site-specific dosimetry at a level that can be used to verify the predictions of modeling experiments in which site-specific dosimetry data are calculated.

ACKNOWLEDGMENTS

This work was supported in part by NIH Grant CA 33541 and U.S. Environmental Protection Agency Cooperative Agreement CR 813113.

REFERENCES

Ahluwalia, B., Browness, G. L., Hales, C., and Kazemi, H. (1981). Regional lung function with nitrogen-13. *Eur. J. Nucl. Med.* **6,** 453–457.

Bates, D. V. (1972). Air pollutants and the human lung. *Am. Rev. Respir. Dis.* **10,** 1–13.

Brain, J. D., Knudsson, D. E., Sorokin, S. P., and Davis, M. A. (1976). Pulmonary distribution of particles given by intratracheal instillation or by aerosol inhalation. *Environ. Res.* **11,** 13–33.

Coleman, R. E., Blinder, R. A., and Jaszczak, R. J. (1986). Single photon emission computed tomography (SPECT): Clinical applications. *Invest. Radiol.* **21,** 1–11.

Foord, N., Black, A., and Walsh, M. (1978). Regional deposition of 2.5–7.5 μm diameter inhaled particles in healthy male nonsmokers. *J. Aerosol Sci.* **9,** 343–357.

Frankel, N., and Coleman, R. E. (1983). New imaging modalities for evaluating pulmonary disease. *Semin. Respir. Med.* **5,** 89–100.

Hughes, J. M. B., Brudin, L. H., Valind, S. O., and Rhodes, C. G. (1985). Positron emission tomography in the lung. *J. Thorac. Imag.* **1,** 78–88.

Jaszczak, R. J., Greer, K. L., Floyd, C. E., Harris, C. C., and Coleman, R. E. (1984). Improved SPECT quantification using compensation for scattered photons. *J. Nucl. Med.* **25,** 893–900.

Jaszczak, R. J., Floyd, C. E., Jr., and Coleman, R. E. (1985). Scatter compensation techniques for SPECT. *IEEE Trans. Nucl. Sci.* **32,** 786–793.

Kay, J., Coates, G., and O'Brodovich, H. (1986). Pulmonary deposition sites on an inhaled radiolabeled submicronic aerosol. *Pediatr. Res.* **20,** 1297–1300.

MacIntyre, N. R., Anderson, H. R., Silver, R. M., Schuler, F. R., and Coleman, R. E. (1983). Pulmonary function in mechanically-ventilated patients during 24 hour use of a hygroscopic condensor humidifier. *Chest* **84,** 560–564.

MacIntyre, N. R., Silver, R. M., Miller, C. W., Schuler, F., and Coleman, R. E. (1985). Aerosol delivery in intubated, mechanically ventilated patients. *Crit. Care Med.* **13,** 81–84.

Madsen, M. T., Hichwa, R. D., and Nichleo, R. J. (1981). An investigation of ^{11}C-methane, ^{13}N-nitrous oxide, and ^{11}C-acetylene as regional cerebral blood flow agents. *Phys. Med. Biol.* **26,** 875–882.

Osborne, D. R., Jaszczak, R. J., Coleman, R. E., and Drayer, B. P. (1981). Single photon emission computed tomography in the canine lung. *J. Comput. Assist. Tomogr.* **5,** 684–689.

Osborne, D., Jaszczak, R. J., Coleman, R. E., Greer, K., and Lischko, M. (1982). *In vivo* regional quantitation of intrathoracic Tc-99m using SPECT: Concise communication. *J. Nucl. Med.* **23,** 446–450.

Osborne, D. R. S., Jaszczak, R., and Coleman, R. E. (1983a). Single photon emission computed tomography and its application in lung. *Radiol. Clin. North Am.* **21,** 789–800.

Osborne, D. R., Jaszczak, R. J., Greer, K., Roggli, V., Lischko, M. S., and Coleman, R. E. (1983b). Detection of pulmonary emboli in dogs: Comparison of single photon emission computed tomography, gamma camera imaging, and angiography. *Radiology* **146,** 493–497.

Osborne, D., Jaszczak, R. J., Greer, K., Lischko, M., and Coleman, R. E. (1985). SPECT quantification of technetium-99m microspheres within the canine lung. *J. Comput. Assist. Tomogr.* **9,** 73–77.

Peacock, A., Alchanat, M., Valind, S., Sopwith, T., Hughes, J. M. B., and Turner Warwick, M. (1984). A non-invasive method for estimation of lung tissue volume in patients with fibrosing lung disease: Comparison with lung density measurement by positron transmission tomography. *Lancet* **2,** 785–787.

Rhodes, C. G., Wollmer, P., Frazio, F., and Jones, T. (1981). Quantitative measurement of regional extravascular lung density using positron emission and transmission tomography. *J. Comput. Assist. Tomogr.* **5,** 783–791.

Schuster, D. P., Mintun, M. A., Green, M. A., and Ter-Pogossian, M. M. (1985). Regional lung water and hematocrit determined by position emission tomography. *J. Appl. Physiol.* **59,** 860–868.

Susskind, H., Bull, A. B., and Harold, W. H. (1986). Quantitative comparison of regional distributions of inhaled Tc-99m DTPA aerosol and Kr-81m gas in coal miner's lung. *Am. J. Physiol. Im.* **1,** 67–76.

Wolf, A. (1981). Special characteristics and potential for radiopharmaceuticals for positron emission tomography. *Semin. Nucl. Med.* **11,** 2–12.

Chapter 19

Dual Laser Doppler System for Real-Time, Simultaneous Characterization of Aerosols by Size and Concentration

M. K. Mazumder
J. D. Wilson
D. L. Wankum
R. Cole
G. M. Northrop
L. T. Neidhardt
Graduate Institute of Technology
University of Arkansas
Little Rock, Arkansas 72204

T. Martonen
Health Effects Research Laboratory
U. S. Environmental Protection Agency
Research Triangle Park, North Carolina 27711

I. SUMMARY

To determine the total lung deposition of inhaled particles, it is necessary to know the periodic respiratory volumetric flow rate, $Q(t)$, and the particle concentration, $C(d_a, t)$, as a function of particle size and time during inspiration and expiration. Problems associated with the measurements of $Q(t)$ and $C(d_a, t)$ for typical aerosol delivery systems used for determining the total lung deposition are examined in this paper. Ideally, experimental studies on lung deposition of aerosol particles require an instrument that measures both $Q(t)$ and $C(d_a, t)$ in real time for all particles within the

Disclaimer: The research described in this article has been reviewed by the Health Effects Research Laboratory, U.S. Environmental Protection Agency, and approved for publication. Approval does not signify that the contents necessarily reflect the views and policies of the Agency nor does mention of trade names or commercial products constitute endorsement or recommendation for use.

given size range, regardless of their chemical composition, physical characteristics, or electrostatic charge.

Details on development and application of a modified single particle aerodynamic relaxation time (SPART) analyzer, with fiber optic transmission of incident and scattered laser radiation, for real-time *in situ* measurement of total lung deposition of inhaled particles are presented. The performance characteristics of the SPART analyzer in determining the aerodynamic size distribution of aerosol particles in the range 0.3 to 20 μm in diameter are discussed. Test aerosols include cigarette smoke, sodium chloride, and therapeutic aerosols. Experimental data on the hygroscopic growth of particles are presented along with a description of the instrument modifications that allow measurements of the particle size and concentration into or out of the mouthpiece in real time.

II. INTRODUCTION

The deposition of inhaled particles in the human respiratory tract is a function of both the size distribution of the particles and the characteristics of individual breathing patterns (Task Group on Lung Dynamics, 1966). For inhaled particles larger than 0.3 μm in aerodynamic diameter, the physical mechanisms involved in the lung deposition process make the aerodynamic size distribution of respirable particles the most significant parameter for determining the deposition fraction. Since physical properties such as size, geometric shape, density, and surface texture of the inhaled particles vary widely, the regional lung deposition is expressed in terms of the equivalent aerodynamic diameter (d_a), which takes into account these factors related to the lung deposition of particles (Task Group on Lung Dynamics, 1966).

Since a large fraction of the inhaled particles are hygroscopic and the rate of growth depends upon particle composition, size, and relative humidity, it is essential to determine the growth-time characteristics of the inhalable hygroscopic particles (Martonen *et al.*, 1982). Some of the particles will continue to grow during their passage through several tracheobronchial (TB) generations. Hygroscopic particles may deposit in a pattern different from that of nonhygroscopic particles of identical preinspired size distribution.

The current methods of instrumental analysis of inhalation exposure systems for laboratory animals and human subjects have been well reviewed in the literature and some of the notable ones are included in the references. In assessing the instrumentation used for characterization and control of exposure aerosols, Phalen (1984) observed, "Aerosol characterizations represent some of the most difficult of all physical measurements." Experimental methods of particle size measurement based on cascade impactors and microscopes can produce basic information on the particle size distribution, but not data in real time. Optical particle counting techniques do provide real-time information, yet they are not readily applicable to determining the aerodynamic size distributions for particles of different composition, shape, density, and surface texture.

Probably the most important advancements in the experimental determination of lung deposition were made by Heyder *et al.* (1980b) who applied light scattering spectroscopy. Using this technique, the particles' size and concentration were measured *in situ* during mouth breathing. Counting and sizing were performed by measuring the number and height of pulses of the scattered light signal generated by each particle passing through a laser sensing volume. This optical scattering method requires calibration relating the measured optical diameter to the equivalent aerodynamic diameter (Marple, 1979). Experiments with light-scattering instruments are often limited to monodisperse, stable particles of known composition for which the relationship between the aerodynamic diameter and the optical diameter is known. The hygroscopic growth of particles may significantly change the physical properties of the particles, including the refractive index; therefore, the method of measurement should be independent of the parameters that are not related to the aerodynamic size (Marple, 1979).

Problems related to the characterization of hygroscopic aerosol particles are discussed by several authors. Gonda *et al.* (1982) observed that direct measurement of particle size distribution at an elevated relative humidity is essential because of the complexity involved in calculating the growth of multicomponent aerosol particles in the lung environment. For multicomponent hygroscopic aerosol particles (Milburn *et al.,* 1957), experimental determination of the growth rate and final size is critically important for estimating the regional lung deposition in terms of aerodynamic diameter. Aerosol growth is very sensitive to small changes of relative humidity at near saturation level. Sensitive temperature measurement is necessary because relative humidity is a function of the ambient temperature. Since particle rate of growth depends upon the relative humidity, the initial particle size, and the hygroscopicity of the particle, it is necessary to determine the particle diameter as a function of the residence time of the inspired particles in the human lung.

This article describes the development of a modified SPART analyzer for real-time *in-situ* measurement of the aerodynamic size distribution of inhaled particles and the applications of the analyzer to the characterization of aerosols entering or exiting the mouthpiece during breathing cycles. The modified SPART analyzer measures $C(d_a, t)$ and d_a on a single particle basis inside the mouthpiece.

III. MEASUREMENT OF TOTAL LUNG DEPOSITION

To determine the total deposition (DE) of the inhaled particles in the lung during steady breathing, one must know the periodic respiratory volumetric flow rate, $Q(t)$, and the particle size distribution, $C(d_a, t)$, as a function of time during inspiration and expiration. It is also necessary that wash-in is completed before the measurements are initiated and the dead space is as small as possible. For stable, nonhygroscopic particles, the total lung deposition can be expressed as (Heyder *et al.*, 1973)

$$\mathrm{DE}(d_a) = 1 - \int_{t_e} C(d_a, t)Q(t)dt \Big/ \int_{t_i} C(d_a, t)Q(t)dt \tag{1}$$

where

d_a = equivalent aerodynamic diameter,
DE = total lung deposition,
$C(d_a, t)$ = particle concentration as a function of d_a and t,
$Q(t)$ = volumetric flow rate,
t_i = inspiration period, and
t_e = expiration period.

A schematic diagram of an inhalation experiment using an aerosol chamber is shown in Fig. 19-1. The aerosol inside the chamber is inhaled through an aerosol delivery tube and a mouthpiece. A microprocessor controlled valve is programmed to cycle automatically when the desired air volume is exchanged. The volumetric flow rate is measured by a pneumotachograph whose output is displayed on an oscilloscope so that the subject can match a standard respiratory pattern previously established for the inhalation studies. The valve-controlling circuitry is programmed to allow exhalation into either a waste bag or into a sample selection bag placed inside the aerosol chamber.

In the chamber, aerosol concentration can be maintained essentially constant at a desired level for particles in the range 0.1 to 1.0 μm in aerodynamic diameter. Sedimentation of larger particles, and diffusion and wall deposition of smaller particles, may cause a significant change in particulate concentration during inhalation experiments.

Figure 19-2 shows an aerosol delivery scheme where the challenge aerosol to be inhaled is drawn from a dynamic aerosol flow system. Figure 19-3

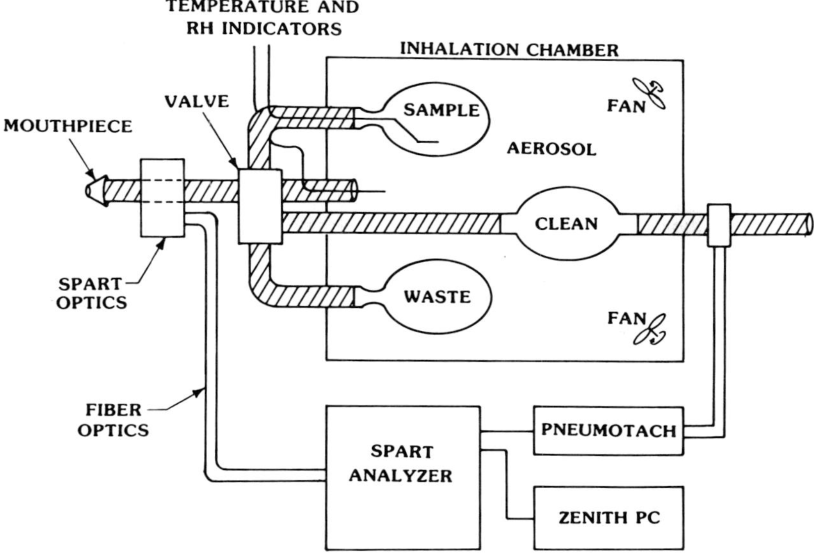

Fig. 19-1 Schematic diagram of the inhalation chamber with microprocessor-controlled valve system for measuring total lung deposition of aerosol particles using a fiber-optic-based SPART analyzer.

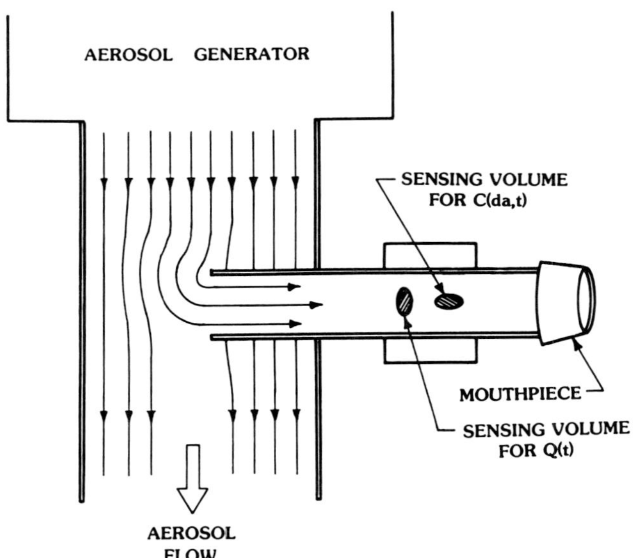

Fig. 19-2 Experimental arrangement showing a dynamic aerosol flow system for determining total lung deposition of aerosol particles using *in-situ* measurements for both particle concentration and volumetric flow rate.

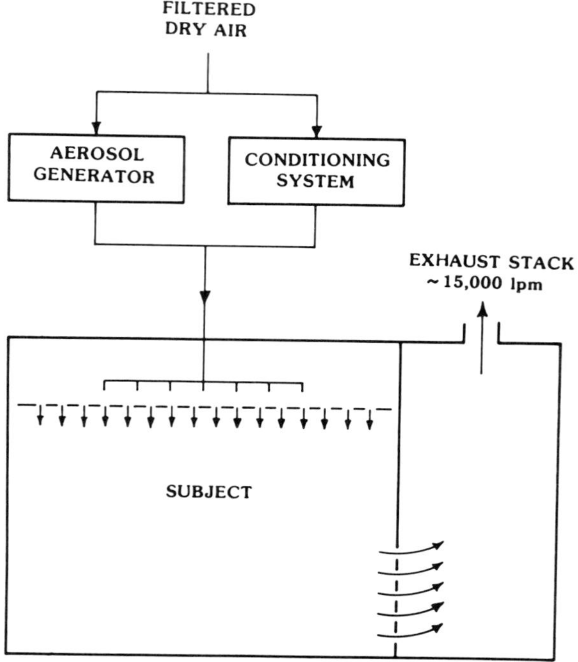

Fig. 19-3 Experimental setup of an exposure chamber where the subject inhales the aerosol delivered in a vertically downward flow field.

shows an exposure chamber where the subject inhales the aerosol delivered in a vertically downward flow field. Whether human subjects, animal models, or surrogate lung models are used, such systems are suitable for aerosol delivery in a wide size distribution for both spherical and non-spherical particles (e.g., coal and cotton dusts, silica, or fibrous particles).

IV. MEASUREMENT OF VOLUMETRIC FLOW RATE $Q(t)$

One of the requirements in the $Q(t)$ measurements for lung deposition studies is that the instrument used must be sensitive to the bidirectional flow field so the times of inhalation and exhalation (t_i, t_e) can be identified. The measurement must be nonintrusive and have characteristics for measuring the flow rates in the aerosol delivery tube.

The volumetric flow rate measurements can be performed by using any one of the following instruments: a pneumotachograph, an ultrasonic flowmeter, or a laser doppler flowmeter.

The pneumotachograph has been widely used for measuring lung functions and determining the total lung deposition of inhaled aerosols. However, it has major limitations because its sensing element must not be contaminated by aerosol deposition, and pressure-to-voltage transducers often require frequent calibration. For similar reasons the hotwire or hot-film anemometers cannot be readily used to measure the volumetric flow rates in lung deposition studies.

The ultrasonic and the laser doppler flow meters are capable of meeting the requirements of measuring the gas flow in a nonintrusive manner and can provide information on both the flow rate and direction.

In an inhalation experiment using an aerosol chamber as shown in Fig. 19-1, a pneumotachograph can be conveniently employed since the inhaled and exhaled air flow is reproduced by the clean air flow into a clean air bag located inside the chamber. However, in experiments where a dynamic flow of aerosol involving particles larger than 2 μm in diameter is needed, as shown in Fig. 19-2 and 19-3, both $Q(t)$ and $C(d_a, t)$ must be measured *in situ.* Heyder *et al.* (1973) employed *in situ* measurement techniques for both the particle size distribution and the volumetric flow rate through the aerosol delivery tube connected to the mouthpiece. The volumetric flow rate was determined by measuring the pressure drop across the flow-stabilizing tubes placed in front of the mouthpiece.

V. MEASUREMENT OF PARTICLE SIZE DISTRIBUTIONS $C(d_a, t)$

The following measurement techniques can be applied to characterize the particles for determining the total lung deposition:

- Microscopes (optical and electron microscopes)
- Inertial separators (including cascade impactors, centrifuges, and elutriators)
- Electrical Aerosol Analyzer (EAA)
- Light-scattering instruments (including optical particle counters and spectroscopes, and photon correlators)
- Laser Doppler Velocimetry-based size analyzers (relaxation time analyzers, LDV-based photon correlators, and fringe visibility measuring devices).

Microscopes, inertial separators, and electrical aerosol analyzers are not applicable for real-time measurements, whereas light scattering and laser Doppler techniques can be applied to real-time particle size and concentration measurements.

A. Sampling versus *in Situ* Measurements

Optical techniques can be used in two configurations: by sampling the aerosol through a tube and then aerodynamically focusing the aerosol stream within a thin sheath of clean air so the aerosol stream passes through the sensing volume in a region of constant intensity distribution, or locating the optical sensing volume within the aerosol delivery tube and measuring the size distribution *in situ.* To determine the total lung deposition, it is desirable to measure the aerodynamic size and concentration of particles in real time during the breathing cycle. Optical techniques provide a nonintrusive measurement of particle concentration and size, as well as $Q(t)$.

If the sampled aerosol stream is focused aerodynamically into a small central zone of light where the intensity distribution of the optical radiation across the zone is nearly constant, the resultant size resolution will be high. In addition, the sheath air greatly reduces the contamination of optical surfaces. Therefore, the intensity of the scattered light can be directly related to the optical equivalent diameter of the particle.

Sampling and aerodynamic focusing require that particles be isokinetically sampled from the aerosol delivery tube. Since both the flow rates and the direction of flow in the aerosol delivery tube change during a breathing cycle, isokinetic sampling and aerodynamic focusing techniques are, in general, incompatible in inhalation experiments. The sampling error becomes significant when the aerodynamic size distributions of aerosol particles are in the supermicron range and the distributions vary from one experiment to another.

Aerosol delivery systems used in inhalation exposure experiments vary widely, depending upon the experimental objectives. However, in most cases, the aerosol is drawn through a sampling port from either a chamber (Fig. 19-1) where the aerosol is stirred or from a flow stream (Fig. 19-2) connected to the output of the aerosol generator. In another scheme, the subject is situated in an exposure chamber where there is a continuous, vertically downward flow of aerosol to the exhaust stack (Fig. 19-3). In all of these arrangements, if the aerosol is sampled in an isokinetic manner, the sampling efficiency for a given aerodynamic diameter will vary with $Q(t)$ and the sampling error will become significant when the particle inertia cannot be neglected.

B. Light-Scattering Photometry

In light-scattering photometry, the particles are passed through the sensing volume of a focused beam of laser radiation or other source of light and a photomultiplier is used to measure the intensity of the scattered light generated by each particle as it passes through the sensing volume. Either the peak height of the scattered light intensity or the integral of the area defined by the intensity versus time curve is used to determine the particle size.

Figure 19-4 shows the variation of light intensity across the sensing volume, using a Gaussian distribution. Particle trajectories at three different points of the Gaussian intensity distribution pattern will result in three

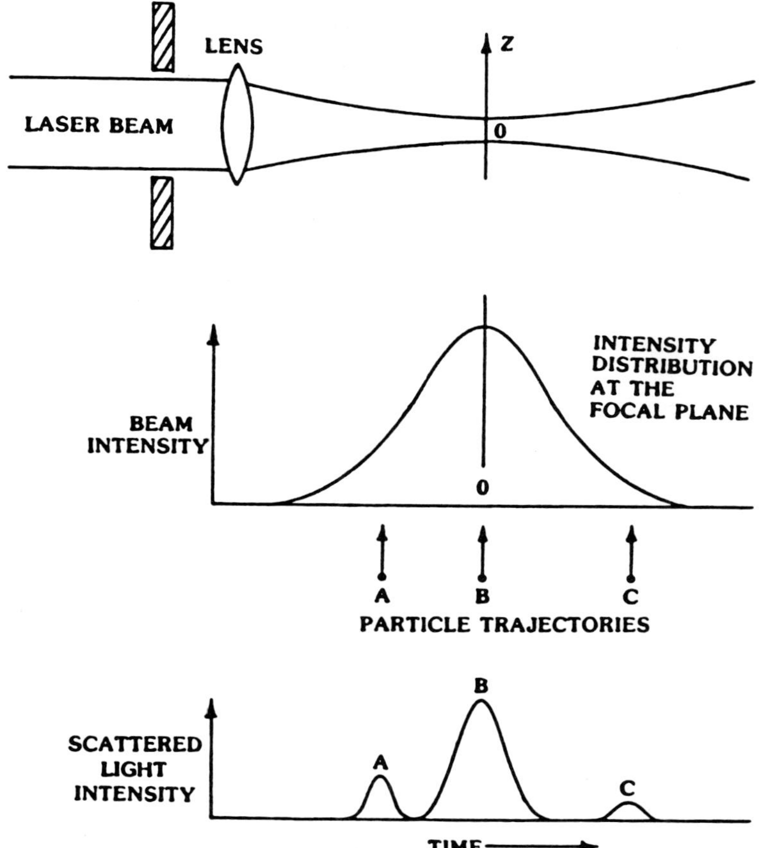

Fig. 19-4 Effect of light intensity variation across the sensing volume on particle size measurement using light-scattering photometry.

different pulse heights as shown in Fig. 19-4. This figure illustrates the effect of intensity variations on particle size measurements when an optical particle counter is used for size discrimination. Even when a constant intensity distribution is achieved, intensity variation exists at the edge of the sensing volume imaged by the receiving optics of the photometer. This is known as an edge effect. Also, intensity varies with the particle shape and refractive index for particles having the same value of d_a. Another complexity in size measurements arises from the contamination of optical surfaces by particles due to the deposition aerosol particles passing through the entire cross-section of the inhalation tube. Contamination of optical surfaces decreases the intensity of the scattered radiation, and the measured size distribution will have a downward shift in the size scale.

In the *in situ* measurements such as the technique used by Heyder *et al.* (1978) and Stahlhofen *et al.* (1980, 1981a), the sensing volume is placed within the aerosol delivery tube leading to the mouthpiece. Since the particles cannot be aerodynamically focused, they pass through a sheet of laser light forming the sensing volume. Contamination of the optical surfaces must be carefully avoided if particle measurements are to be performed with a good size resolution. The equivalent optical diameter is calculated and can be used to determine aerodynamic size distribution for spherical particles of known composition.

Ideally, the *in situ* instruments should measure flow rate $Q(t)$ and particle size distribution $C(d_a, t)$ in real time for all particles within a given size range, regardless of the chemical composition and physical characteristics of the particles.

C. SPART Measurement of Aerodynamic Size Distributions

SPART is an acronym for single particle aerodynamic relaxation time analyzer. Using a laser, the analyzer measures the aerodynamic size distribution of particles suspended in air. The principle of operation and performance characteristics of this analyzer have been discussed in the literature (Mazumder *et al.,* 1979). Only a brief description of the instrument related to the current application for *in-situ* measurement of particle size and concentration using fiber optic coupling is discussed in this chapter.

The SPART analyzer has three basic components:

1. A dual-beam, frequency-biased laser Doppler velocimeter (LDV) with fiber optic coupling of the laser beam between the instrument and the sensing volume;
2. An acoustic relaxation cell that is mounted circumferentially around the aerosol delivery tube; and
3. An electronic signal and data processing system coupled with a personal computer.

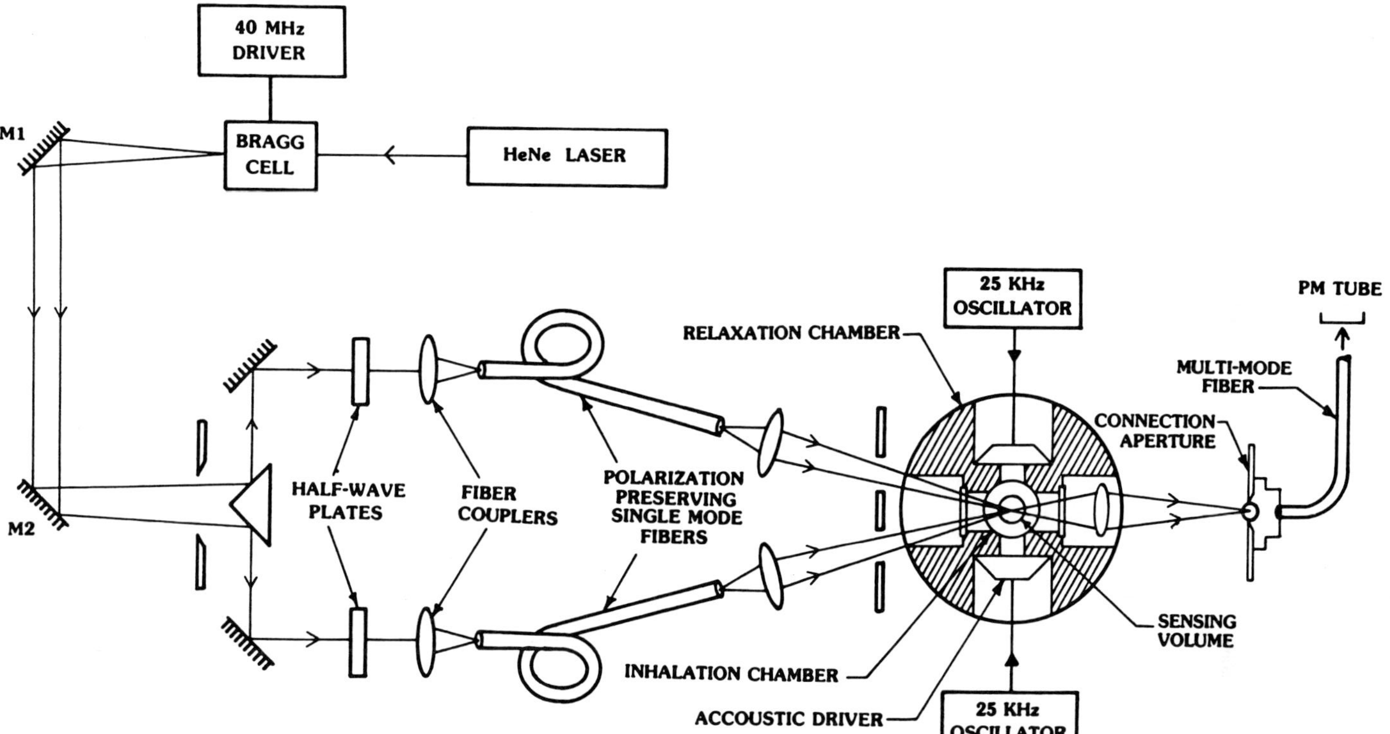

Fig. 19-5 Block diagram of the SPART analyzer incorporating fiber optics for *in situ* measurement of aerodynamic size distribution of particles during individual breathing cycles.

The experimental arrangements are shown in Fig. 19-5, 19-6, and 19-7. Figure 19-5 is a block diagram of the SPART analyzer incorporating fiber optics. A helium–neon laser is used as the monochromatic light source operating at the radiation wavelength of 633 nm and at an output power of 10 mW. The laser beam is first passed through a Bragg cell (acousto-optic deflector) producing two laser beams of equal power, one shifted in frequency by 40 MHz and the other unshifted. The two beams are then reflected by two mirrors, M1 and M2, and then are incident on a right-angle prism mirrored on the front surface which separates the two beams. The two half-wave plates are used to rotate the polarization of the two beams so that the electric vectors of both shifted and unshifted beams are coplanar. The two beams are then coupled to two polarization-preserving single mode fibers. Each fiber has a core diameter of 4 μm and is approximately 1.5 m long. The fibers are connected to the relaxation cell as shown in Fig. 19-5. The laser beams emerging out of these fibers are then collimated and focused at the sensing volume where the two beams intersect each other. Light scattered from aerosol particles transiting the sensing volume is captured by receiving lenses and focused to a multimode fiber which is coupled to a photomultiplier tube. The photomultiplier output signal is then processed to obtain the particle size information in the following manner.

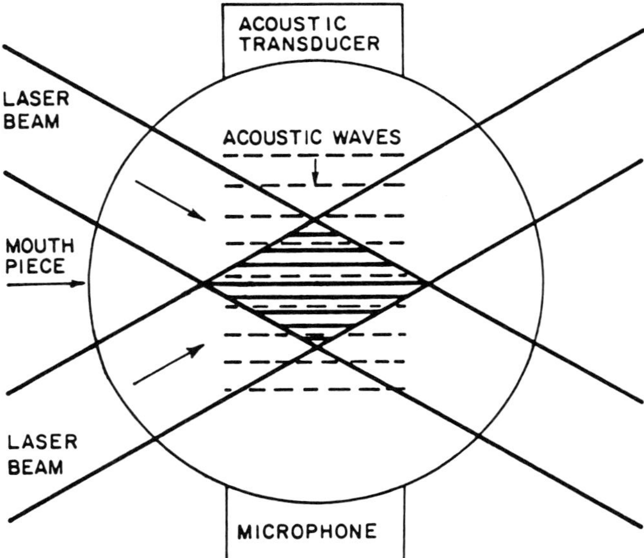

Fig. 19-6 Sensing volume of the SPART analyzer.

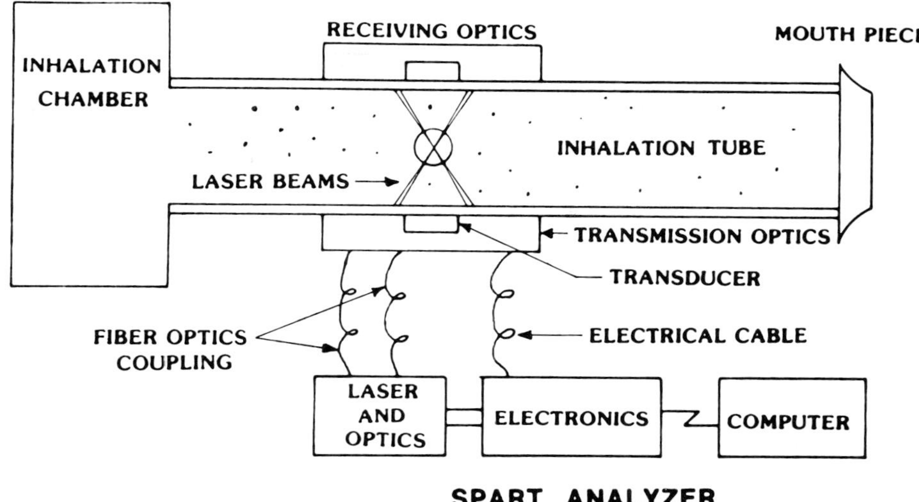

Fig. 19-7 Schematic of the lung deposition analyzer showing the sensing volume of the fiber-optic SPART analyzer inside the inhalation tube located in front of the mouthpiece.

In the SPART analyzer, the LDV measures the particle velocity in its sensing volume as shown in Fig. 19-6. The sensing volume, defined by the intersection of the two incident laser beams and the view volume of the receiving optics, is located inside the aerosol delivery tube in front of the mouthpiece (Fig. 19-7). Figure 19-6 illustrates the cross-section of the aerosol delivery tube showing the laser beams intersecting at the center of the tube. The direction of aerosol flow is perpendicular to the plane containing the two laser beams. As the particle passes through the LDV sensing volume, it experiences an acoustic excitation that causes the particle to oscillate in the direction that is parallel to the plane containing the two intersecting laser beams and, therefore, is in the plane of LDV velocity measurements. The acoustic excitation is generated and measured by a pair of acoustic transducers that are symmetrically located across the LDV sensing volume. To operate the analyzer in the size range 0.3 to 20.0 μm in diameter, it is necessary to use two acoustic frequencies, one approximately 25 kHz and the other approximately 1.0 kHz, operating sequentially to cover this entire range.

The LDV signal burst generated at the output of the photomultiplier is increased with an RF amplifier, passed through a limiter, and applied to a delay line-type FM discriminator to demodulate the signal that represents the particle motion inside the sensing volume. The filtered output is applied to a high-speed, zero crossing detector to generate digital pulses for measuring the relative phase shift of the particle motion with respect to the acoustic reference signals. The phase measurement is performed digitally using the direct memory access of a personal computer. The phase data acquisition and processing is performed in real time. The computer is also programmed to take data from other sources, such as the instrument for measuring the volumetric flow rate $Q(t)$, to compute the total aerosol intake during the inhalation and the amount of particles exhaled during the exhalation cycle. The basic scheme for computation of the total lung deposition is shown in Fig. 19-8.

If the count rate is given by $N_t(d_a)$, then the concentration $C(d_a)$ of aerosol can be expressed as

$$N_t(d_a) = A_s\, U(t)C(d_a) \tag{2}$$

where A_s is the cross-sectional area of the sensing volume projected in the direction of aerosol flow, $U(t)$ is the instantaneous local velocity of the aerosol stream across the sensing volume, and $C(d_a)$ is the concentration of particles per unit volume.

The cross-sectional area A_s is a function of the light scattering properties of the particles transiting through the sensing volume, therefore, the SPART analyzer is calibrated for calculating $C(d_a)$ from the measured value of $N_t(d_a)$. A typical upper limit of particle concentration measurements by a SPART analyzer without using any dilution device is 10^6 particles/cm^3.

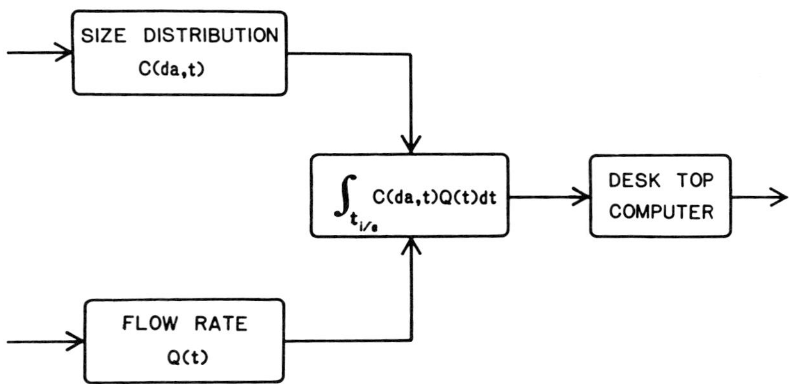

Fig. 19-8 Personal computer-controlled SPART data processing system for determining total lung deposition of aerosol particles versus aerodynamic diameter in individual breathing cycle.

D. Measurement of Particulate Concentrations during Breathing Cycles

If the SPART count is $\Delta N(d_a)$ in a time interval Δt, then, from Equation (2)

$$\Delta N(d_a) = C(d_a)\Delta t Q_s(t) \tag{3}$$

where

$$Q_s(t) = A_s U(t)$$
$$Q_s(t) = F[Q(t)]$$

$Q_s(t)$ is the aerosol flow rate through the SPART sensing volume. The pneumotachograph measures the total volumetric flow rate $Q(t)$. If the quantity $Q_s(t)$ is integrated over the entire cross-sectional area of the aerosol delivery tube, one will obtain the total volumetric flow rate $Q(t)$ in plug flow conditions. During the inhalation process, the total number of particles N_i entering into the mouthpiece can be expressed by Equations (1) and (3), as

$$N_i(d_a) = \int_{t_i} C(d_a)Q(t)dt = \sum_{j=0}^{t_i} \Delta N_j(d_a)Q(t)/F[Q(t)] \tag{4}$$

Similarly, during the exhalation process, the total number of particles exhaled for the size d_a can be expressed as

$$N_e(d_a) = \int_{t_e} C(d_a)Q(t)dt = \sum_{j=0}^{t_e} \Delta N_j(d_a)Q(t)/F[Q(t)] \tag{5}$$

and the deposition fraction, DE, can be expressed from Equation (1) as

$$\mathrm{DE}(d_a) = 1 - N_e(d_a)/N_i(d_a) \tag{6}$$

Equation (6) is useful for describing the deposition fraction of nonhygroscopic, polydisperse aerosol particles. When hygroscopic aerosol particles are inhaled, the size distribution of the exhaled particles will be different and, therefore, Equation (6) cannot be directly used. If we consider a monodisperse aerosol containing hygroscopic particles of diameter d_a during inhalation and diameter d'_a as the particles are exhaled, Equation (6) can be written as

$$\mathrm{DE}(d_a) = 1 - N_e(d'_a)/N_i(d_a) \tag{7}$$

The hygroscopic growth characteristics for a polydisperse aerosol particle can be experimentally determined and Equation (7) can be used for estimating lung deposition fraction.

Earlier studies (Hiller *et al.,* 1982a, b; Wilson *et al.,* 1985) on the application of the SPART analyzer for measuring total lung deposition were performed by inspecting the aerosol from an aerosol chamber representing the inspired aerosol and from a sample bag which contained the expired aerosol. The experimental arrangement used was similar to the one shown in Fig. 19-1, in which measurements of total lung deposition of aerosol particles were performed by testing the aerosol sample through the SPART analyzer. The basic difference between the experimental arrangement previously used and the experiments which are now in progress with the fiber-optic SPART is that the aerosol concentrations are now measured *in situ,* in front of the mouthpiece, to determine the lung deposition. In such a process, it is not necessary to assume uniform concentration of aerosol particles both in the aerosol chamber and in the sample bag to determine the lung deposition in each breathing cycle.

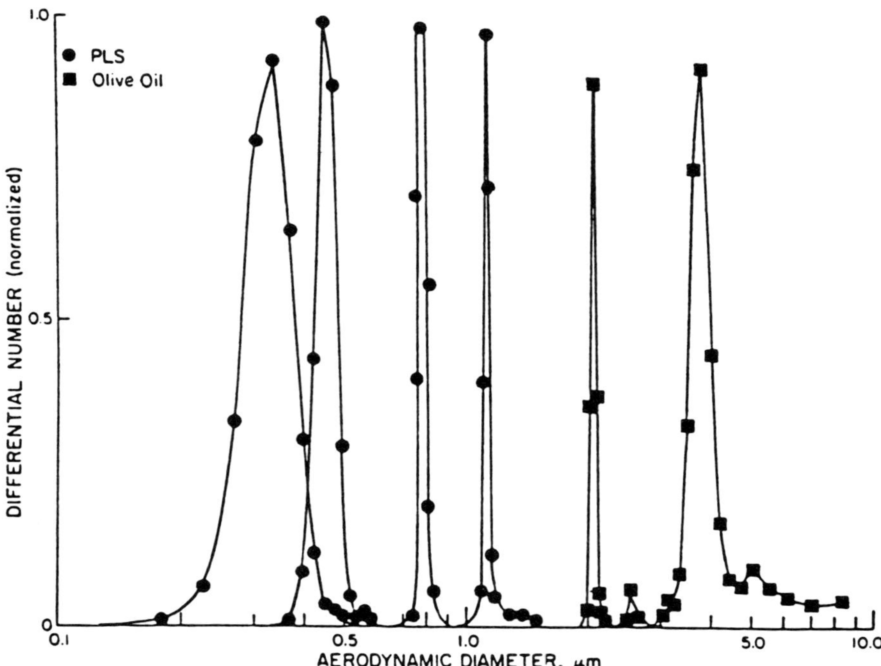

Fig. 19-9 Normalized size distribution of several monodisperse aerosols as measured by the SPART analyzer and plotted in the same graph. The known diameters of particles used in this study are 0.312-, 0.481-, 0.804- and, 1.101-μm diameter for PLS particles and 2.0- and 3.8-μm diameter for olive oil droplets.

E. Size Range and Resolution

Using two excitation frequencies, 1 kHz and 25 kHz, particle size in the range 0.3 to 20 μm in aerodynamic diameter can be measured by the SPART analyzer. Figure 19-9 shows the size distributions of several monodisperse particles measured by the SPART analyzer. In this case, the instrument was operated at an acoustic excitation frequency of 25 kHz. Figure 19-10 shows the size distribution of bisethylhexyl sebacate (BES) droplets measured by the SPART analyzer from 2.5 μm to 20.0 μm in diameter. The analyzer was operated at a 1-kHz acoustic frequency. Figure 19-11 shows the size resolution of the SPART analyzer in measuring aerosols containing 0.726 and 0.822 μm polystyrene latex sphere (PLS) particles, while the analyzer was operating at 25 kHz. Figures 19-9, 19-10, and 19-11 are experimental data obtained from laminar, unidirectional aerosol flow.

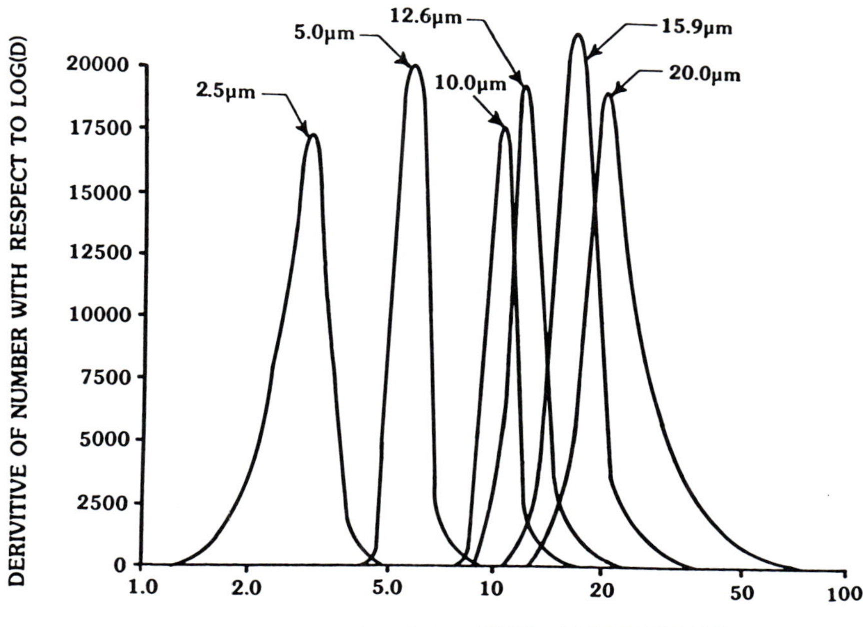

Fig. 19-10 Normalized size distribution of several aerosols containing BES droplets of known diameter (as shown) measured by the SPART analyzer and plotted in the same graph.

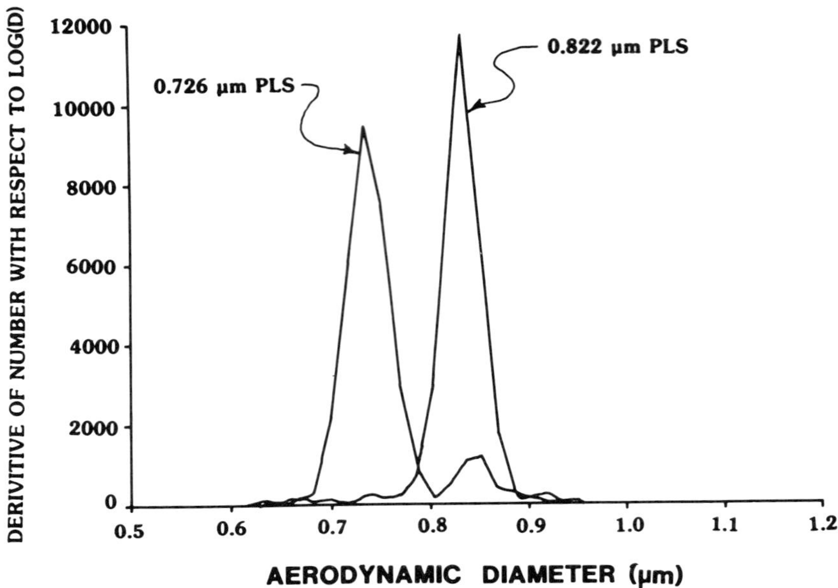

Fig. 19-11 Size distribution of aerosol particles containing two known sizes, 0.726- and 0.822-μm diameter PLS, as measured by the SPART analyzer. The distribution shows the resolution of the analyzer operation at 25 kHz.

VI. MEASUREMENT OF PARTICLE GROWTH IN A HUMID ENVIRONMENT

As the aerosol particle enters through a mouthpiece and travels through the upper conducting airways to the pulmonary region, it will experience increasing relative humidity. Similarly, particles inhaled, but not deposited in the lung, will be experiencing a gradual decrease in relative humidity as the particles are exhaled from the pulmonary region through the upper conducting airways back to the mouthpiece. If the RH values for different generations of the pulmonary tree are known, the rate of condensation and evaporation for an isolated particle during inhalation and exhalation can be calculated if its initial diameter and chemical composition are known. For a single spherical particle with geometric diameter, (d_p), the rate of change in diameter can be expressed as

$$d(d_p)/dt = 4D_vM(P - P_s)/[RT\rho_p d_p] \qquad (8)$$

where D_v is the diffusion coefficient of the water vapor, M is the molecular weight of the water molecules, P is the partial pressure of water vapor surrounding the particle, P_s is the partial vapor pressure at the particle surface, R is the gas constant, T is the temperature in degrees Kelvin, and ρ_p is the particle density. The condensation and evaporation of water vapor are generally estimated for isolated particles far removed from the influence of other particles. If the aerosol contains particles of different sizes and compositions and if the particle concentration is relatively high, the growth and decay processes may be more complex than are estimated from an isolated particle. For example, since the growth rate is inversely proportional to the particle diameter, small particles present at a high concentration will grow more rapidly than the larger particles. The growth and decay processes for highly hygroscopic particles are more significant in the RH range from 90 to 100% than in any other regions for equal change of RH. For example, the particle of NaCl of initial diameter 0.21 μm at RH 10% will form a droplet of 0.38 μm in diameter at 75% RH and will reach 1.0 μm in diameter at 99.5% RH.

Figure 19-12 shows the calculated rate of growth for NaCl particles of different initial diameters at a relative humidity of 98.5%. Within the first

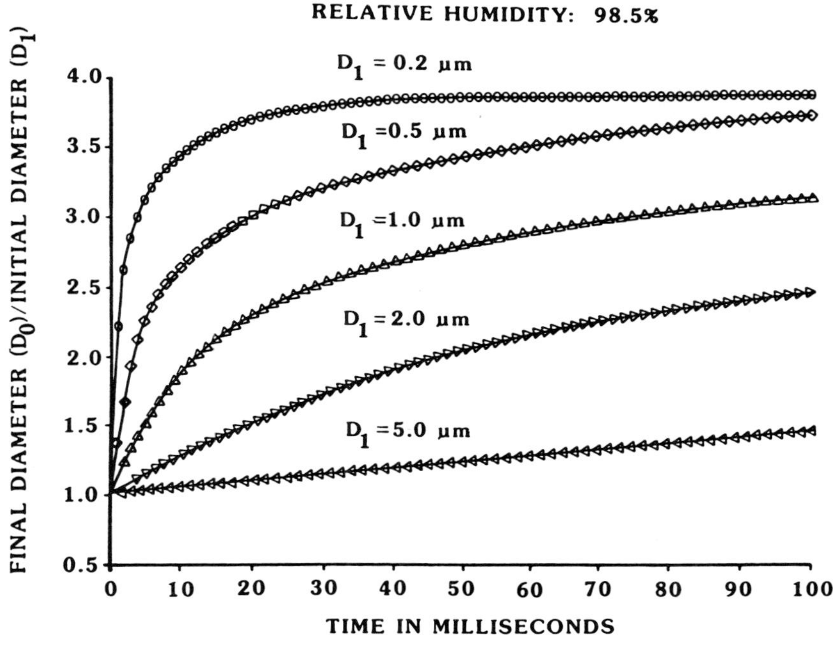

Fig. 19-12 Hygroscopic growth of NaCl particles of different initial diameters plotted as the ratio of the final diameter (D_0) to an initial diameter (D_1) as a function of time showing their rate of growth.

100 msec, particles smaller than 0.5 μm in diameter will attain their near-equilibrium size, while particles greater than 0.5 μm will continue to grow as they enter into the respiratory system. For particles larger than 5 μm in diameter, the growth process may continue during the entire inhalation cycle in a normal breathing period (which is typically 4 sec). Figure 19-13 shows the calculated growth of a 0.2-μm NaCl particle as a function of increasing relative humidity at 37°C. Figures 19-12 and 19-13 are based on Equation (7). Since the SPART analyzer measures d_a for each particle in 0.5 msec, the instrument can be used to measure both the growth rate and the final equilibrium size for particles of unknown composition in inhalation experiments. The SPART analyzer was used to measure the growth of aerosol particles in an atmosphere simulating the humid environment of a human lung. In this experiment the aerosol was allowed to grow in a humid chamber maintained at 37°C with a relative humidity that could be varied from 16 to 95% or higher. The tests included sodium chloride, cigarette smoke, and therapeutic aerosols. Figures 19-14 and 19-15 provide experimental data showing the growth as measured by the SPART analyzer for NaCl aerosols and for cigarette smoke.

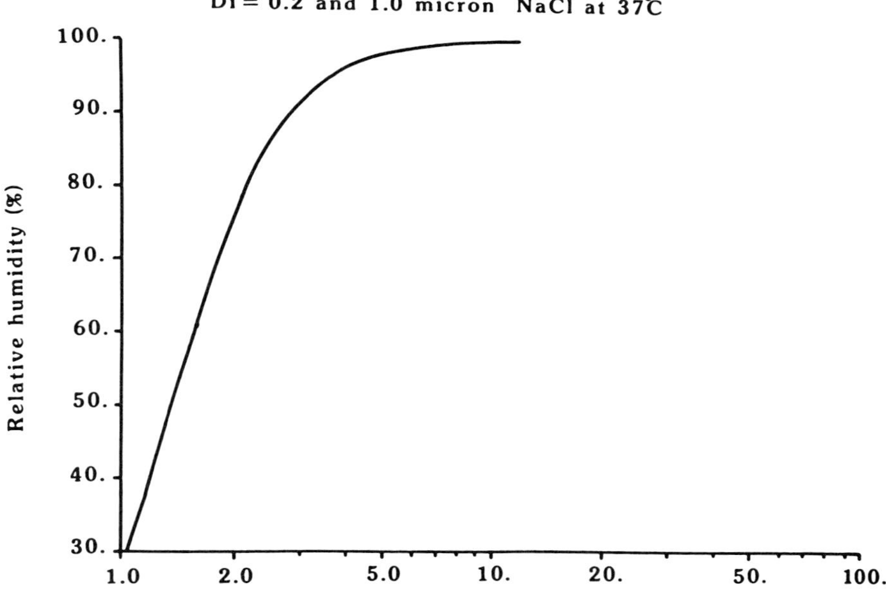

Fig. 19-13 Hygroscopic growth of 0.2-μm diameter NaCl particles as a function of relative humidity showing the variation of particle size in the human tracheobronchial tree.

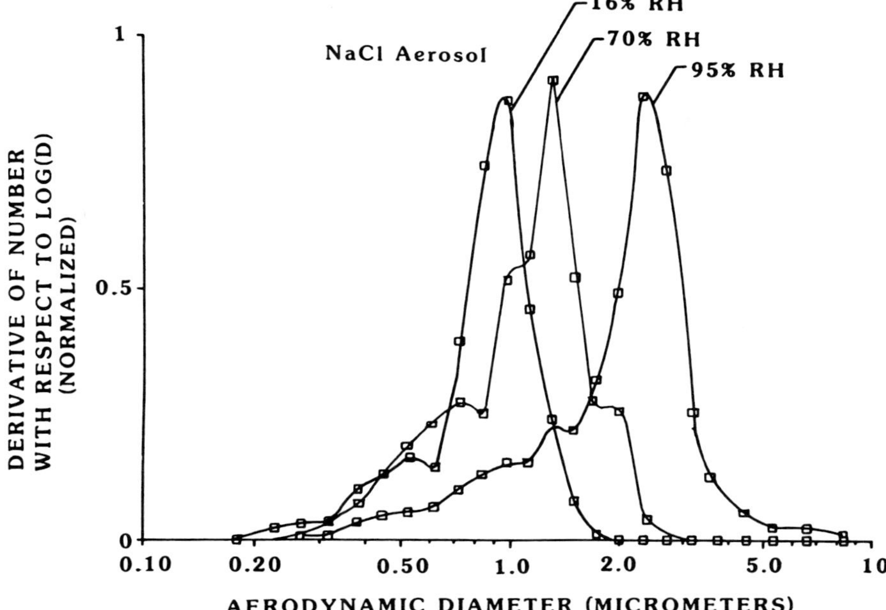

Fig. 19-14 Aerodynamic size distributions of NaCl particles measured at three relative humidities by the SPART analyzer, showing the change in the size distribution for the same aerosol particles at different relative humidities.

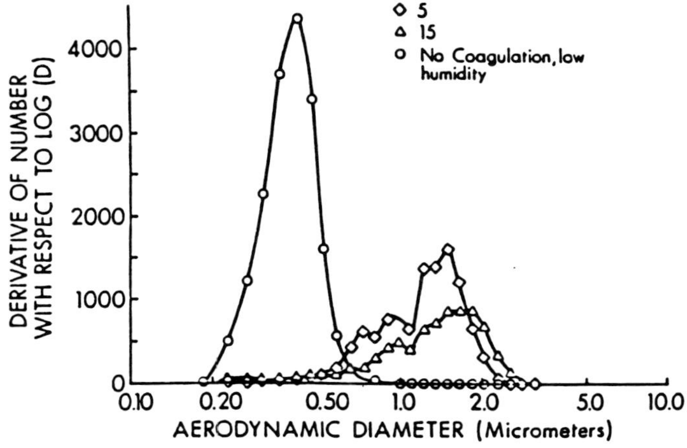

Fig. 19-15 Aerodynamic size distribution of cigarette smoke particles measured by the SPART analyzer at three different conditions: smoke particles diluted immediately after the cigarette smoke left the butt end (○); smoke particles allowed to coagulate for 5 sec before dilution (◇); and smoke particles allowed to coagulate 15 sec before dilution (△). Coagulation took place at an ambient RH of 98%. (The data were taken by K. McCusker.)

VII. CONCLUSIONS

Both theoretical analysis and experimental data obtained to date show that the fiber-optic-based SPART analyzer can be used for determining the total lung deposition of hygroscopic and nonhygroscopic polydispersed aerosols of different chemical compositions. The unique characteristics of the analyzer are:

1. Aerodynamic size distribution measurements are made directly;
2. There is no sampling loss, because measurement is performed *in situ* and in real time; and
3. The hygroscopic growth of the particles is measured in terms of aerodynamic diameter.

Aerodynamic size measurements are performed on a single-particle basis at a maximum count rate of 100 particles/sec. The fiber optic SPART analyzer permits measurements of total lung deposition of inhaled aerosol particles during individual breathing cycles. The measurement process is noninvasive and can provide a better correlation of the particle deposition with physiological parameters such as inspiration and expiration, tidal volume, and breath-holding. In addition, the process can be applied to both inhalation toxicology and aerosol therapy.

ACKNOWLEDGMENTS

This work is funded in part by U.S. EPA Grant CR 812675 and in part by a grant from the Arkansas Science and Technology Authority. The authors are grateful to many colleagues who made many significant contributions, particularly to K. Tennal, A. Eisner, I. Lau, and K. McCusker. The authors are also grateful to D. Belk, D. Watson, P. Archer, and S. Turner for their help in the preparation of the manuscript, and to the reviewers of the manuscript for many valuable suggestions.

REFERENCES

Altshuler, B., Yarmus, L., Palmes, E. D., and Nelson, N. (1957). Aerosol deposition in the human respiratory tract. *AMA Arch. Ind. Health* **15,** 293–303.

Camner, R., Hellstrom, P. A., and Lundborg, M. (1973). Coating 5 μm particles with carbon and metals for lung clearance studies. *Arch. Environ. Health* **27,** 331–333.

Camner, P., Hellstrom, P. A., and Lundborg, M. (1976). Lung clearance of 4 μm particles coated with silver, carbon, or beryllium. *Arch. Environ. Health* **32,** 58–62.

Dasgupta, P. K., Raabe, O. G., Duvall, T. R., and Tarkington, B. K. (1980). Generation and characterization of sodium sulfate aerosols for application in inhalation toxicologic research. *Am. Ind. Hyg. Assoc. J.* **41,** 660–667.

Davies, C. N., Heyder, J., and Subba Ramu, M. C. (1972). Breathing of half-micron aerosols. I. Experimental. *J. Appl. Physiol.* **32,** 591–600.

Gonda, I., Kayes, J. B., Groom, C. V., and Fildes, F. J. T. (1982). Characterisation of hygroscopic inhalation aerosols. *Particle Size Conf., Loughborough Univ. Technol. 22–25 Sept., 1981* pp. 31–43.

Heyder, J., Gebhart, J., Heigner, G., Roth, C., and Stahlhofen, W. (1973). Experimental studies of the total lung deposition of aerosol particles in the human respiratory tract. *J. Aerosol Sci.* **4,** 191–208.

Heyder, J., Gebhart, J., Roth, C., Stahlhofen, W., and Stuck, B. (1978). Intercomposition of lung deposition data for aerosol particles. *J. Aerosol Sci.* **9,** 147–155.

Heyder, J., Gebhart, J., Rudolf, G., and Stahlhofen, W. (1980a). Physical factors determining particle deposition in the human respiratory tract. *J. Aerosol Sci.* **11,** 505–515.

Heyder, J., Gebhart, J., and Stahlhofen, W. (1980b). Inhalation of aerosols—Particle deposition and retention. *In* "Generation of Aerosols" (K. Willeke, ed.), pp. 65–103. Ann Arbor Science Publ., Ann Arbor, Michigan.

Hiller, F. C., McCusker, K. T., Mazumder, M. K., Wilson, J. D., and Bone, R. C. (1982a). Quantitative deposition of sidestream cigarette smoke in the human respiratory tract. *Am. Rev. Respir. Dis.* **125,** 406–408.

Hiller, F. C., Mazumder, M. K., Wilson, J. D., McLeod, P. C., and Bone, R. C. (1982b). Human respiratory tract deposition using multimodal aerosols. *J. Aerosol Sci.* **13,** 337–343.

Landahl, H. D. (1972). The effect of gravity, hygroscopicity, and particle size on the amount and site of deposition of inhaled particles with particular reference to hazard due to airborne viruses. *In* "Assessment of Airborne Particles" (T. T. Mercer, P. E. Morrow, and W. Stober, eds.), pp. 421–428. Thomas, Springfield, Illinois.

Landahl, H. D., Tracewell, T. N., and Lassen, W. H. (1951). On the retention of airborne particulates in the human lung. *AMA Arch. Ind. Hyg. Occup. Med.* **3,** 359–366.

Landahl, H. D., Tracewell, T. N., and Lassen, W. H. (1952). Retention of airborne particulates in the human lung. III. *AMA Arch. Ind. Hyg. Occup. Med.* **6,** 508–511.

Lippmann, M. (1977). Regional deposition of particles in the human respiratory tract. *In* "Handbook of Physiology, Section 9: Reactions to Environmental Agents" (D. H. K. Lee, H. L. Falk, S. F. Murphy, and S. R. Geiger, eds.), pp. 213–232. American Physiological Society, Bethesda, Maryland.

Marple, V. A. (1979). The aerodynamic size calibration of optical particle counters by inertial impactors. *In* "Aerosol Measurement" (D. A. Lundgren, F. S. Harris, Jr., W. H. Marlow, M. Lippmann, W. E. Clark, and M. D. Durham, eds.), pp. 207–215. Univ. Press of Florida, Gainesville.

Martonen, T. B. (1984). Measurements of hygroscopic growth rates of medical aerosols. *In* "Aerosols" (B. Y. H. Liu, Y. H. Pui, and H. J. Fissan, eds.), pp. 1003–1006. Elsevier, New York.

Martonen, T. B., Bell, K. A., Phalen, R. F., Wilson, A. F., and Ho. A. (1982). Growth rate measurements and deposition modelling of hygroscopic aerosols in human tracheobronchial models. *In* "Inhaled Particles V" (W. H. Walton, ed.), pp. 93–108. Pergamon, Oxford.

Mazumder, M. K., Ware, R. E., Wilson, J. D., Renninger, R. G., Hiller, F. C., McLeod, P. C., Raible, R. W., and Testerman, M. L. (1979). SPART analyzer: Its application to aerodynamic size distribution measurement. *J. Aerosol Sci.* **10,** 561–569.

Mercer, T. T. (1973). "Aerosol Technology in Hazard Evaluation." Academic Press, New York.

Milburn, R. H., Crider, W. C., and Morton, S. D. (1957). The retention of hygroscopic dusts in the human lungs. *AMA Arch. Ind. Health* **15,** 59–62.

Morrow, P. E., Mehrhof, E., Casarett, L. J., and Morken, D. A. (1958). An experimental study of aerosol deposition in human subjects. *AMA Arch. Ind. Health* **18,** 292–298.

Patterson, R. K., and Wagman, J. (1977). Mass and composition of an urban aerosol as a function of particle size for several visibility levels. *J. Aerosol Sci.* **8,** 269–279.

Phalen, R. F. (1984). "Inhalation Studies: Foundations and Techniques." CRC Press, Boca Raton, Florida.

Stahlhofen, W., Gebhart, J., and Heyder, J. (1980). Experimental determination of the regional deposition of aerosol particles in the human respiratory tract. *Am. Ind. Hyg. Assoc. J.* **41,** 385–398.

Stahlhofen, W., Gebhart, J., and Heyder, J. (1981a). Biological variability of regional deposition of aerosol particles in the human respiratory tract. *Am. Ind. Hyg. Assoc. J.* **42,** 348–353.

Stahlhofen, W., Gebhart, J., Heyder, J., Philipson, K., and Comner, P. (1981b). Intercomparison of regional deposition of aerosol particles in the human respiratory tract and their long-term elimination. *Exp. Lung Res.* **2,** 131–139.

Swift, D. L., Shanty, F., and O'Neill, J. T. (1977). Human respiratory tract deposition of nuclei particles and health implications. *In* "Airborne Radioactivity," pp. 183–189. American Nuclear Society, La Grange Park, Illinois.

Tang, I. N. (1980). Deliquescence properties and particle size change of hygroscopic aerosols. *In* "Generation of Aerosols" (K. Willeke, ed.), pp. 153–167. Ann Arbor Science Publ., Ann Arbor, Michigan.

Task Group on Lung Dynamics. (1966). Deposition and retention models for internal dosimetry of the human respiratory tract. *Health Phys.* **12,** 173–207.

Wilson, F. J., Hiller, F. C., Wilson, J. D., and Bone, R. C. (1985). Quantitative deposition of ultrafine stable particles in the human respiratory tract. *J. Appl. Physiol.* **58,** 223–229.

Yu, C. P. (1978). Exact analysis of aerosol deposition during steady breathing. *Powder Technol.* **21,** 55–62.

Chapter 20

Dosimetry of Inhaled Particles by Means of Light Scattering

J. Gebhart
Gesellschaft für Strahlen- und Umweltforschung mbH
Institut für Biophysikalische Strahlenforschung
D-6000 Frankfurt/M., Paul-Ehrlich-Str. 20, Federal Republic of Germany

I. INTRODUCTION

Dosimetry of inhaled particles involves determining the fraction of inhaled particles deposited in the respiratory tract. For many years, light-scattering photometers combined with monodisperse test aerosols have been used in inhalation studies to measure the deposited fraction of aerosol particles in the human respiratory tract, as well as to study gas-mixing processes inside the airways (Altshuler *et al.,* 1957, 1959; Muir and Davies, 1967; Muir, 1967; Davies *et al.,* 1972; Giacomelli-Maltoni *et al.,* 1972; Heyder *et al.,* 1973, 1975, 1986, 1988; Gebhart *et al.,* 1981). The photometer technique permits continuous recording of the aerosol concentration close to the mouth during the whole course of a breathing cycle. From these records and simultaneous measurements of respiratory volumes and flow rates, the amount of aerosol in successive fractions of inspired and expired air can be evaluated.

Different devices based on the principle of light scattering have been developed to evaluate the number of particles in inspired and expired air of a subject. Photometric arrangements which measure the aerosol concentration in the main stream of inspired and expired air immediately at the

Extrapolation of Dosimetric Relationships for Inhaled Particles and Gases

entrance of the respiratory tract are called on-line systems. One advantage of these systems is that particle losses in connecting tubes and valves do not influence the results.

II. ON-LINE INHALATION APPARATUSES

A. Analog System

The analog system has been introduced in aerosol inhalation by Heyder *et al.* (1975, 1980). Its principle of operation is illustrated in Fig. 20-1. A continuous stream of a monodisperse aerosol passes through a mixing chamber to obtain a uniform particle concentration over the cross-section of the inhalation–exhalation channel. The volumetric flow rate of the aerosol stream always exceeds the respiratory flow rate so that the exhaled aerosol is constantly removed from the mixer. From continuous recordings of flow rate, $\dot{V}(t)$, with a pneumotachograph and of particle concentration, $c(t)$, with a photometer close to the mouth, the ratio of the number of expired to inspired particles per breath, N_e/N_i, is calculated by the equation

$$\frac{N_e}{N_i} = \frac{\int_{t_e} \dot{V}(t)c(t)dt}{\int_{t_i} \dot{V}(t)c(t)dt} \tag{1}$$

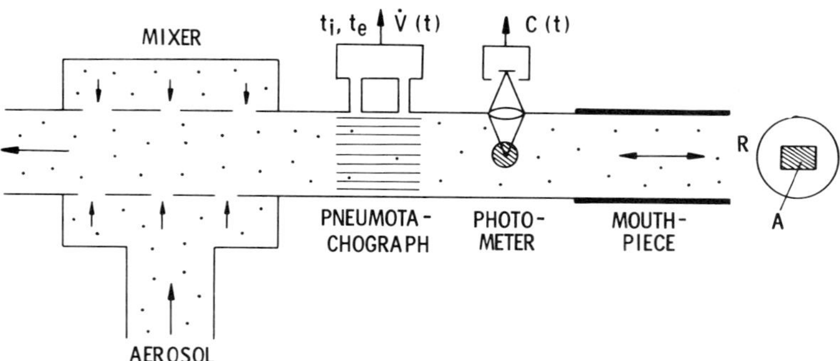

Fig. 20-1 Schematic diagram of an on-line apparatus operated in the analog mode.

where $t_i(t_e)$ is the inspiratory (expiratory) period. For this purpose, the electrical analog signals of the photometer and the pneumotachograph are fed to a computer for multiplication and integration. At the end of an inhalation experiment, the inspired and expired volumes and the number of inspired and expired particles are printed for each breath. The on-line analog system allows a high degree of automation, but needs careful calibration and adjustment of the photometer and the pneumotachograph.

B. Counting System

An aerosol photometer which counts individual particles in the respiratory flow immediately at the entrance of the respiratory tract has been developed by Gebhart *et al.* (1980) and used by Heyder *et al.* (1980) for measurements of total deposition. Its principle of operation is demonstrated in Fig. 20-2. The system is supplied with aerosol (monodisperse or polydisperse) via a mixing chamber like that in Fig. 20-1. The sensitive area of the photometer is formed by a sheet of laser-light, which covers almost the whole cross-section of the inhalation–exhalation channel and is inclined at a certain angle to the direction of the aerosol flow. A particle traversing the light sheet produces a flash of scattered light which is received by a photomultiplier and converted into an electrical pulse. The time signals controlling the period of inspiration, t_i, and expiration, t_e, and the volumetric flow rate, $\dot{V}(t)$, are obtained from a pneumotachograph. A computer counts the number of signals during inspiration and expiration

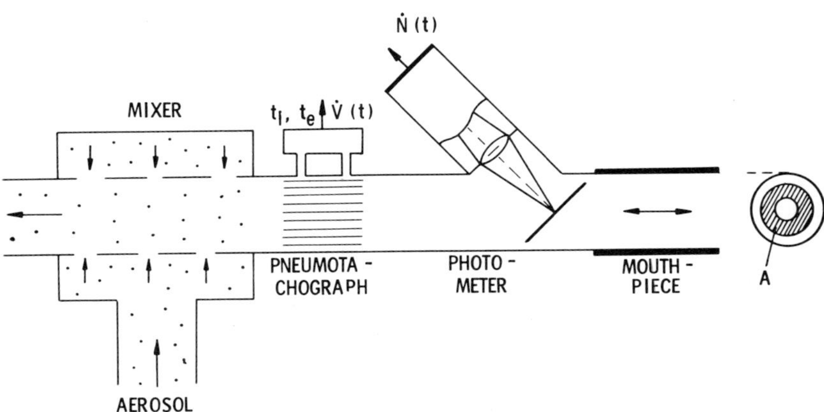

Fig. 20-2 Schematic diagram of an on-line apparatus operated in the counting mode.

and calculates the ratio of the number of expired and inspired particles per breath according to the equation

$$\frac{N_e}{N_i} = \frac{\int_{t_e} \dot{N}(t)dt}{\int_{t_i} \dot{N}(t)dt} \tag{2}$$

where $\dot{N}(t)$ is the particle stream (count rate) monitored by the photometer. The inspired and expired volumes, V_i and V_e, respectively, are achieved by integrating the volumetric flow rates, $\dot{V}(t)$. At the end of a breathing experiment the quantities, V_i, V_e, N_i, N_e, N_e/N_i, and $1 - N_e/N_i$, are printed with their means ± S.D. for each breath. Since the sensitive area of the photometer does not cover the whole cross-section of the inhalation–exhalation channel but only a representative fraction, N_i and N_e are relative numbers. Nevertheless, the ratio N_e/N_i as expressed by Equation (2) is measured correctly. The on-line counting system works in a straightforward manner and can be combined directly with modern computer techniques.

III. REQUIREMENTS

For a correct measurement of the ratio of expired to inspired particle number, N_e/N_i, some requirements concerning the aerosol supply and the flow pattern in the respiratory channel must be fulfilled. They are explained and summarized in Fig. 20-3. In the analog method, where the particles are illuminated by a pencil of light, the sensing volume forms a cylinder and its cross-sectional area A perpendicular to the direction of flow is nearly rectangular. In the counting method, where the particles are illuminated by a sheet of light, the sensing volume forms a flat circular disk and its cross-sectional area A perpendicular to the direction of flow has an elliptical shape. In the analog mode, the instantaneous particle concentration, $c(t)$, of a monodisperse aerosol is derived from light scattering on an assembly of particles. In this case, the number concentration of the particles passing the sensitive area A has to be representative for the whole cross-section, πR^2, of the channel. In the counting mode the instantaneous particle stream, $\dot{N}(t)$, of the aerosol is obtained from direct counts of single particles. This technique also requires a homogeneous particle concentration over the cross-section of the aerosol channel. In addition, the flow velocity, $u(x, y, t)$, through the sensitive area A has to be representative for the whole cross-section, πR^2, of the channel. This is achieved by a disklike sensing volume which covers a large part of the channel.

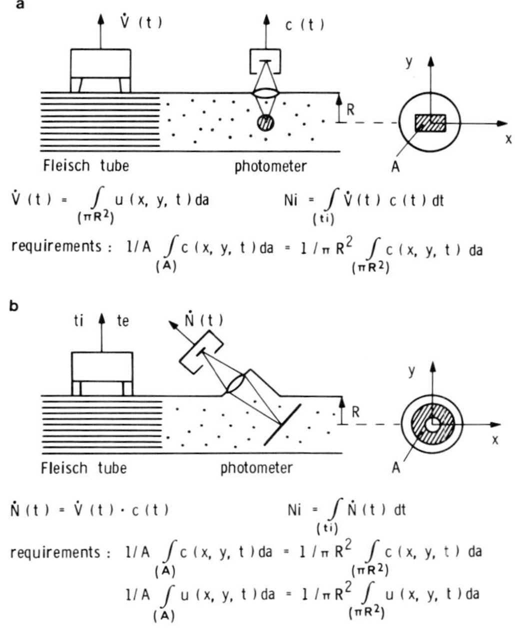

Fig. 20-3 Requirements for the on-line measurement of particle number in the respiratory flow (a) Analog method, (b) counting method.

Other sources of error in inhalation photometry can arise from the light scattering properties of small particles. Let an aerosol with particle number concentration, c, and particle size distribution function, $f(D)$, pass through a photometer with sensitive volume, V_m, and illumination intensity, I_0. The flux of the light, P, scattered by the assembly of particles into the receiver aperture, $\Delta\Omega$, is then given by the equation

$$P = I_0 V_m c \int_0^\infty f(D) S(D, m, \lambda, \Delta\Omega) dD \tag{3}$$

where $S(D, m, \lambda, \Delta\Omega)$ is the partial scattering cross-section of a spherical particle with diameter, D, and refractive index, m, at the wavelength, λ. The sensitive volume, V_m, the receiver aperture, $\Delta\Omega$, and the wavelength, λ, are fixed by the photometer design. If the reading of the photometer now varies, one can generally not distinguish whether the number concentration, c, the function, $f(D)$, or the optical constants, m, of the particle material have changed. Since only the change in the number concentration, c, is to be measured, the size distribution function, $f(D)$, and the refractive index, m, of the particles have to be the same during inspiration and expiration. Therefore, monodisperse aerosols consisting of nonhygroscopic particles must be used in combination with the analog method. To be sure that the function, $f(D)$, is not altered by the filter characteristic of the respiratory tract, the geometrical standard deviation of these aerosols should be below 1.15. Nonhygroscopic particles are required for the analog mode, since otherwise the uptake of water in the lungs leads to a change of the size and refractive index of the particles in expired air. In the counting mode changes of light scattering properties have no effect on the evaluation of particle number. In this case, all particles above a certain detection limit are counted regardless of their optical properties and size distribution, so the ratio N_e/N_i is measured correctly even for polydisperse aerosols which differ in their size distribution, $f(D)$, between inhalation and exhalation.

Finally, differences in temperature and moisture content between inspired and expired aerosols may also contribute to erroneous measurements of particle number, N. Such differences change the volume of an aerosol which results in a corresponding variation of its concentration. Applying the equations for ideal gases, the relative change of the aerosol volume, $\Delta V/V$, can be estimated according to the equation (Muir and Davies, 1967)

$$\frac{\Delta V}{V} = \frac{(760 - \eta 47)310}{(760 - 47)(273 + \zeta)} - 1 \tag{4}$$

Equation (4) assumes that the expired aerosol entering the photometer has a temperature of 37°C and is saturated with water vapor at pressure p_s (37°C) = 47 mm Hg. During inspiration, it is assumed that the aerosol has a temperature ζ(°C) and a water vapor pressure ηp_s (37°C). From Equation (4) it is evident that the diluting effect can be neglected only if the aerosol offered for inhalation is warmed to body temperature and humidified above 80% relative humidity. To avoid supersaturation inside the photometer during expiration, the whole unit must be kept slightly above body temperature.

IV. PERFORMANCE OF A 2-MODE PHOTOMETER

In the last few years the photometer described in Fig. 20-2 has been reconstructed as a 2-mode-photometer for aerosol inhalation studies (Gebhart *et al.*, 1980, 1983). It can be operated in the analog mode above 100 particles/cm^3 and in the counting and sizing mode below 10 particles/cm^3. With this device, total deposition in human subjects can be measured at very low number concentrations. Under certain assumptions the photometer can also be used as a particle size spectrometer which classifies aerosol particles in the main stream of inspired and expired air immediately at the entrance of the respiratory tract.

A. Opto-Mechanical Set-Up

The parallel beam of an argon ion laser (2 Watts) traverses a system of diaphragms and cylindrical lenses to form, within the sensitive volume, a sheet of light of 80 μm thickness and 15 mm height. The sheet of light is inclined at an angle of 60° to the direction of the aerosol flow and covers almost the whole cross-section of the aerosol channel ($\phi = 16$ mm). The sheetlike laser beam enters the metallic housing of the photometer through a glass window, crosses the aerosol channel, and disappears in a light trap. The light flashes scattered from single particles crossing the light sheet are received by a microscope objective ($f = 25$ mm; aperture angle 12°) under a mean scattering angle of 90°, whereby the sensitive volume is imaged into the plane of a field stop (Fig. 20-4). The field stop confines the sensitive area of the photometer and keeps stray light from the photomultiplier. The size of the field stop allows about half the number of particles traversing the total cross-section of the channel to be seen by the photomultiplier. Since the sheet of light coincides with the object plane, only those particles which are imaged sharply into the field stop are illuminated. In this way, more than 97% of the counted particles can be classified without being influenced by the edge of the stop. A field lens images the aperture of the microscope objective onto the cathode of the photomultiplier, so all light flashes originating from single particles strike the same area of the cathode regardless of their position inside the sensitive volume.

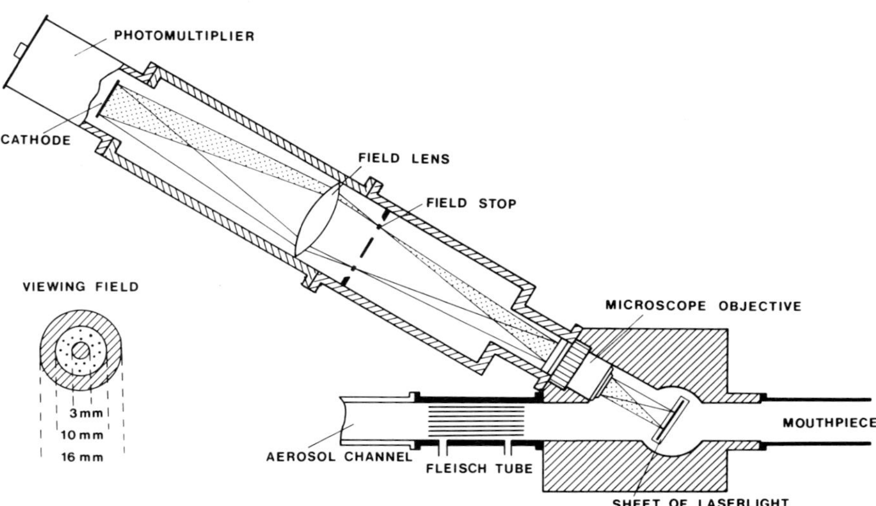

Fig. 20-4 Receiver unit of the 2-mode-photometer.

B. Electronic Equipment

The time signals for the onset of inspiration and expiration are supplied from the differential pressure transducer of the pneumotachograph. The analog signal of the pneumotachograph represents the respiratory volumetric flow rate, $\dot{V}(t)$, which after integration yields the inspired and expired volumes. In the analog mode the instantaneous aerosol concentration, $c(t)$, is represented by an analog voltage which is fed to an analog-digital-converter (ADC) and stored in terms of binary numbers as a function of inspired and expired volumes in steps $\Delta V = 0.6\ \text{cm}^3$.

In the counting mode all electrical pulses of the photomultiplier above a certain pulse height are counted. During normal breathing, pulse durations between 50 and about 500 μsec are achieved. These pulses are fed to the ADC of a pulse-height-analyzer, classified according to their height, and stored separately during inspiration and expiration. All pulses being selected by a comparator are counted in electronic counters to obtain N_e/N_i.

C. Measurement of Total Deposition

A typical experiment which illustrates the determination of total deposition in the counting mode is given in Fig. 20-5. The figure contains a specification of the breathing maneuver, the particle size distributions measured in inspired and expired air, and a print-out of computer calculated data. The subject inhaled polystyrene spheres of 2.02-μm diameter under standardized breathing conditions. The printed data represent four

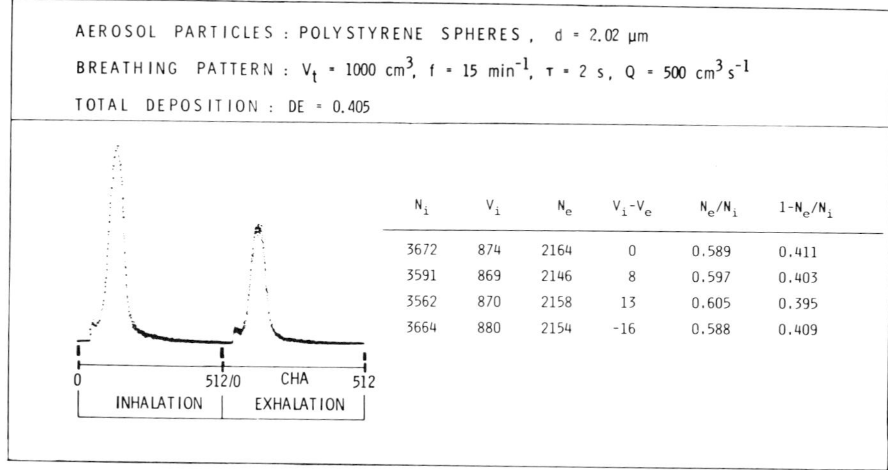

N_i	V_i	N_e	V_i-V_e	N_e/N_i	1-N_e/N_i
3672	874	2164	0	0.589	0.411
3591	869	2146	8	0.597	0.403
3562	870	2158	13	0.605	0.395
3664	880	2154	-16	0.588	0.409

Fig. 20-5 Measurement of total deposition in the counting mode.

successive breaths under steady state conditions after the completion of the wash-in. For each breath the number of counts during inspiration and expiration, N_i and N_e; the inspired volume, V_i; the difference between inspired and expired volume, $V_i - V_e$; the ratio, N_e/N_i; and the deposition per breath, $1 - N_e/N_i$, are listed. Since about half of the total number of particles passing through the aerosol channel are counted, the subject inhaled about 7000 particles per breath and per liter, which corresponds to a concentration of only 7 cm^{-3}. Human deposition data evaluated with the analog method at relatively high concentrations and with the counting method after extreme dilution of the aerosol have been found to be in good agreement (Heyder *et al.*, 1980). The counting method has the advantage that the lung burden originating from particulate matter can be kept extremely low.

D. Growth Studies of Hygroscopic Particles

In the counting and sizing mode the photometer can be used as a sensitive monitor for hygroscopic properties of aerosol particles, since any change of the scattering behaviour of the particles is indicated immediately. For a quantitative analysis 150-cm^3 boluses of monodisperse sodium chloride aerosols have been injected into the inspiratory flow of different subjects, transported into a defined lung depth, and expired again with the same flow rate as during inhalation (Anselm *et al.*, 1986). To obtain the diameter of the expired aqueous NaCl droplets from their scattering signals, theoretical response functions of the photometer have been calculated for various refractive indices ranging from pure water ($m = 1.334$) to a saturated aqueous NaCl solution with $m = 1.381$. The curves derived from Lorenz-Mie theory have been matched with experimental data of test standards. To evaluate the diameter of a solution droplet, it was first assumed that the expired droplets consist of pure water. Then, starting from the known NaCl mass solved in the droplet, an iteration process was initiated until the best fit for the droplet diameter was achieved. Figure 20-6 compares growth factors measured for 0.7-μm NaCl particles with theoretical predictions when exposing these particles at time zero to different relative humidities (Ferron, 1977). It has been found that 0.7-μm NaCl particles are expired from the alveolar region as aqueous NaCl droplets, with a mean diameter of 3.85 μm which corresponds to a growth factor of 5.5.

High-resolution size spectrometry based on the intensity of scattered light requires particles of spherical shape and a known refractive index. These conditions are met well by salt particles which are converted into homogeneous solution droplets inside the respiratory tract. No knowledge of the optical properties and shape of the particles is required if the single particle aerodynamic relaxation time analyzer (SPART) is used for size

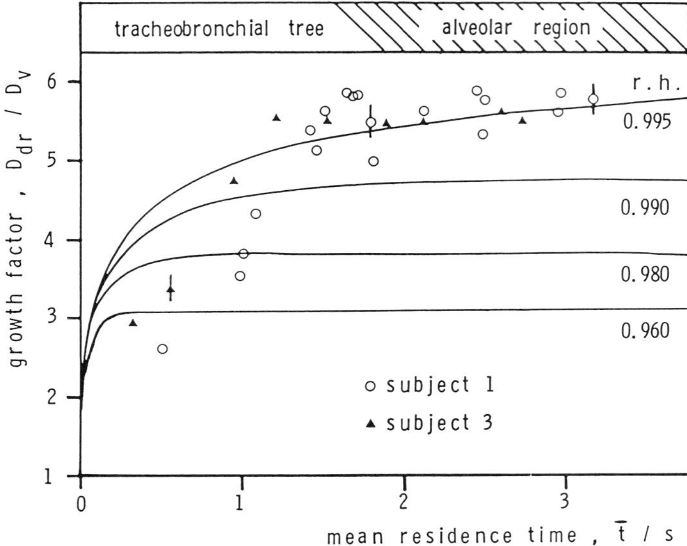

Fig. 20-6 Growth factor of expired aqueous NaCl droplets as function of mean residence time in the lungs after inhalation of 0.7-μm NaCl particles in comparison to theoretical predictions for different relative humidities (r. h.). (After Ferron, 1977.)

analysis in the respiratory flow (Mazumder *et al.*, Chapter 19, this volume). When using the SPART analyzer for on-line measurements of total deposition, care must be taken by control circuits to ensure that the particle stream through the small sensing area of the device is representative for the entire cross-section of the respiratory channel.

V. CONCLUSIONS

Light-scattering photometers in combination with pneumotachography and well-defined test aerosols can be used to measure the deposited fraction of inspired aerosol particles in the human respiratory tract as well as to study convective gas transport and deposition mechanisms inside the airways. These techniques permit continuous recordings of the aerosol concentration close to the entrance of the respiratory tract during the whole course of a breathing cycle and allow a high degree of automation. With an intense laser beam for the illumination, a 2-mode-photometer which operates in the photometric mode (assembly of particles) at high number concentrations and counts single particles at concentrations below about 10 cm^{-3} has been constructed. In the counting mode this instrument can also be used as a particle size spectrometer which classifies aerosol particles in the main stream of the respiratory flow. Modern photometer techniques in association with model aerosols offer new possibilities in lung function tests which should be introduced not only in aerosol laboratories but also in the field of clinical research.

REFERENCES

Altshuler, B., Yarmus, L., Palmes, E. D., and Nelson, N. (1957). Aerosol deposition in the human respiratory tract. *A.M.A. Arch. Ind. Health* **15,** 293–303.

Altshuler, B., Palmes, E. D., Yarmus, L., and Nelson, N. (1959). Intrapulmonary mixing of gases studied with aerosols. *J. Appl. Physiol.* **14,** 321–327.

Anselm, A., Gebhart, J., Heyder, J., and Ferron, G. (1986). Human inhalation studies of growth of hygroscopic particles in the respiratory tract. *Proc. Int. Aerosol Conf., 2nd, Berlin* pp. 252–255.

Davies, C. N., Heyder, J., and Subba Ramu, M. C. (1972). Breathing of half-micron aerosols. I. Experimental. *J. Appl. Physiol.* **32,** 591–600.

Ferron, G. (1977). The size of soluble aerosol particles as a function of the humidity of the air. Application to the human respiratory tract. *J. Aerosol Sci.* **8,** 151–167.

Gebhart, J., Heigwer, G., Roth, C., and Stahlhofen, W. (1980). A new device for aerosol and gas inhalation studies and its application in lung investigations. *J. Aerosol Sci.* **11,** 237–239.

Gebhart, J., Heyder, J., and Stahlhofen, W. (1981). Use of aerosols to estimate pulmonary air-space dimensions. *J. Appl. Physiol.* **51,** 465–476.

Gebhart, J., Heigwer, G., and Stahlhofen, W. (1983). Method and apparatus for determining the deposition of particles in the human respiratory tract and/or for checking the function of the respiratory tract. US Patent No. 4, 370,986; Appl. Ser. No. 06/191,821.

Giacomelli-Maltoni, G., Melandri, C., Prodi, V., and Tarroni, G. (1972). Deposition efficiency of monodisperse particles in human respiratory tract. *Am. Ind. Hyg. Assoc. J.* **33,** 603–610.

Heyder, J., Gebhart, J., Heigwer, G., Roth, C., and Stahlhofen, W. (1973). Experimental studies of the total deposition of aerosol particles in the human respiratory tract. *J. Aerosol Sci.* **4,** 191–208.

Heyder, J., Armbruster, L., Gebhart, J., Grein, E., and Stahlhofen, W. (1975). Total deposition of aerosol particles in the human respiratory tract for nose- and mouth-breathing. *J. Aerosol Sci.* **6,** 311–328.

Heyder, J., Gebhart, J., and Stahlhofen, W. (1980). Inhalation of aerosols: Particle deposition and retention. *In* "Generation of Aerosols—Their Characteristics, Facilities and Exposure Experiments" (K. Willeke, ed.), pp. 65–103. Ann Arbor Science Publ., Ann Arbor, Michigan.

Heyder, J., Gebhart, J., Rudolf, G., Schiller, C. F., and Stahlhofen, W. (1986). Deposition of particles in the human respiratory tract in the size range 0.005–15 μm. *J. Aerosol Sci.* **17,** 811–825.

Heyder, J., Blanchard, J. D., Feldman, H. A., and Brain, J. D. (1988). Convective mixing in human respiratory tract: Estimates with aerosol boli. *J. Appl. Physiol.* **64,** 1273–1278.

Muir, D. C. F. (1967). Distribution of aerosol particles in exhaled air. *J. Appl. Physiol.* **23,** 210–214.

Muir, D. C. F., and Davies, C. N. (1967). The deposition of 0.5 μm diameter aerosols in the lung of man. *Ann. Occup. Hyg.* **10,** 161–174.

Part Five

Modeling Approaches to Predicting Dosimetry of Inhaled Particles and Gases

Chapter 21

Modeling Approaches: Introduction

W. Stöber
Fraunhofer Institute of Toxicology and Aerosol Research
Hannover, West Germany

I. RATIONALE

The presence and final retention of inhaled foreign gases and particulates in the lung could, and frequently do, cause certain health effects in humans if the foreign matter has toxic properties or is deposited in excessive amounts. In view of this potential threat to human health and in order to facilitate an assessment of the effective dose and the associated health risk, it is desirable to obtain reliable information regarding the deposition, dose distribution, accumulation, clearance, and final removal or retention of inhaled toxic trace gases and aerosols in the human lung. Efforts to acquire this information have led to the development of a variety of physical, anatomical, and physiological lung models. These models utilize verified relevant data, scientifically sound assumptions, plausible hypotheses, and certain simplifications for a theoretical simulation of the distribution of doses, lung burdens, clearance rates, other effects, and biological responses. The resulting simulation data are supposed to represent the actual body burden of the inhaled material and to predict the potential human health risk.

Of course, the complexity of biological mechanisms and the stochastic nature of biological cause–effect relationships make the predictions of any biological model less precise than those of purely physical processes. Undoubtedly, the data of these simulation models are not suited for deterministic individual predictions and can only be considered as a statistical probability.

Extrapolation of Dosimetric Relationships for Inhaled Particles and Gases

Certainly, the modeling approaches apply a basic principle of science: the search for the essential variables common to a set of known phenomena and data in order to predict their quantitative influence on a merely qualitatively known or expected effect. Many of the great successes of the physical sciences have developed along these lines. But in biomedical modeling of human risk, it is not always possible to continue along the chain of prediction, subsequent testing, theoretical improvement, and final verification of the model. For obvious ethical reasons, there is a restrictive limit to acquiring experimental human data on health risk, particularly if irreversible health effects could be encountered. To avoid this danger, substitute studies with laboratory animals or other biological systems become necessary. Unfortunately these systems introduce additional uncertainties for the assessment of risk.

Because of statistical limitations and the impossibility of direct verification, there is uncertainty in risk assessment by biomathematical models. This requires the introduction of margins of safety and the use of conservative estimates when results of toxicity studies with animals under severe exposure conditions are to be extrapolated to people living under relatively light occupational or environmental exposures.

Occasionally, such substitute procedures for risk assessment are disputed, and some regulatory decisions based on extrapolation of animal data are resented or praised by the public that is divided in special interest groups. In reverence to a North Carolinian genius loci, Arthur C. Stern, the early pioneer of air pollution abatement, a quote is offered here. He gave a proper description of the dilemma of this problem.

> "There is a range of ambiguity in our health criteria data. In this range there is disagreement among experts as to its validity and interpretation. Thus from the same body of health effects data, one could adopt an air quality standard on the high side of the range of ambiguity or on the low side. Much soul-searching is required before accepting the results of questionable human health effects research and being accused of imposing large costs on the public by so doing; or of rejecting these results, and being accused of subjecting the public to potential damage to human health."

In this situation, it is important that modeling approaches be as precise as possible and that they do not unnecessarily increase the range of ambiguity of the associated health criteria. Model simplifications should not be accepted if they significantly increase the computed risk or its upper-bound confidence level. Disputed model hypotheses or vague assumptions should be identified and taken into account when the associated computed risk is considered.

These caveats do not imply that risk assessment by biomathematical modeling is always in a desolate situation. But not all of the models used at the present time are of the same sophistication and quality. Models are most advanced if they can rely predominantly on well-known quantitative physical or physicochemical relationships. For instance, the models on deposition and deposit distribution of trace gases and aerosols in the respiratory tract have excellent precision and suffer merely from the imperfections of the adopted anatomical model of the respiratory tract. Thus, none of the subsequent chapters address the basics of the kinetic models of the trace gas diffusion and absorption to the walls of the airways or the models of the particle dynamics in the respiratory tract. There is little need if any for improvement.

However, the great complexity and the strong variations between individual human lungs, with regard to both the conducting airways and the pulmonary region, require certain simplifications and standardizations which may not always be acceptable. This leads to the chapters outlined below.

II. CURRENT MODELING EFFORTS

A review by Ménache and Graham (Ch. 22) evaluates published data on dimensions of developing lungs of children which, of course, must be at variance with the standard model data. The investigators' present conclusion is that data of sufficient detail which could serve as a basis for a reliable model of the developing lung are not yet available. However, the need for such a model is demonstrated by Hofmann *et al.* in Chapter 26 which discusses an age-dependent lung dosimetry model for inhaled radon daughters. According to their model, the deposition of radon progeny inhaled by children may be greater than for adults. Thus, children may be at a higher risk by receiving a higher radiation dose.

Another shortcoming of symmetrical anatomical standard lung models with typical and uniform pathways from the trachea to the alveoli is described by Overton *et al.* in Chapter 23, dealing with the influence of the actual variability of airway paths and air-flow rates on the model predictions of trace gas absorption in rat lungs. For rats inhaling ozone, the authors show that the dose to the first alveolated airways in the pulmonary region may vary by a factor of three, depending on whether a standard anatomical model or calculation along the actual airway path was used. However, for the model assumptions made, the authors conclude that the use of the standard typical pathway model in conjunction with the common equivalent paths dosimetry model is still sufficient for a prediction of the ozone dose to the lower respiratory tract for the rats.

A practical aspect of aerosol inhalation and its importance for modeling is pointed out in Chapter 25 by Martonen *et al.* This chapter deals with the influence of aerosol hygroscopicity on aerosol deposition patterns. According to their calculations, they conclude that by ignoring the hygroscopicity of sulfuric acid droplets, their net deposition would be underestimated by a factor of two, while the fractional deposition for submicron particles would be overestimated by 50%.

Accounting for the above ad hoc modifications will certainly further improve the existing models on deposition and dose distribution of inhaled substances. However, the variations of the deposition patterns and dose distributions in individuals are at least of the order of the correction factors discussed so far. The associated change of health risk will probably vary by the same factor. But these changes are not impressive when compared to increases of the ambiguity range of the predictions by the high margins of safety which have to be introduced when animal data must be used to have a tangible dose–effect relationship with which to work.

A good example of the great interspecies variability that may be encountered in inhalation studies is given by Cuddihy *et al.* in Chapter 24 on predictions of respiratory tract clearance in humans. A comparison of experimental human data with results of animal studies indicates that the daily fractional pulmonary clearance rate for mice and rats is up to 50 times greater than for dogs and guinea pigs, with the human data somewhere in between. Thus, Cuddihy and co-authors stay away from interspecies extrapolations and confine their paper to the prediction of clearance in humans by an empirical model.

In contrast, Chapter 27 by Oberdörster on deposition and retention modeling of inhaled cadmium involves the quantitative extrapolation of effects in rats to the effects in people. It is a typical example of an attempted risk assessment by extrapolating observed health effects in animals and converting them into predicted health effects in humans.

In general, at a first stage of such an assessment, the mathematical models of lung deposition and retention of inhaled particles are to be utilized for an interpretation and generalization of the empirical data obtained by the aerosol inhalation studies with laboratory animals. As discussed in part above, the models can be adjusted to specified anatomical, physiological, and biochemical conditions which vary from species to species and in accordance with the experimental arrangements. In this way, experimental data of laboratory studies can be evaluated and condensed to species-dependent inhalation parameters for a variety of lung compartments, animals, and exposure conditions.

At the second stage, the species-dependent parameters may be used as yardsticks for interspecies comparisons. It would be instructive, for instance, to determine the actual lung burden variations between different animal species that were exposed to the same aerosol concentration. Or, one could establish the changes in species with regard to their susceptibility for certain respiratory diseases by evaluating the differences in the corresponding dose–response relationships. Or, the question could be settled as to what actually represents the proper measure of effective dose of a deposited harmful substance.

In a final stage, after contrasting data of different species, the comparisons must then be extended to humans and include available human model parameters and exposure conditions. Ultimately, the experimental animal health hazards have to be translated into an extrapolated potential human risk which normally requires the introduction of a safety margin. In the reported cadmium study, the author avoided the incorporation of a safety margin in his assessment by the plausible, but intuitive, assumption that equal amounts of accumulated cadmium deposits per g of lung tissue have the same carcinogenic potential in the peripheral lung in rat and humans.

Fortunately, the conservative risk assessment of the Oberdörster study (Ch. 27) by a conventional linear extrapolation from high- to low-exposure concentrations showed that the human risk of lung cancer from cadmium under environmental conditions is rather small. Of course, in case of a low risk, nobody is overly concerned about the precision of the risk estimate as long as it is reasonably conservative. However, if the model estimate would indicate a cadmium lung cancer risk of significant proportions, it would be necessary to evaluate the model to determine whether its uncertainties have inadvertently increased the upper bound of the 95% confidence level of the risk assessment.

III. REQUIREMENTS FOR IMPROVED MODELING OF PARTICLE CLEARANCE AND RETENTION

Whatever influence a model refinement with regard to particle translocation and removal may have on the risk assessment, there is no doubt that there are major differences in quality between the model of the deposition of the aerosol particles and the model of clearance and retention. On one hand, there are the sophisticated computations of particle dynamics and deposition of respirable aerosols, and on the other hand, there is a rather simple conventional assumption for the clearance and the retention of particles in the airways. This is also apparent from the calculations of Cuddihy *et al.* in Chapter 24.

Up to the present time, all clearance and retention data have merely been characterized by applying a simple mathematical model of first-order reaction kinetics to the various clearance processes that remove particulate deposits from the lung. The model provides a convenient parameter which relates the lung clearance to an apparent biological half-life of a particle mass deposit in the lung. However, because the model does not specify any underlying clearance mechanism it is possible to apply this purely mathematical concept to both bronchial and alveolar lung clearance, although the actual clearance mechanisms in these two regions are entirely different.

Accordingly, all acquired biological half-life values of particle deposits have been taken as empirical entities. In general, there have been no attempts so far to deduce half-life values from physiological or biochemical species-dependent variables, and there are indications for macrophage-mediated pulmonary clearance that the notion of an existence of true half-life values may even be conceptually incorrect.

Typical experimental half-life values range from less than an hour for fast bronchial clearance processes to many weeks or even years for alveolar clearance. The relatively fast bronchial clearance is primarily determined by the motion of the bronchial mucous layer as mediated by the ciliated epithelium of the conductive airways. However, no mechanistic model which would relate to the empirically well-established first-order kinetics has yet been developed.

For the assessment of potential health hazards, the empirically observed slow pulmonary clearance is of special importance, because the slow removal process permits the accumulation of sizeable particle deposits in the alveolar region before a steady state equilibrium may be reached by a balance of the rates for deposition and clearance. This steady state is a mathematical consequence of the first-order reaction kinetics model of clearance. Undoubtedly, this state must be very delicate for slow clearance processes or high deposition rates, and may not even exist under these circumstances. It is well known that the simple retention model which predicts a steady state at some point in time is easily impaired when cytotoxicity of particles is encountered or when the deposition rate is so high that the slow clearance process is overwhelmed. In these cases, the particulate lung burden will increase steadily in time and does not reach a steady state.

The relatively slow clearance rate in the alveolar region is usually explained by the slow, macrophage-mediated removal process which involves the phagocytosis of the particles by the macrophages and a transport onto the mucous layer of the conductive airways. Unfortunately, no established mechanistic model is available for this process. A review of alveolar macrophage data concerning their population, turnover rate, and phagocytic activity seems to indicate that the removal of particle deposits should be considerably faster if the majority of the macrophages would leave the alveolar region via the conductive airways. In this case, the clearance half-time should not exceed the residence time of the macrophages on the alveolar surface, which is given as no more than 35 days. In contrast, the actual clearance half-times are certainly multiples of this figure. This may indicate that the macrophages are rather poorly adapted for the transport of particles onto the mucociliary escalator.

It may be useful to recall that the primary purpose of macrophages is to find and to incorporate living foreign biological particles in order to kill and lyse them rather than to transport them away from the alveolar surface. In fact, the macrophage-mediated removal of inorganic particles is so slow that a low solubility of the deposited particles may well contribute to or compete with the removal by transport.

It will probably be a long time until it is possible to develop a suitable model for alveolar clearance and retention which could take care of species differences. Beforehand, detailed information will be needed which is presently, at best, incompletely available and sometimes inconsistent. A suitable model may require data on the macrophage population per alveolar surface area, the macrophage residence time in the alveolar region, the effective size of the macrophages, their migratory velocity with and without chemotactic influences, their phagocytic efficiency, and their volume capacity for foreign particles.

IV. CONCLUSIONS

The previous section of this introduction has described some of the improvements that are needed in developing appropriate models of inhalation risk assessment, and some of the requirements have been delineated for an improved data base. This is probably not a complete list of requirements, but it indicates how much empirical information for a mechanistic pulmonary clearance model may be needed. Unless such data become available, the conventional approach by determining nominal half-lives which do not lend themselves to mechanistic interpretation will continue for some time to come, and meanwhile, risk assessment near the critical limits will continue to suffer from the uncertainties of the entirely empirical approach to modeling retention of particles in the lungs.

Chapter 22

Issues That Must Be Addressed When Constructing Anatomical Models of the Developing Lung

M. G. Ménache
R. C. Graham
NSI Technology Services Corporation
Research Triangle Park, North Carolina 27709

I. INTRODUCTION

To understand the effects of inhaled toxicants on human health, the dose delivered to the lung and other target organs must be estimated. Theoretical considerations have been coupled with results from experimental studies to develop mathematical dosimetry models capable of simulating deposition of inhaled gases and particles in the human adult lung. Models of the developing lung are needed to assess chronic and lifetime exposures to radionuclides and other compounds as well as to estimate delivered doses of gases and particles in children.

A description of lung dimensions, including airway lengths, diameters, branching angles, and inclinations to gravity for the conducting airways, is needed for these dosimetry models. Branching angles in the upper airways

Disclaimer: This study was supported by funds provided by the U.S. Environmental Protection Agency under contract No. 68-02-4450. This report has been reviewed by the Health Effects Research Laboratory, U.S. Environmental Protection Agency, and approved for publication. Approval does not signify that the contents necessarily reflect the views and policies of the U.S. Environmental Protection Agency, nor does mention of trade names or commercial products constitute endorsement or recommendation for use.

contribute to impaction and also affect turbulence, which alters the probability of deposition at the airway bifurcations. The length and diameter of an airway affect the probability of deposition due to sedimentation and diffusion within that airway. Some data are available on these parameters in the human adult lung, but very little information has been reported for children. As a result, growth of the airways in the tracheobronchial (TB) region as a function of age is based primarily on simple regression models using the available empirical studies.

Data on the anatomical characteristics of the acinar region are even scarcer. Estimates of the number of alveoli vary significantly, and there is incomplete evidence on the age at which this number stabilizes. Incorporation of dimensions of the respiratory airways in a dosimetry model, therefore, relies heavily on assumptions made concerning the number of generations and number of airways per generation as a function of age. Differences in the airway dimensions, structure, and growth of this region used in dosimetry models are based on assumptions that can not be empirically validated because of lack of data.

In this chapter, published data on airway dimensions in the developing human lung, including estimated age-dependent changes as well as experimental data, have been summarized to illustrate the variability and gaps in the data used to develop anatomical models for deposition calculations. A comparison of the morphological dimensions resulting from various assumptions about the volume distribution of pulmonary airways is also presented. Although deposition is not examined here, Hofmann *et al.* (Chapter 26, this volume) have shown twofold differences due to various geometries in the TB region alone and, more importantly, a reversal in predicted deposition depending on age. That is, one model does not consistently predict higher deposition than the other. Further investigation of lung dimensions is necessary to reduce the variability in reported measurements, thereby improving the data base necessary for risk assessment.

II. METHODS

The literature was reviewed to find studies with data on dimensions in the developing lung. The data presented in the figures were taken directly from published tables or digitized from published plots. Some authors presented empirically or theoretically derived equations which were used to provide continuous curves of age-related dimension changes when empirical data were not reported. It was not possible to include all of the data in the figures; however, Tables 22-1 to 22-5 provide a detailed summary of the studies reviewed and the nature of the data presented in these papers. Studies on the development of the prenatal lung are not considered in this chapter.

Table 22-1 Trachea

Study	Number of Subjects	Age (yrs)	Diameter	Length	Volume	Presentation of Data	Comments
Griscom and Wohl (1985)	100	<20	X	X	X	Regressions, graphs	
Kawakami *et al.* (1986)	34	16	X			Tables of averages	Genetic factors examined
Morgan and Steward (1982)	50	3–16		X		Regressions, tables, graphs	Measurements made to calculate optimal lengths for tracheal tubes
Johnson *et al.* (1978)	NR[a]	0–15	X	X		Table	Developmental physiology

[a] NR, not reported.

Table 22-2 TB Region

Study	Number of Subjects	Age (yrs)	Airway Dimensions Given	Structure Described	Presentation of Data	Comments
Cudmore *et al.* (1962)	20	0–73			Graphs	Number of generations are independent of age
Hislop *et al.* (1972)	12	0–adult	X		Graphs	
Hogg *et al.* (1970)	16	0.25–50	X		Graphs	Diameters and conductance
Huizinga (1933)	85	2–adult	X		Tables	Trachea and main bronchi diameters
Phalen *et al.* (1985)	20	0.1–21	X	X	Regressions, graphs	One path per lung measured

Table 22-3 Pulmonary Region[a]

Study	Number of Subjects	Age (yrs)	N_a	V_a	S_a	V_{duct}	Presentation of Data	Comments
Angus and Thurlbeck (1972)	32	adult	X				Graph, table	
Boyden (1967)	4	0.1–6.7					Drawings	Describes structure
Davies and Reid (1970)	8	0–11	X			X	Table	
Dunnill (1962)	10	0–8	X	X	X	X	Tables	
Emery and Mithal (1960)	309	0–12.5	X				Averages for groups	No totals for whole lung
Emery and Wilcock (1966)	165	0–15	X					No totals for whole lung
Hislop and Reid (1974)	12	0–adult					Drawings	Describes structure
Hogg *et al.* (1970)	16	0.25–50					Graphs	Distance between alveoli
Langston *et al.* (1984)	42	−0.3–0.1	X	X		X	Regressions	Fit to crown–rump length
Thurlbeck (1982)	56	0.1–14	X		X		Regressions, graphs	

[a] N_a, number of alveoli; V_a, alveolar volume; S_a, alveolar surface area; V_{duct}, duct volume.

Table 22-4 Age-Dependent Lung Models

Study	Age (yrs)	Extra-Thoracic Region	TB Region Dimensions/Structure	Pulmonary Region Dimensions/Structure
Crawford (1981)	0–adult	ICRP	ICRP/Weibel, Findeisen-Landahl	Follows TB growth/Weibel
Hofmann (1982)	0–adult	None	Curve fit of several authors/Weibel	Curve fit of several authors/Weibel
Martonen *et al.* (1989)	0.6–adult	None	Phalen/Yeh and Schum	Dunnill/Weibel
Overton and Graham (1989)	0–adult	Proportional to TB volume	Scaled Phalen/Yeh and Schum	Dunnill/Hansen and Ampaya
Phalen *et al.* (1985)	0–adult	None	Phalen/Yeh and Schum	
Thomas and Healy (1985)	0.1–adult	50 ml in adult	Handbook of Respiration/Landahl	Dunnill and others/Landahl
Xu and Yu (1986)		Proportional to trachea	Follows Hofmann	Follows Hofmann

Table 22-5 Other Studies

Study	Age (yrs)	Presentation of Data	Comments
Altman and Dittmer (1971)	all	Regressions, graphs	Compilation of volumes, V_t, f from several authors
Altman and Dittmer (1972)	all	Regressions, graphs	
Hart *et al.* (1963)	4–42	Regressions, graphs	
Wood *et al.* (1971)	7–40	Regressions, graphs	

III. THE CONDUCTING AIRWAYS

Measurements of conducting airways have been made in living subjects with bronchograms, radiograms, and computed tomography as well as using the lung itself either by dissection or creation of a solid cast. There are, of course, sources of measurement error associated with the technique selected and differences due to uncontrollable genetic and environmental variability in the subjects. In addition, differences in experimental protocols may contribute to variation. The lung is an elastic organ and changes shape during respiration. The pressure at which a lung is inflated or the point in the respiration cycle at which an image is made will affect the reported size of the airways. Although airways are commonly considered as a series of connecting tubes, they are neither perfectly round nor of a single diameter throughout their length. Definitions of length and diameter are different from study to study and also contribute to variability.

Age-dependent measurements of airway lengths and diameters are found

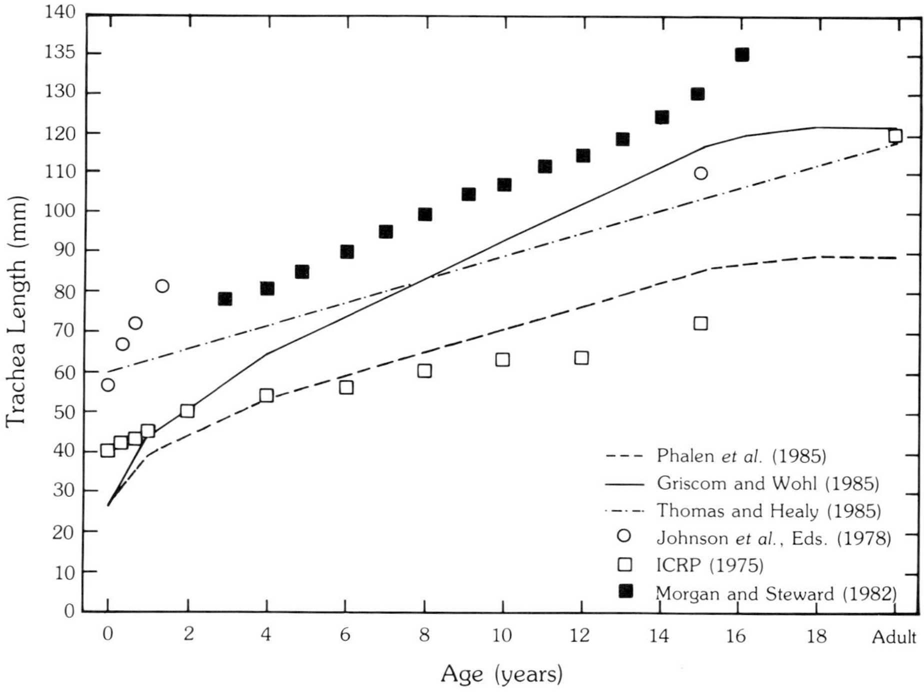

Fig. 22-1 Trachea length growth from birth to young adulthood. All data are taken from published reports as indicated on the figure. Curves are used to indicate predicted lengths; symbols represent reported measurements.

in the literature. Although branching and gravity angles are discussed in some studies, data are presented in only one study (Walker *et al.,* 1987). The most commonly measured airway was the trachea. Length and diameter measurements were found for the terminal bronchioles, while a few investigators measured all airways along a single, or very few, paths. Some measurements for airways in the other generations were reported, but are not reproduced here due to differences in the definition of a generation.

Age-related changes in tracheal length are shown in Fig. 22-1. Wide variation in length measurements is seen, with more than twofold differences found at some ages. In general, it was not possible to determine how the trachea length was measured from information in the articles. A common definition of length involves calculating the intersections of the axes of the two daughter airways with the central axis of the parent airway (Raabe *et al.,* 1976). This would define the endpoint of the trachea but not the initial point. A common definition for the beginning of the trachea is the first tracheal ring. Morgan and Steward (1982), whose reported measurements are greater than all others, defined tracheal length to begin with

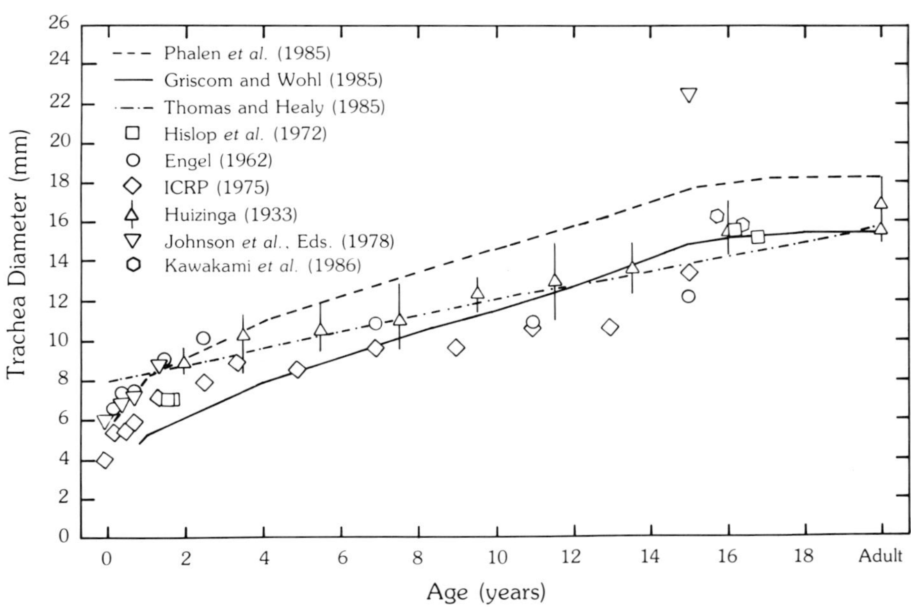

Fig. 22-2 The growth of the tracheal diameter as a function of age. Lines are used to indicate predicted changes in diameter; symbols represent reported measurements. The data of Huizinga (1933) are presented as the range of measurements and the means for groups of 6–11 subjects.

the vocal cords and end at the carina. Their data should, therefore, define an upper bound on length since both of their endpoints are more extreme than the commonly used definitions.

Data on age-dependent tracheal diameters (Fig. 22-2) exhibit slightly less variability than that observed in the lengths. There are never greater than twofold differences in measured diameters, and only one measurement, for a 15-year-old (Johnson *et al.,* 1978), seems inconsistent with the other data. Such agreement is rather surprising, as these measurements were made using a variety of techniques at different points in the respiratory cycle and using diverse measures of diameter (hydraulic, transverse, and antero-posterior). A possible explanation is that the stiff structure of the trachea due to the cartilaginous rings reduces diameter variability during measurement.

For the 2-year-old, it was possible to compare predicted diameters of airways along a single path in the TB region with empirical data. Figure 22-3 shows that there is substantial absolute variability between estimated and actual diameters in generations five through eight, but better agreement is found in the other airways. Twofold or smaller differences are

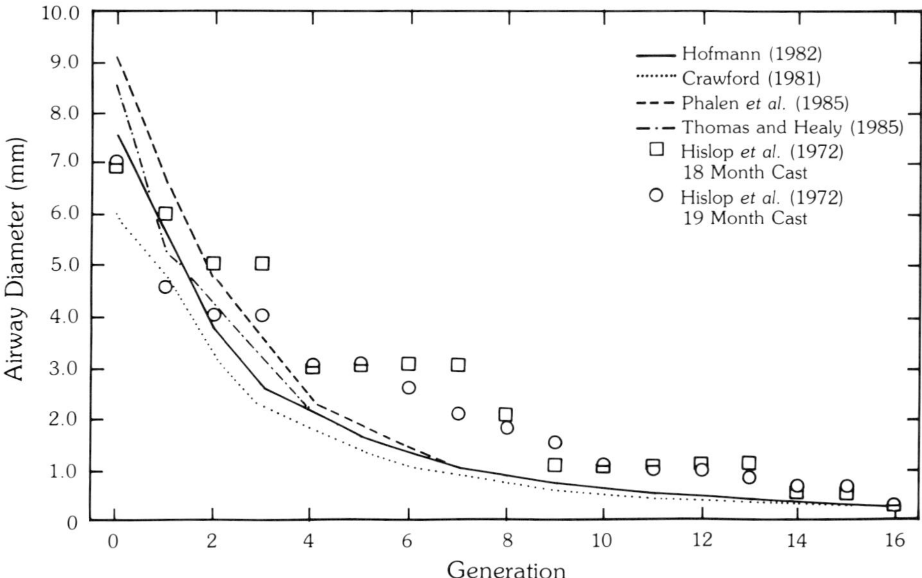

Fig. 22-3 Estimated airway diameters in the TB region (from the trachea to terminal bronchiole) of a 2-year-old child are compared to measurements made in casts from children of a similar age. Curves are used to indicate predicted diameters; symbols represent reported measurements.

observed in most generations, with approximately threefold differences seen in generations five to eight. These variations seem to be due to an apparent plateau in the measured diameters, which is not modeled by the continuously declining estimated diameters. Also, the paths used in the estimated equations are single paths that are adjusted to be representative of the whole lung, while the experimentally measured airways are from a single, actual path, which might not be characteristic of the entire lung.

Terminal bronchiole diameter growth is plotted in Fig. 22-4. The data of Hislop *et al.* (1972), which were used by Hofmann (1982) in the development of his theoretical model, suggest a significant increase in terminal bronchiole diameter with age. The terminal bronchiole diameters reported by Engel (1962) appear to increase slightly with age, whereas Phalen *et al.* (1985) reported no growth with age.

Tracheobronchial volume is plotted (Fig. 22-5) as a function of age to assess the overall effect of airway sizes in all conducting generations. With the exception of Langston *et al.* (1984), the volumes were estimated. Hart *et al.* (1963) measured dead space in subjects ranging from 4 years old to adult. Their estimates of dead space are included as an upper bound for the TB volumes.

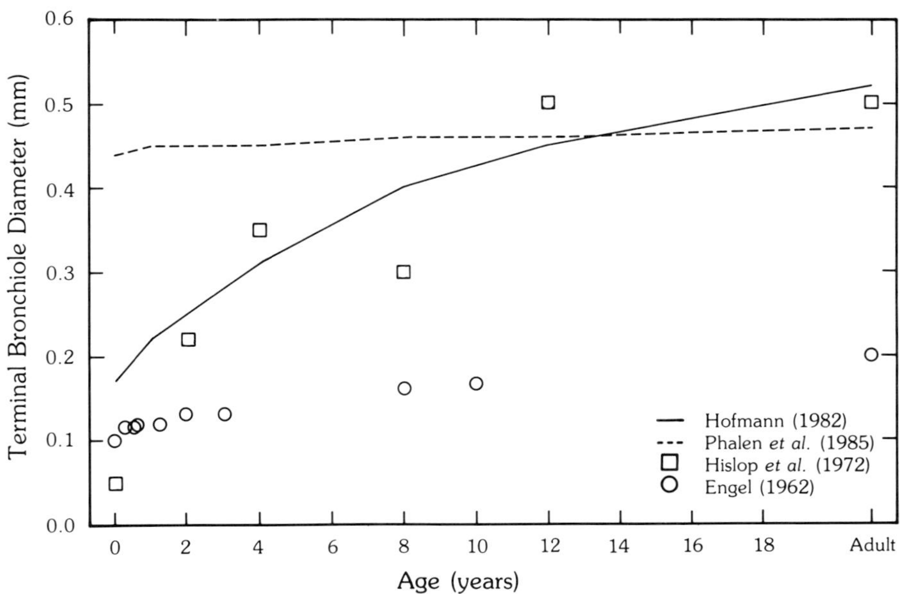

Fig. 22-4 Measurements of terminal bronchiole diameters and diameters estimated as a function of age. Curves are used to indicate predicted diameters; symbols represent reported measurements.

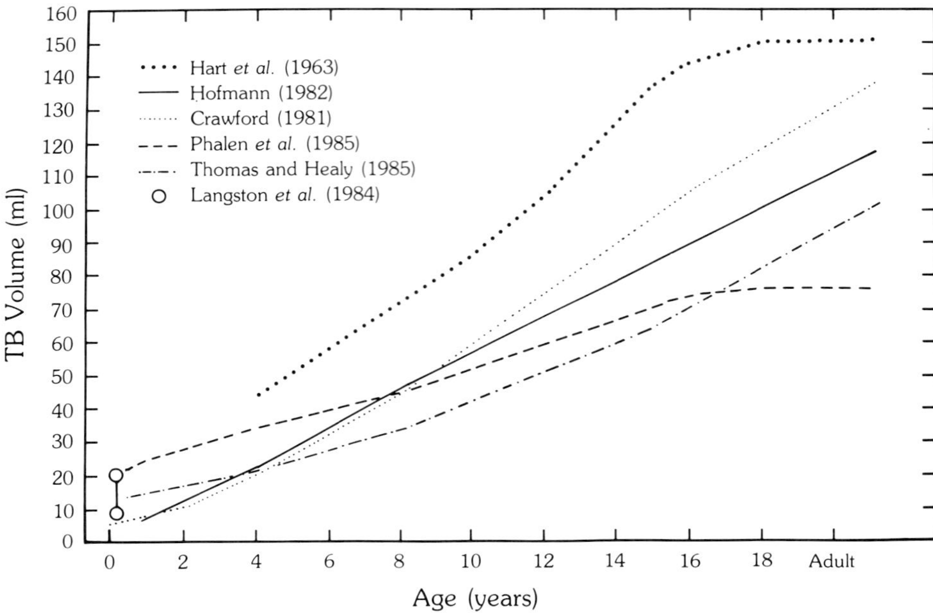

Fig. 22-5 Estimates of TB volume. The curve of Hart *et al.* (1963) provides an estimate of dead space and should serve as an upper bound. Langston *et al.* (1984) measured TB volume in newborn children. The range of their measurements is indicated.

IV. THE PULMONARY REGION

Measurements of the pulmonary airways in the developing lung present different problems from measurements of the conducting airways for a number of reasons. While the conducting airways are all present at birth and develop by growth only (Dunnill, 1962; Hislop *et al.*, 1972), the pulmonary region is incomplete at birth (Dunnill, 1962; Horsfield, 1978). Development thus includes an increase in the number of airways and alveoli and a possible change in function of some airways as well as growth of the existing airways. Measurements of pulmonary airway dimensions must be made from a lung directly or from a cast, so it is impossible to use living subjects. Because the pulmonary airways can double in volume during respiration, inflation pressure can significantly affect measured airway dimensions. Finally, the right circular cylinder approximation used to describe the airways is even less applicable in this region than in the TB region. Despite the interest in describing the development of the acinar region of the lung, very few quantitative studies have been published.

Although two fairly complete studies exist (Dunnill, 1962; Thurlbeck, 1982), their measurements do not always agree, and neither study made an attempt to measure changes in dimensions with age.

Several studies were found in which an attempt was made to quantitate the number of alveoli as a function of age (Fig. 22-6). The data of Dunnill (1962) and Davies and Reid (1970) are similar and show an exponential increase in the number of alveoli that appears to stabilize at about 8 years of age. The data of Thurlbeck (1982), however, exhibit a wide scatter with no obvious trend.

Dosimetry models require information on the structure of the alveolar region; however, no measurements are available on airway dimensions, number of pulmonary generations, or alveoli distribution among the generations as a function of age. Estimation of pulmonary parameters such as duct diameters, therefore, depends entirely on assumptions. Because of the number of variables involved, even if two models are based on the same age-dependent pulmonary volumes, estimates of the duct diameters may differ due to the distribution of the volume among the theoretically determined generations. This problem is illustrated in Fig. 22-7, which shows

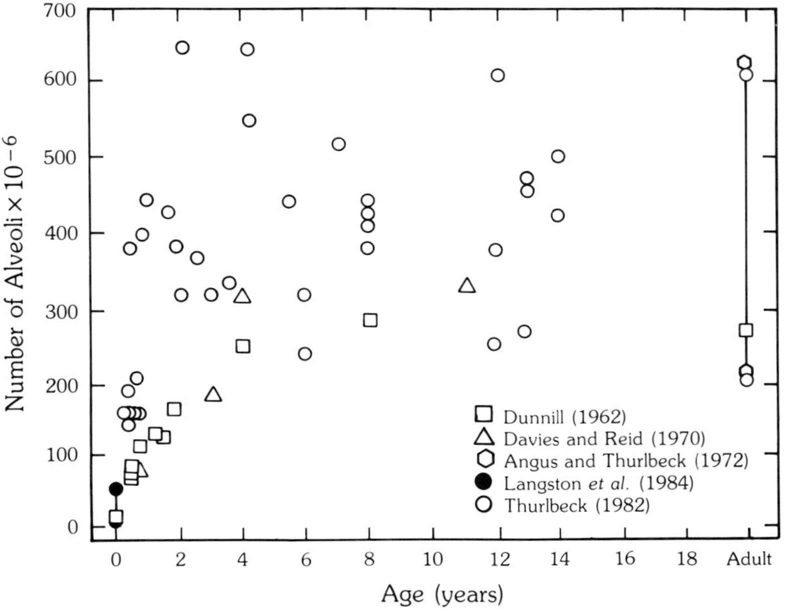

Fig. 22-6 The number of alveoli as a function of age. Ranges of measurements are shown for Langston *et al.* (1984) who made numerous measurements in newborns and for Angus and Thurlbeck (1972) who made numerous measurements in adults.

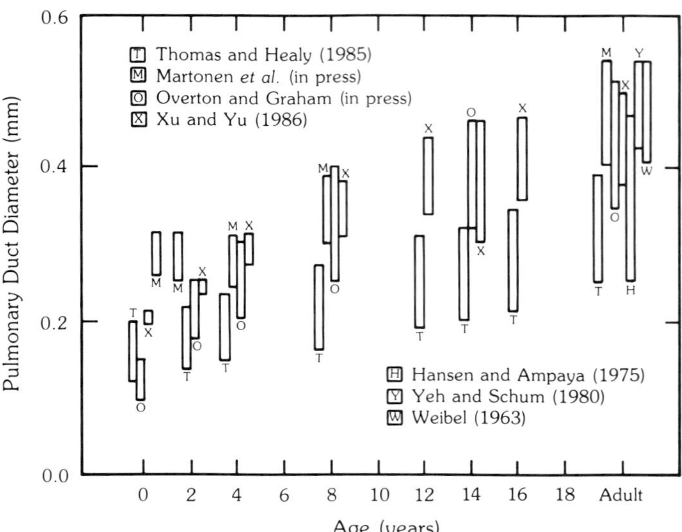

Fig. 22-7 Estimates of the range of pulmonary duct diameters as a function of age based on the indicated models. Adjacent vertical bars are for the same age; they have been separated horizontally for legibility.

ranges of pulmonary duct diameters for various models. For example, Martonen *et al.* (1989) and Overton and Graham (1989) both used volumes reported by Dunnill (1962) for various ages, but the duct diameters are different, especially below 8 years of age, because they assumed different structures for the pulmonary region. The effect of the assumed structure on diameter is also illustrated by the comparison of predicted diameters for the adult, based on the model of Hansen and Ampaya (1975) with that of Yeh and Schum (1980). By reducing several adult models to the same volume, Yu and Diu (1982) have shown that this difference is not due to inflation.

V. DISCUSSION AND CONCLUSIONS

Published data on dimensions of the developing lung report numerous measurements for the length and diameter of the trachea. Some authors have presented measurements for other conducting airways. Terminal bronchiole diameters were reported by Hislop *et al.* (1972) and Engel (1962). Measurements of length and diameter for the main bronchi have been summarized elsewhere (ICRP, 1975). Phalen *et al.* (1985) graphed observed values of length and diameter for the third generation. Phalen *et al.* (1985), Hofmann (1982), and Crawford (1981) presented equations to estimate lengths and diameters in all conducting airway generations, while Hislop *et al.* (1972) measured diameters for a single path. Few authors, on the other hand, have reported measurements for the alveolar region in the adult lung, and no one has presented data in enough detail to define models of developing lungs for simulation purposes.

Considerable variability in many of the measurements and estimates is due to a number of factors. A variety of experimental techniques were used involving different fixation methods and pressures in the casts and different breathing protocols in humans. Variations exist in the definition of generation, length, and diameter. While it was not possible to examine genetic or environmental differences using these data, it has been shown that significant intersubject variability exists and affects airway dimension (Soong *et al.,* 1979; Nikiforov and Schlesinger, 1985). Measurement of pulmonary airways poses particular problems, which have not yet been resolved. It is likely that three-dimensional reconstructions are needed to make accurate measurements, and it has been suggested that this is the best way to trace small pathways and the most accurate method to determine shapes (Hansen and Ampaya, 1974).

Variability in anatomical dimensions used in dosimetry can have a significant impact on estimated deposition fractions. Yu and Diu (1982) investigated the effects of the assumptions of five different models of the human adult lung on aerosol deposition and found that regional deposition varied significantly. For example, for mouth breathing, the TB deposition of 1- to 4-μm particles for the Yeh and Schum (1980) model was half the deposition of that for the Weibel (1963) model, while the deposition profiles within the TB region showed no similarity. The conclusions from the Yu and Diu (1982) study point to the importance of lung dimensions in adult models of deposition and suggest that lung geometry should be studied carefully in models of the developing lung.

Why dimensions are important to deposition can be seen by simplifying the equations for deposition fractions (Table 22-6) due to impaction (F_i) and sedimentation (F_s) in aerosol models and for steady state absorption in the trachea (F_t) and the pulmonary region (F_p) in gas models. Simplified deposition fractions were the same based on the theoretical models of several authors (Martonen, 1982; Yeh and Schum, 1980; Yu and Xu, 1987). Simplified absorption was based on the theoretical model of Overton *et al.* (1987) and is in agreement with the experimental data of Schiller *et al.* (1986).

As shown in Table 22-6, F_i, F_s, and F_t are functions of length and/or diameter. The factor of two to three in estimates of airway lengths will have

Table 22-6 Relation Between Deposition and Lung Geometry[a]

Fraction of aerosol, F_i, deposited by impaction

$$F_i = \frac{K_i \theta V_E}{nD^3}$$

Fraction of aerosol, F_s, deposited by sedimentation

$$F_s = \frac{K_s nLD}{V_E}$$

Fraction of gas, F_t, absorbed in the trachea

$$F_t = \frac{K_t LD}{V_E}$$

Fraction of gas, F_p, absorbed in the pulmonary region

$$F_p = \frac{K_p V_a}{V_E} = \frac{K_p(V_T - V_D)}{V_T}$$

[a] θ = branching angle; n = number of airways; D = diameter; L = length; V_E = minute volume; V_a = alveolar ventilation; V_T = tidal volume; V_D = dead space; K_i, K_s, K_t, K_p = constants that do not depend upon dimensions.

the same effect on the deposition fraction. Similarly, F_s and F_t will increase or decrease in the same proportion as the diameter. F_i can be significantly over- or underestimated because it is a function of diameter to the third power. F_p is a function of dead space. As seen from the age-dependent estimates of TB volume (Fig. 22-5), an important component of dead space, predictions vary by as much as a factor of two. Again, this can have a significant effect on estimates of F_p, particularly at low tidal volumes.

Knowledge of the dimensions of the developing lung is incomplete and highly variable. Based on other studies (Yu and Diu, 1982) and an analysis of the simplified form of deposition equations, however, it is clear that geometry is an important factor in calculating deposition fractions. Because these deposition fractions are needed to calculate delivered dose and to relate dose to health effects for risk assessment purposes, it is important to provide more data to improve the models of lung structure being used in the simulation models.

ACKNOWLEDGMENT

The authors would like to thank John Overton for his valuable comments on this paper.

REFERENCES

Altman, P. L., and Dittmer, D. S., eds. (1971). "Respiration and Circulation." Fed. of Am. Soc. for Exp. Biol., Bethesda, Maryland.

Altman, P. L., and Dittmer, D. S., eds. (1972). "Biology Data Book." Fed. of Am. Soc. for Exp. Biol., Bethesda, Maryland.

Angus, G. E., and Thurlbeck, W. M. (1972). Number of alveoli in the human lung. *J. Appl. Physiol.* **32,** 483–485.

Boyden, E. A. (1967). Notes on the development of the lung in infancy and early childhood. *Am. J. Anat.* **121,** 749–762.

Crawford, D. J. (1981). "A Generalized Age-Dependent Lung Model with Applications to Radiation Standards." U.S. Nuclear Regulatory Commission, Office of Nuclear Regulatory Research, Washington, D.C.

Cudmore, R. E., Emery, J. L., and Mithal, A. (1962). Postnatal growth of bronchi and bronchioles. *Arch. Dis. Child.* **37,** 481–484.

Davies, G., and Reid, L. (1970). Growth of the alveoli and pulmonary arteries in childhood. *Thorax* **25,** 669–681.

Dunnill, M. S. (1962). Postnatal growth of the lung. *Thorax* **17,** 329–333.

Emery, J. L., and Mithal, A. (1960). The number of alveoli in the terminal respiratory unit of man during late intrauterine life and childhood. *Arch. Dis. Child.* **35,** 544–547.

Emery, L., and Wilcock, P. F. (1966). The postnatal development of the lung. *Acta Anat.* **65,** 10–29.

Engel, S. (1962). "Lung Structure." Thomas, Springfield, Illinois.

Griscom, N. T., and Wohl, M. E. (1985). Dimensions of the growing trachea related to body height. *Am. Rev. Respir. Dis.* **131,** 840–844.

Hansen, J. E., and Ampaya, E. P. (1974). Lung morphometry: A fallacy in the use of the counting principle. *J. Appl. Physiol.* **37,** 951–954.

Hansen, J. E., and Ampaya, E. P. (1975). Human air space shapes, sizes, areas, and volumes. *J. Appl. Physiol.* **38,** 990–995.

Hart, M. C., Orzalesi, M. M., and Cook, C. D. (1963). Relation between anatomic respiratory dead space and body size and lung volume. *J. Appl. Physiol.* **18,** 519–522.

Hislop, A., and Reid, L. (1974). Development of the acinus in the human lung. *Thorax* **29,** 90–94.

Hislop, A., Muir, D. C. F., Jacobsen, M., Simon, G., and Reid, L. (1972). Postnatal growth and function of pre-acinar airways. *Thorax* **27,** 265–274.

Hofmann, W. (1982). Mathematical model for the postnatal growth of the human lung. *Respir. Physiol.* **49,** 115–129.

Hogg, J. C., Williams, J., Richardson, J. B., Macklem, P. T., and Thurlbeck, W. M. (1970). Age as a factor in the distribution of lower-airway conductance and in the pathologic anatomy of obstructive lung disease. *N. Engl. J. Med.* **282,** 1283–1287.

Horsfield, K. (1978). Quantitative morphology and structure: Functional correlations in the lung. *In* "The Lung: Structure, Function, and Disease" (W. M. Thurlbeck and M. R. Abell, eds.). Williams & Wilkins, Baltimore.

Huizinga, E. (1933). Ueber die Weite und das Wachstum des Bronchialbaumes. *Z. Hals Nasen Ohren Heilkd.* **33,** 546–558.

(ICRP) International Commission on Radiological Protection. (1975). "Report of the Task Group on Reference Man." ICRP Publication 23. Pergamon, Oxford.

Johnson, T. R., Moore, W. M., and Jeffries, J. E., eds. (1978). "Children are Different: Developmental Physiology." Ross Laboratories, Columbus, Ohio.

Kawakami, Y., Kusaba, H., Nishimura, M., and Abe, S. (1986). Trachea and lung dimensions in nonsmoking twins: Morphological and functional studies. *J. Appl. Physiol.* **61,** 495–499.

Langston, C., Kida, K., and Reed, M. (1984). Human lung growth in late gestation and in the neonate. *Am. Rev. Respir. Dis.* **129,** 607–613.

Martonen, T. B. (1982). Analytic model of hygroscopic particle behavior in human airways. *Bull. Math. Biol.* **44,** 425–442.

Martonen, T. B., Graham, R. C., and Hofmann, W. (1989). Human subject age and activity level: Factors addressed in a biomathematical deposition program for extrapolation modeling. *Health Phys.* In press.

Morgan, G. A. R., and Steward, D. J. (1982). Linear airway dimensions in children, including those with cleft palate. *Can. Anesth. Soc. J.* **29,** 1–8.

Nikiforov, A. I., and Schlesinger, R. B. (1985). Morphometric variability of the human upper bronchial tree. *Respir. Physiol.* **59,** 289–299.

Overton, J. H., and Graham, R. C. (1989). Predicting the uptake and distribution of ozone in the lower respiratory tract of children. *Health Phys.* In press.

Overton, J. H., Graham, R. C., and Miller, F. J. (1987). A model of the regional uptake of gaseous pollutants in the lung. II. The sensitivity of ozone uptake in laboratory animal lungs to anatomical and ventilatory parameters. *Toxicol. Appl. Pharmacol.* **88,** 418–432.

Phalen, R. F., Oldham, M. J., Beaucage, C. B., Crocker, T. T., and Mortensen, J. D. (1985). Postnatal enlargement of human tracheobronchial airways and implications for particle deposition. *Anat. Rec.* **212,** 368–380.

Raabe, O. G., Yeh, H. C., Schum, G. M., and Phalen, R. F. (1976). "Tracheobronchial Geometry: Human, Dog, Rat, Hamster—A Compilation of Selected Data from the Project Respiratory Tract Deposition Models." U.S. Govt. Printing Office, LF-53, 4C-48.

Schiller, C. F., Gebhart, J., Rudolf, G., and Stahlhofen, W. (1986). Factors influencing total deposition of ultrafine aerosol particles in the human respiratory tract. *J. Aerosol Sci.* **17,** 328–332.

Soong, T. T., Nicolaides, P., Yu, C. P., and Soong, S. C. (1979). A statistical description of the human tracheobronchial tree geometry. *Respir. Physiol.* **37,** 161–172.

Thomas, R. G., and Healy, J. W. (1985). "Models of the Respiratory Tract for Children Aged 1 Month to Adult." Los Alamos National Laboratory, Los Alamos, New Mexico.

Thurlbeck, W. M. (1982). Postnatal human lung growth. *Thorax* **37,** 564–571.

Walker, J. T., Nelson, N. S., and Mortensen, JD (1987). Age dependent changes in gravity angles of conducting airways in the human lung. *In* "Deposition and Clearance of Aerosols in the Human Respiratory Tract" (W. Hofmann, ed.), pp. 129–134. Facultas, Vienna.

Weibel, E. R. (1963). "Morphometry of the Human Lung." Academic Press, New York.

Wood, L. D. H., Prichard, S., Weng, T. R., Kruger, K., Bryan, A. C., and Levison, H. (1971). Relationship between anatomic dead space and body size in health, asthma, and cystic fibrosis. *Am. Rev. Respir. Dis.* **104,** 215–222.

Xu, G. B., and Yu, C. P. (1986). Effects of age on deposition of inhaled aerosols in the human lung. *Aerosol Sci. Technol.* **5,** 349–357.

Yeh, H., and Schum, G. M. (1980). Models of human lung airways and their application to inhaled particle deposition. *Bull. Math. Biol.* **42,** 461–480.

Yu, C. P., and Diu, C. K. (1982). A comparative study of aerosol deposition in different lung models. *Am. Ind. Hyg. Assoc. J.* **43,** 54–65.

Yu, C. P., and Xu, G. B. (1987). "Predictive Models for Deposition of Inhaled Diesel Exhaust Particles in Humans and Laboratory Species." Health Effects Institute, Cambridge, Massachusetts.

Chapter 23

Significances of the Variability of Tracheobronchial Airway Paths and Their Air Flow Rates to Dosimetry Model Predictions of the Absorption of Gases

J. H. Overton
Inhalation Toxicology Division
U.S. Environmental Protection Agency
Research Triangle Park, North Carolina 27711

A. E. Barnett
R. C. Graham
NSI Technology Services Corporation
Research Triangle Park, North Carolina 27709

I. INTRODUCTION

Health scientists must be able to estimate dose at specific lower respiratory tract (LRT) sites in order to extrapolate LRT health effects from laboratory animals to humans. Such dose predictions will allow estimation of the human exposure concentrations that would produce equivalent doses predicted or observed in animals. For this purpose a mathematical dosimetry model has been developed (Miller *et al.*, 1985; Overton *et al.*, 1987) that simulates the uptake and distribution of irreversibly reacting gases in the LRT of humans and laboratory animals.

The model was originally developed to simulate the absorption of ozone (O_3) in the LRT of humans and laboratory animals in conjunction with

Disclaimer: The research described in this article has been reviewed by the Health Effects Research Laboratory, U.S. Environmental Protection Agency, and approved for publication. Approval does not signify that the contents necessarily reflect the views and policies of the Agency nor does mention of trade names or commercial products constitute endorsement or recommendation for use.

typical path whole lung models (TPWLM) or symmetric lung models of the type developed by Weibel (1963) for humans and Yeh *et al.* (1979) for rats. These models summarize, approximate, and simplify the very complex geometry and structure of the airways and airspaces of animal and human lungs. The numerous varying pathways and airway dimensions of real lungs are approximated by many identical or equivalent pathways from the trachea to the terminal bronchioles or pulmonary sacs. All lung model pathways have the same number of generations and all airways and airspaces in the same generation have the same dimensions. In addition, for dosimetry modeling, the air flow rates (i.e., volume per unit time) at equivalent locations along each path are usually assumed to be the same. As a result, the development of mathematical dosimetry models is greatly simplified, because the treatment of transport and absorption can be the same along all paths, no matter how many. We call a mathematical dosimetry model that takes advantage of this simplicity an equivalent paths dosimetry model.

In contrast, actual paths in a respiratory tract (RT) can be very different from each other and from the model path of a symmetric lung model or TPWLM. In Fig. 23-1 the rectangles represent airways; the width and height are to scale and correspond to airway diameter and length, respectively. Figure 23-1A is a schematic of cast data from Raabe *et al.* (1976) for the first four generations of the tracheobronchial region (TB) of a rat lung. Figure 23-1B represents the first four generations of the TPWLM of Yeh *et al.* (1979), which is based on the same cast data represented in Fig. 23-1A. Only the tracheas have common dimensions, and the number of airways is different in generation 2 and in generation 3 (see Yeh *et al.*, 1979, for the rationale behind the number of airways in each generation of their TPWLM). Obviously, in Fig. 23-1A, all paths traced from the trachea to equivalent but distinct locations are different. However, all paths are equivalent in the TPWLM. For dosimetry modeling, the air-flow rates in each airway of the same generation of the TPWLM are usually assumed to be equal. How the flow is split at bifurcations in the real lung, as symbolized in Fig. 23-1A, is not known in general. Even if air flow rates were known, a reasonable assumption is that they would be different at equivalent locations along different paths.

The absorption of gases along a particular sequence of airways depends, among other things, on the dimensions of the airways and the rate of air flow through the airway. Thus, we can expect different doses in various airways of the same generation in a real lung as well as different doses at equivalent morphological locations. (Here, dose is a quantitative measure of the amount of pollutant reacting with the biochemical constituents at the given location; e.g., μg of O_3 absorbed by the liquid lining in a given generation.) This suggests the possibility that these locations may be at

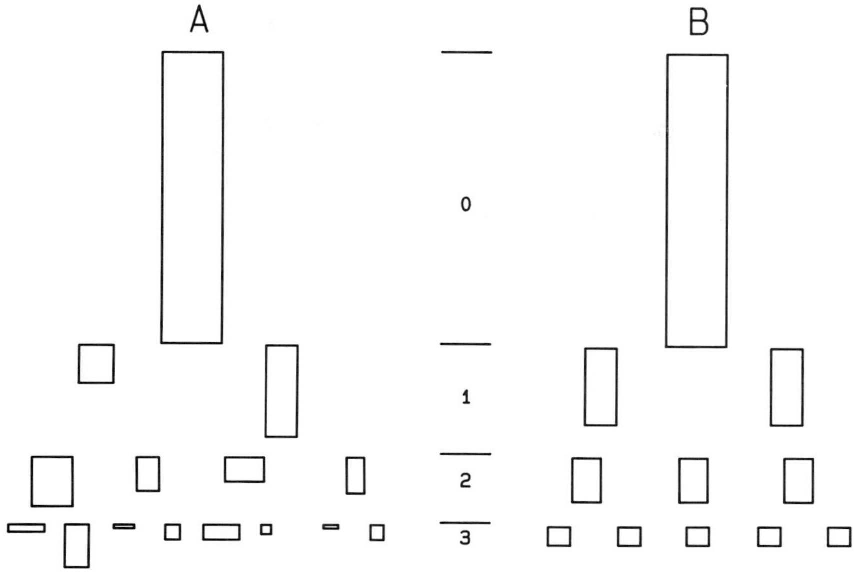

Fig. 23-1 Schematic of two models of the first four generations of the rat tracheobronchial (TB) region. Drawn to the same scale, the rectangles represent the airways, with height and width corresponding to airway length and diameter, respectively; A, the actual airways based on the Lovelace Foundation cast data (Raabe *et al.*, 1976); B, the airways of the Typical Path Whole Lung Model (Yeh *et al.*, 1979).

different toxicological risks within the same lung. Gaseous dosimetry modeling used in conjunction with TPWLMs cannot tell us the importance of accounting for actual lung paths and flow rates and may over- or underestimate dose in locations of toxicological importance.

In order to investigate the importance of (1) taking into account the variability of lung paths, and (2) the role of air-flow rates along these paths, a procedure has been developed that makes use of the computer simulation program developed to solve the equations of the equivalent paths dosimetry model described by Overton *et al.* (1987). This procedure is intuitive and is considered a first approximation to the very complex problem of simulating for a real lung the uptake of a highly reactive but moderately soluble gas. In this paper we describe the procedure and present results of using it to investigate the significance to dosimetry modeling of (1) and (2). Also considered are the implications of (1) and (2) to interpreting the predictions of the equivalent paths dosimetry model used in conjunction with TPWLMs.

Presented first is a discussion of the procedure for using the equivalent paths dosimetry model computer simulation program in conjunction with airway cast data to simulate absorption along one actual or real airway path and for predicting total LRT dose. (An actual or real airway path is a linear sequence of specific airways and airspaces and their dimensions as they really occur in a lung or cast data.) Next, the results of the simulation of the uptake along two actual airway paths, using the procedure, and along the path of a TPWLM are presented and compared. Following this, the results of predicting a total LRT dose using the TPWLM and actual airway data are presented and compared. Finally, the results of the investigation are summarized and discussed.

II. METHODS

Although the equivalent paths dosimetry model computer simulation program was developed to be used in conjunction with lung models of the TPWLM type, the program can also be employed to simulate absorption along one arbitrarily defined path, such as a real or actual airway path defined by data from lung casts. This requires (at least) airway dimensions along the path and the axial air-flow rate in each generation along the path. Given this information, uptake along the actual path can be estimated using the present simulation program in conjunction with a special lung model which is created so that all model airway paths are equivalent to the defined path and the air-flow rates in each model airway of a given generation are the same as defined for the actual airway. We call this type of lung model an actual path lung-like model, or APLLM. If the dimensions of the sequence of airways used to construct the APLLM are based on real airway data and on air flow rates along the path, then the equivalent paths dosimetry model computer program in conjunction with the APLLM can be used to estimate uptake along the actual path.

Although an APLLM can be used for the simulation of uptakes along only one airway path, an APLLM can be constructed for each airway path in the real lungs. Using the dosimetry model simulation program in conjunction with each of the APLLMs will then provide estimates of uptakes along all real airway paths in the LRT which can be used to predict total LRT uptake. The implementation of this, preceded by a brief discussion of the equivalent paths dosimetry model and the construction of APLLMs, is now addressed.

A. Mathematical Model

The equivalent paths dosimetry model used for this work is described in Miller *et al.* (1985) and Overton *et al.* (1987). The model was developed for ozone, a highly reactive but moderately soluble gas; however, the model can be used for any gas that has a first-order irreversible reaction with the biochemical constituents of RT tissues and fluids.

In this model the equation of transport for the airway lumens or air spaces of each generation (Overton *et al.*, 1987) is

$$\frac{\partial C}{\partial t} + U\frac{\partial C}{\partial X} = D\frac{\partial^2 C}{\partial X^2} - \frac{S}{V}KC \tag{1}$$

where C is the cross-sectional average concentration which is a function of axial distance (X) and time (t). The average velocity in the X direction is U, and D is the dispersion coefficient. K is the overall mass transfer coefficient for transfer to the surface of the airway or alveoli. S and V are the surface area and volume, respectively, of the airway or airspace. For a discussion of these parameters and how they are defined, see Overton *et al.* (1987).

B. The Development of Actual Path Lung-Like Models

To construct an APLLM, the air-flow rates in each airway of the LRT are needed. The rates in each airway are assumed to be proportional to the air-flow rates in the trachea. The constant of proportionality for an airway is assumed independent of time and is determined from assumptions concerning flow, not only along the airway's path, but in all airways in the TB region (i.e., the air-flow rates at all bifurcations are conserved). In principle, the constants of proportionality could be experimentally derived or based on the results of modeling air flow, however, such data are unavailable. The method for choosing the constants for this work is discussed later.

Given the airway dependent constant (which is the fraction of the tracheal air flow through the airway) for each real airway along the path, the APLLM is constructed so that the number of airways in a given generation of the APLLM is the inverse of the constant for that generation and real airway. The resulting APLLM is the same form or type as the TPWLM or symmetric lung model in which all airway paths from the trachea to the pulmonary sacs are equivalent. Furthermore, when an APLLM is used in conjunction with the equivalent paths dosimetry model simulation program, the air flow rates in each created airway of the same generation are equal, giving the required air flow rates for the actual airway in each generation along the chosen path.

For this investigation, the rat lung (Raabe *et al.,* 1976) was chosen because TB airway data available on this species were the most complete. The TPWLM is that described by Yeh *et al.* (1979), who based the TB region part of their LRT model on the TB airway data provided by the Lovelace Foundation (Rabbe *et al.,* 1976). The airway dimensions of the TB region of each APLLM are also based on the Lovelace data. The

APLLM pulmonary region dimensions are based on those of the TPWLM of Yeh *et al.* (1979). Since data are available to specify variability only in the TB region, airway variability is considered only for this region.

The volume of the LRT at functional residual capacity (FRC) was assumed to be half the volume defined by the sum of the Lovelace TB airways and the Yeh *et al.* (1979) pulmonary region (Overton *et al.*, 1987). The decrease in dimensions was performed by reducing all linear dimensions of the airways of the TPWLM, the Lovelace data, and the pulmonary region model by the factor of $.5^{1/3}$. We infer from Hoppin *et al.*, 1977, that the extent of this decrease in TB airway dimensions is justified. After the reduction, the APLLMs were constructed.

C. Using Many APLLM Simulations to Estimate Total LRT Dose

In an ideal LRT dosimetry model, absorption in all the airspaces and airways of an animal would be simulated simultaneously. That is, a multiple-path dosimetry model would be used in conjunction with a lung model composed of data on all airways. In principle, the development of such a model is possible, but not presently practical due to the large number of LRT airways and ducts. However, the LRT dose of an ideal multiple-path dosimetry model (and that of a breathing animal) can be estimated by constructing an APLLM for each airway path, performing a simulation using each of the APLLMs, and appropriately averaging the resulting doses.

In order to understand the use of APLLMs in estimating LRT doses, we first consider the ideal LRT dosimetry model or absorption in the real lung. Here, the dose, D, is simply the sum of the doses of all the airways, ducts, and sacs in the lung:

$$D = \sum_{i,g} a_{ig} \tag{2}$$

where a_{ig} is the mass of pollutant per minute removed from the air by the ith airway wall in the gth generation. For simplicity, and unless stated otherwise, the meaning of airway includes ducts and sacs.

The definition of LRT dose based on using the equivalent paths dosimetry model in conjunction with APLLMs is similar to the above, but the fact that the results of many simulations are combined must be taken into account. Dose estimated in this way is expressed as

$$\overline{D} = \sum_{i,g} \overline{a}_{ig} \tag{3}$$

where $\overline{D}$ is the estimated LRT dose and $\overline{a}_{ig}$ is the average of the mass per min estimated to have been absorbed by the ith actual airway in the gth generation.

The average, $\overline{a}_{ig}$, is calculated by averaging the (i,g) airway dose value for each simulation in which the (i,g) airway occurs. In theory, the total number of simulations necessary to compute an $\overline{a}_{ig}$ depends on the assumed air-flow scheme. For example, if the air-flow rates are assumed to be proportional to the number of distal terminal bronchioles, then the number of simulations to be carried out is the same as the number of distinct paths from the trachea to the terminal bronchioles. Since the trachea is a member of all the APLLMs (one for each path), there will be as many values of tracheal dose as there are paths (one dose for each possible path simulation). The predicted tracheal dose is defined as the average of these values. When only one path passes through an airway, there is only one value of calculated dose for the airway, which is the estimated or predicted dose for that airway.

In the Lovelace rat data, there are about 2400 terminal bronchioles. Performing this many simulations to obtain an estimate of total LRT dose for only one set of ventilatory parameters is not feasible. Fortunately, there is a family of air-flow schemes that allows considerable simplification in calculating the LRT dose based on averaging the results of APLLMs. These schemes have the following properties:

1. A set of airway or pulmonary units distal to the trachea is defined such that each unique path from the trachea to the pulmonary sacs goes through or into one and only one unit; the air flow rates into or through the units are assumed equal (we call each one of these units an equal flow unit, EFU). Depending on anatomical and physiological reasonableness, examples of sets include the terminal bronchioles, all airways of a specific diameter, the alveolar sacs, the first completely alveolated generation, etc.
2. The air flow rate in any airway proximal to EFUs is assumed to be proportional to the number of EFUs distal to the airway.
3. The lung structures distal to the EFUs are equivalent.

In this family of flow schemes, the average of all the total doses computed using all the APLLMs is equivalent to summing over the average of the estimated doses of all the actual airways. Specifically

$$\overline{D} = \frac{1}{N} \sum_{p} D_p = \sum_{i,g} \overline{a}_{ig} \tag{4}$$

where D_p is the LRT dose computed using the equivalent paths dosimetry model in conjunction with the APLLM for the pth path and N is the total number of paths. In this context, and unless otherwise indicated, a path is defined as a unique tracing or trajectory through a specific set of sequential airways from the trachea (or portal of entry) to a specific EFU; there is one and only one path for each EFU. A proof of the relationship in Equation (4) is outlined in the appendix.

As defined, a property of this family of air-flow schemes is that the regions of the lung distal to the EFUs for each of the APLLMs corresponding to a distinct path are exactly alike in all respects (i.e., volumes, surface area, the number of airways per generation, the number of acini, etc., are the same). The only difference between the APLLMs for two paths is in the LRT region proximal to the EFUs where the volume, surface area, number of generations, number of airways per generation, etc., are path dependent. Constructed this way, the volume (including the upper respiratory tract, URT) of an APLLM proximal to its EFUs is equal to the quantity of air that must be inhaled in order that fresh air first reaches the analogous EFU in the real lung. We call this path-dependent volume the critical volume (V_{crit}) for the path. Since surface area is approximately proportional to volume to the power of 2/3, and the URT surface area is independent of the path, then the surface areas proximal to the EFUs of two APLLMs with the same V_{crit} are approximately the same. Thus, the equivalent paths dosimetry model simulation program used in conjunction with APLLMs is expected to predict approximately equal total doses for two distinct paths with same V_{crit}.

For the flow scheme in which the air-flow rate is proportional to the number of distal terminal bronchioles (i.e., terminal bronchioles chosen as EFUs), simulations not presented here were carried out validating this conclusion and showing that the predicted total LRT dose using either a TPWLM or an APLLM is highly correlated with the alveolar ventilation as defined by $f(V_T - V_{crit})$, where V_T is the tidal volume, f is the breathing frequency, and V_{crit} is the volume proximal to the EFUs (in this case, terminal bronchioles). This result suggested a way of estimating the actual total LRT dose without carrying out as many APLLM simulations as there are paths, which would be prohibitive. The paths are grouped according to their V_{crit} (the groups are defined by a range of V_{crit}). One path from each group is chosen for a simulation and results in a dose, D_p. The estimated LRT dose for the real lung [as defined in Equation (4)] is then calculated by the weighted (by the frequency of occurrence of paths in the groups) average of the APLLMs doses.

III. RESULTS

For the simulation results illustrated, the air-flow rate in an airway is assumed to be proportional to the number of terminal bronchioles distal to the airway. This air-flow scheme is equivalent to assuming that all acini expand equally and uniformly during the breathing cycle, which is discussed at length by Horsfield and Cumming (1968). The scheme is used to define the APLLMs which are used in conjunction with the equivalent paths dosimetry model to estimate dose along actual airway paths as well as to predict the LRT dose by averaging over the doses indicated by APLLMs.

A. Predicted Dose along Different Airway Paths

Figure 23-2 is an example of the results using the above method for simulating the absorption of a gas along two actual paths and the path of the TPWLM. On the figure are plots of net dose versus generation for three simulations, one for the longest actual path, one for the shortest actual path, and one for the typical path of the TPWLM of Yeh *et al.* (1979). Net dose is the quantity of gas absorbed in a generation per min per area per ambient gas concentration. Long and short refer to the actual path with the largest and smallest critical volume, respectively. These two critical volumes are also equivalent to the paths with the longest and shortest time of flight of a particle moving by convection along the specific path from the nose to the first alveolated generation along that path.

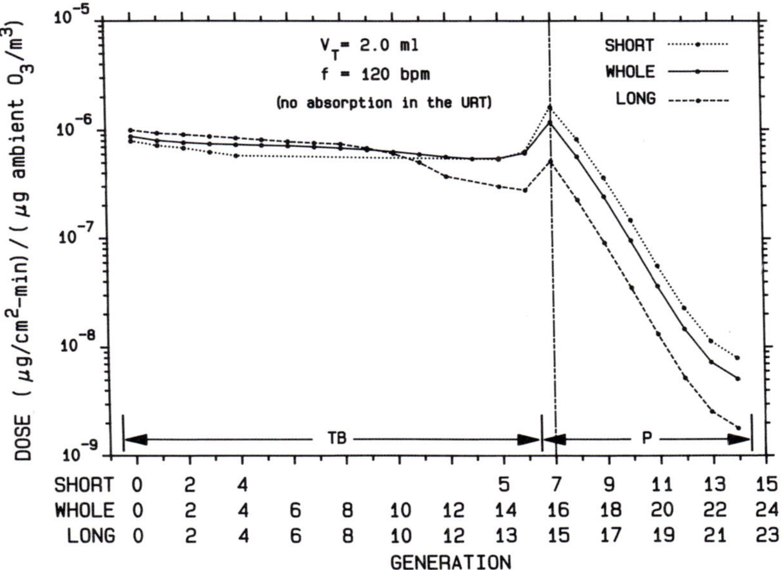

Fig. 23-2 Net dose versus generation for two actual paths and for the Typical Path Whole Lung Model (TPWLM) path. Whole refers to the TPWLM path and short and long indicate the simulation results using two airway paths from the cast data. The dose per generation is the O_3 flux to the airway wall in the given generation. TB and P indicate the tracheobronchial and pulmonary regions, respectively. V_T and f are the tidal volume and breathing frequency. URT stands for upper respiratory tract.

In Fig. 23-2, a generation axis is given for each path because the three paths do not have the same number of generations. This results in the breaks or gaps in the long and short path generation axis and in the plotted data. Where possible, however, morphologically equivalent generations are associated with the same tic mark. Because the model of the pulmonary region is the same for all three paths and because there is a one-to-one correspondence between the terminal bronchioles in all three paths, the corresponding generations in the model acini are morphologically equivalent. For the short (7 TB generations) and long (15 TB generations) paths, the first four generations (0–3) are made up of the same airways. Except for the trachea (generation 0), and the terminal bronchioles, there is no such correspondence (by generation) of these two paths with the TPWLM path in the TB region. Also, the first generations proximal to the terminal bronchioles are associated on the plots, however, we have no information that they are morphologically similar. On the figure, the predicted values are plotted as small solid circles. For a given simulation, these darkened circles are joined by a connecting line for ease of identification.

Interestingly, the number of TB generations in the whole lung model (TPWLM) does not lie between the number of short and long TB generations; the number of whole lung TB generations is one more than the number of long TB generations. This is because the longest path is defined in terms of the largest V_{crit}, not the number of generations. For these lungs the number of TB generations that correspond to the largest V_{crit} is one less than the median generation number used by Yeh *et al.* (1979) to define the number of TB generations in their TPWLM.

The general shapes of the dose curves in Fig. 23-2 are the same. Net dose decreases slowly in the TB region from the trachea to the terminal bronchioles. The doses then increase to a local maximum in the next generation, the first alveolated duct. Proceeding distally in the pulmonary region, the doses decrease rapidly.

The doses in the same TB airways (long and short, generations 0–3) are different, whereas, they should be the same. The lack of agreement is a result of estimating dose along each path in separate simulations. This method does not take into account the effect of the other paths, except through the specification of air-flow rates. Nevertheless, at the end of inhalation, the doses of common airways are predicted to be the same. The discrepancy occurs as a result of the exhalation phase. During inhalation, the dose of an airway is mainly influenced by the proximal gas concentration, which is independent of the actual path being simulated. On exhalation, the quantity of gas distal to a location influences the local concentration and dose. This quantity is path dependent and the amount of gas that flows back over a location during exhalation can vary considerably from one path to another, causing different predicted doses in the same airway.

The effect is evident in Fig. 23-2 where the dose in the first four common airways is larger for the long than for the short path. This is because less gas was removed distal to the common airway on inhalation during the long path simulation than during the short path simulation. Thus, on exhalation for the long path simulation, more O_3 was available to pass back over the common airway than was available to pass over the same airway during the short path simulation. In the pulmonary region where the paths are modeled as the same, the effect is probably less significant. However, in real lungs pulmonary paths are not the same.

The predicted net dose in the first alveolated generation of the short path is 3.1 times as large as in the morphologically equivalent airway of the long path. Proceeding distally this factor increases to 4.4 in the alveolar sacs (last generation). Dose values at equivalent pulmonary locations along other paths are expected to lie between the values predicted by the long and short paths. This is because pulmonary region doses decrease with increasing critical volume and the paths chosen correspond to the minimum and maximum critical volumes. Interestingly, the predictions for doses in the acinus of the TPWLM are more similar to those in the short path than in the long path. The short path predictions of pulmonary region generational dose are about 1.5 times as large as the whole lung predictions. To better understand the significance of the relationship of whole lung predictions to the many centriacinar tissue doses, a distribution of doses would be helpful; however, this is beyond the scope of this chapter.

B. LRT Dose Predictions as a Function of Ventilatory Parameters

1. Use of the TPWLM in Conjunction with the Dosimetry Model to Predict LRT Dose

Figure 23-3 is a plot of predicted LRT doses versus tidal volume from simulation results using the equivalent paths dosimetry model in conjunction with the TPWLM. Nine sets of ventilatory parameters (V_T and f) were used, giving nine values of LRT dose for three different values of V_T and three values of f. For each frequency the results fall on a line of constant slope and all the data can be summarized by the simple regression formula given in the figure (multiple $R^2 = .999$). The regression result is valid for a large range of physiologically reasonable ventilatory parameters. However, dose is positive and must deviate from the regression lines (plotted straight lines) as V_T becomes sufficiently small. Nevertheless, over a physiologically reasonable range of V_T and f, the equation can be used to estimate the LRT dose (in absence of URT absorption) that would be computed by the equivalent paths dosimetry model used in conjunction with the TPWLM.

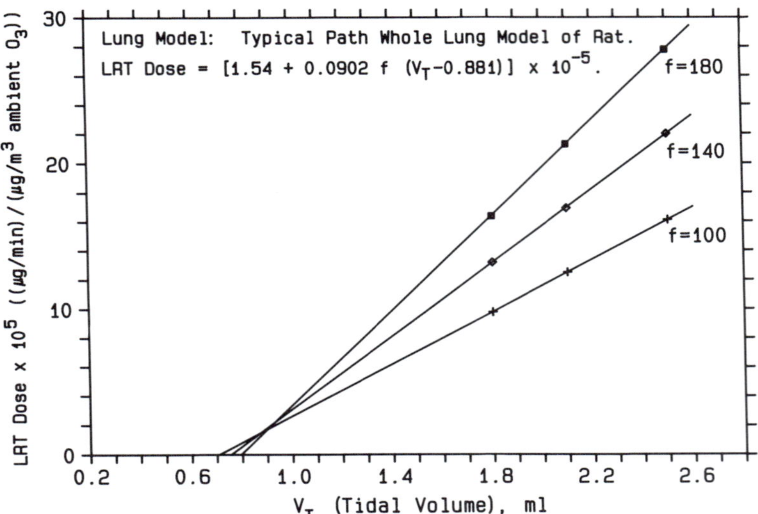

Fig. 23-3 Dose to the lower respiratory tract (LRT) versus tidal volume for three breathing frequencies. Predictions were made using the Typical Path Whole Lung Model of Yeh *et al.* (1979) in conjunction with the O_3 dosimetry model. The values of the nine predicted doses (indicated by the symbols on the lines of constant slope) are summarized by the equation on the figure. V_T and f are tidal volume and breathing frequency, respectively.

2. Use of APLLMs in Conjunction with the Dosimetry Model to Predict LRT Dose

Figure 23-4 shows the frequency histogram of V_{crit} for the flow scheme in which flow in an airway is proportional to the number of distal terminal bronchioles. Also plotted for three (out of five) sets of ventilatory parameters are the APLLM LRT doses (D_p) predicted for eight actual paths. Using the information in this figure, the estimated LRT dose for the real LRT was calculated by the method of weighted averages as previously described.

These results were tabulated and compared to corresponding LRT dose estimates based on use of the TPWLM (see Table 23-1). Except for the estimates corresponding to the lowest minute volume, both ways of calculating LRT doses agree to within less than 1% and the worst-case estimates agree to within less than 2.5%.

C. Other Air-Flow Schemes

Figure 23-5 illustrates density functions of V_{crit} for four air-flow schemes. The solid dark line is the distribution for the scheme identified as #PTB[1],

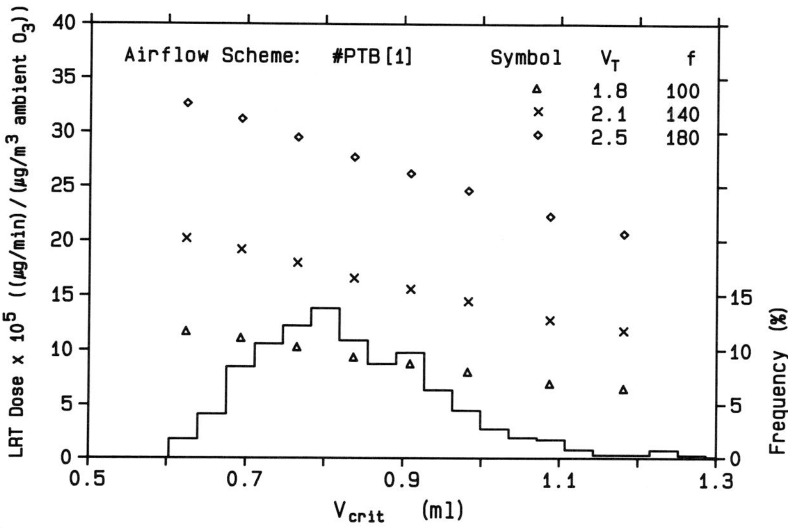

Fig. 23-4 Two sets of data are plotted: lower respiratory tract dose of actual path lung-like models (left vertical axis) for three sets of ventilatory parameters versus critical volume (V_{crit}) and a frequency histogram of critical volumes. The histogram gives the percentage (right vertical axis) of the airway paths in a given range of critical volume. V_T and f indicate tidal volume and breathing frequency, respectively. The air-flow scheme is designated as #PTB[1]: the air flow rate in an airway is proportional to the number of distal terminal bronchioles.

Table 23-1 Comparison of the Results of the Two Approaches to Estimating Dose

V_T (ml)	f (breaths/min)	$\dot{V}_E$ (ml/min)	DOSE[a] TPWLM	DOSE[a] ACTUAL PATHS
1.8	100	180	9.82	9.58
1.8	180	324	16.40	16.43
2.1	140	294	16.90	16.84
2.5	100	250	16.10	15.97
2.5	180	450	27.80	27.90

[a] ((10^{-5} μg/min)/(μg/m^3 ambient O_3))

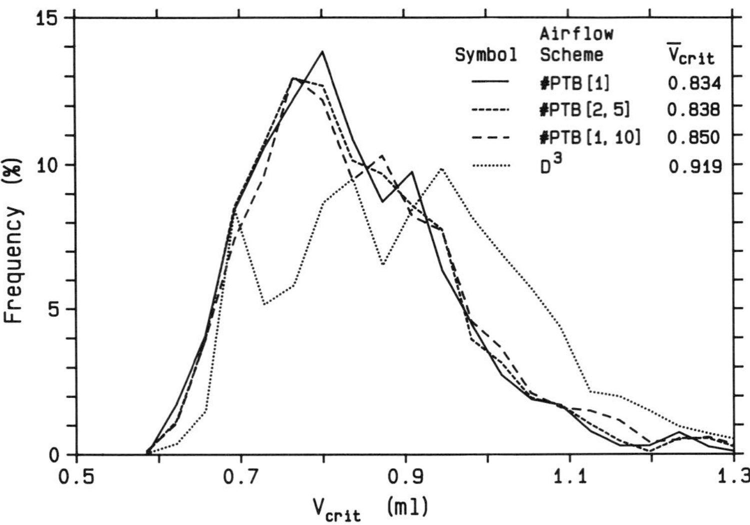

Fig. 23-5 Frequency density function of critical volume (V_{crit}) for four air-flow schemes. The flow schemes and their average critical volume are indicated. The numbers in the brackets indicate the number of equal flow units distal to the terminal bronchioles: [1] indicates one EFU; [I,J] indicates I, I + 1, . . . , J − 1, or J units distal to the terminal bronchioles. D^3 indicates the flow scheme in which the air-flow rate in a tracheobronchial airway (TB) is proportional to the TB airway diameter cubed.

which was used for the above simulations in which the air flow rate in an airway was assumed to be proportional to the number of distal terminal bronchioles. (The part of the scheme indicator, #PTB, means the number of pulmonary units distal to the terminal bronchioles.) In the schemes indicated by #PTB[2,5] and #PTB[1,10], a random number of equally expanding units or volumes is attached distally to each terminal bronchiole. In the #PTB[2,5] scheme, the number of units is either 2, 3, 4, or 5, with equal probability. The average number of units is 3.5, which is approximately the same as the 3.6 ventilatory units distal to the terminal bronchiole reported by Mercer and Crapo (1987). Conceptually, two to five ventilatory units have been attached to each terminal bronchiole and all are assumed to expand and contract equally during the breathing cycle. For the #PTB[1,10] scheme, 1 to 10 volumes that expand equally are assumed distal to the terminal bronchioles. In the D^3 scheme, air flow at a bifurcation is assumed to split in proportion to the cube of the diameter of the daughter airway. This choice of flow scheme was inspired by the work of Horsfield and Thurlbeck (1981), who discuss flow splitting in proportion to airway diameters taken to powers ranging from 2.3 to 3.

The first three schemes have very similar V_{crit} frequency density functions and the average critical volumes differ by less than 2%. On the other hand, the D^3 flow scheme density function is shaped differently and has an average V_{crit} that is 8 to 10% larger than the average V_{crit} of the first three schemes. We speculate that the more similar the V_{crit} distributions, the more similar the predictions of LRT dose. No matter how similar predictions are of LRT doses, the distribution of absorbed gases along the same actual paths for two flow schemes could be very dissimilar. Unfortunately, further investigation is beyond the scope of the present work.

IV. DISCUSSION AND SUMMARY

An equivalent paths dosimetry model has been used to investigate the distribution of an absorbed gas along different airway paths in the LRT of the rat, to predict LRT dose by estimating dose to all airways of the rat, and to compare these results to predictions using a TPWLM of the same species. The investigation was limited to one gas (O_3), one animal species, and the assumption that air-flow rate in an airway was proportional to the number of distal terminal bronchioles.

The results indicate that the shape of the dose versus generation curves is independent of airway path from the trachea to an arbitrary terminal bronchiole. From the trachea to the generation proximal to the terminal bronchioles, the net dose (flux to airway wall of a given generation) decreases slowly and then increases rapidly to a local maximum at the first alveolated generation. The dose then decreases rapidly. On the other hand, the value of the predicted dose in morphologically equivalent generations depends on the airway path (Fig. 23-2). The O_3 doses of the first alveolated airways of the rat pulmonary region are predicted to differ by a factor of up to 3.

Although the shapes of the dose curves for the actual and TPWLM paths are similar, only one dose is predicted for morphologically equivalent locations when the TPWLM is used in conjunction with the dosimetry model. Thus, the use of the traditional lung model of the TPWLM type cannot be expected to result in correctly predicted doses at specific locations; the actual path to the location must be considered. In the present study, the use of the TPWLM underestimates the maximum predicted centriacinar dose by 30% and overestimates the minimum centriacinar dose by 100%.

A method for estimating LRT dose by computing the dose of actual airways (as defined by cast data) was discussed and illustrated by predicting

LRT doses for several sets of breathing parameters. The results were compared to predictions of LRT dose using the TPWLM. There were no significant differences in the two approaches. Thus, for the present assumptions, the use of a TPWLM in conjunction with an equivalent paths dosimetry model is a sufficient method as well as the more economical way of predicting LRT dose.

One of the major reasons that the two ways of predicting total LRT dose agree is that the dead space volume of the TPWLM (.86 ml) is approximately the same as the average critical volume for the flow scheme used (.834 ml). The importance of the approximate equality can be inferred from information in Fig. 23-3 and from Fig. 23-4 by interpolation to estimate the APLLM dose for the critical volume equal to the TPWLM dead space volume. The interpolated doses for $V_T = 2.1$ and 2.5 ml differ from the corresponding TPWLM doses (Fig. 23-3) by less than 4%. For $V_T = 1.8$ ml, the difference is 7%. More importantly, APLLM simulations for which the path critical volume is equal to the average critical volume predict LRT doses that differ from the actual path results (Table 23-1, last column) by less than 2%. These agreements in different ways of estimating LRT dose suggest that the dead space volume of a TPWLM should be constructed approximately, if not exactly equal, to the critical volume of the actual TB regions.

Another important factor that affects absorption is the surface area of the TB region. Lung models defined in terms of sequential sets of cylindrical airways (as they are in the traditional TPWLM) cannot accurately reflect the correct surface area (as defined by the surface of the cylinders) along the model airway path. This is because airways are not cylinders and there are many paths of different lengths, numbers of generations, and airway dimensions within a generation. These aspects make it impossible to construct lung models with reasonable airway dimensions out of sequential sets of cylinders for which the model surface area and volume are consistent with the actual surface areas and volumes. The surface areas and volumes should be specified independently along with airway length and diameter and the number of airways per generation.

The above conclusions depend on many assumptions, including those concerning the air-flow rates in airways, the physical and chemical properties of O_3, the biochemical compounds that react with O_3, and the anatomical and physiological characteristics of one animal species, the rat. The conclusions may not apply for different conditions or assumptions, an aspect that needs exploring. For example, the various flow schemes briefly discussed in conjunction with Fig. 23-5 could be used to test the conclusions within the framework of the present method of defining air-flow schemes. Other aspects that could be considered for future work include nonuniform filling of different regions of the lung, various absorption rates in these regions, gases that react reversibly, the use of tidal volumes and breathing frequencies outside the ranges used in the present work, and other animal species.

Probably, one of the major drawbacks to the method presented is that gas exchange between different paths at bifurcations is not possible; for this, concurrent simulations along different paths would be required. This has been done theoretically for inert, insoluble gases in asymmetric models of the human acinus (e.g., Paiva and Engel, 1984, among many). Because of the differences in the gases of interest, the results of these works may not be applicable to reactive gases such as O_3.

Knowledge of the variability of doses at equivalent morphological locations can help to interpret the results using a lung model of the TPWLM type or of a toxicological experiment. For example, APLLMs could be used in conjunction with dosimetry models to determine the range or even the distribution of dose at equivalent morphological locations of toxicological interest. Experimental data (on toxic effects or dose) from LRT locations chosen at random will have more meaning when compared to a predicted range of doses than to one value of a predicted dose.

Because the simulations were carried out for a specific gas, O_3, several comments relative to the predictions about this toxic pollutant are appropriate. According to Menzel (1984), maximal O_3 damage occurs in the vicinity of the first alveolated airways in the rat and other laboratory animals. Not shown in Fig. 23-2 are the predicted tissue doses. The tracheal tissue dose for each of the three paths graphed is more than two and one-half orders in magnitude smaller than the net dose in the trachea. The tissue dose for each path increases distally to a value nearly the same as the net dose in the first alveolated duct. From here and distally, the tissue dose and net doses of each path are predicted to be essentially the same. Based on the tissue doses, and in agreement with experiments, the worst possible case for toxic effects is predicted to occur in the first alveolated airway of any path. Furthermore, the toxicological risks at centriacinar locations are predicted to be different. This is in agreement with the observations of Schwarz *et al.* (1976) and Boorman *et al.* (1980) of variable responses in the centriacinar areas of the same rat.

ACKNOWLEDGMENTS

The authors appreciate the technical assistance of Dorothy E. Pugh. We also thank the reviewers for helpful suggestions.

APPENDIX

The following definitions are needed to prove the relationship in Equation (4):

1. N = the total number of paths from the trachea to the EFUs.
2. N_{ig} = the number of paths through the ith airway in the gth generation of the real lungs.
3. n_{ig} = the number of airways in the gth generation of an APLLM for the path in the real lungs that includes the (i,g) real airway.
4. a_{pig} = dose of an airway in the gth generation of an APLLM for the pth path in the real lungs that includes the (i,g) real airway. $a_{pig} = 0$, if the pth path p does not go through the (i,g) real airway.
5. $\bar{a}_{ig}$ = the average of the doses for the (i,g) real airway, estimated using the APLLMs; thus,

$$\bar{a}_{ig} = \frac{1}{N_{ig}} \sum_{p} a_{pig}. \tag{A1}$$

6. D_p = total dose computed using the APLLM of path p from the real lungs.

The assumption that the air-flow rate is proportional to the number of distal EFUs is equivalent to the rate being proportional to N_{ig}. Thus,

$$n_{ig} = N/N_{ig} \qquad \text{(note that } n_{ig} \text{ need not be an integer).} \tag{A2}$$

Using definitions 3, 4, and 6,

$$D_p = \sum_{i,g} n_{ig} a_{pig}. \tag{A3}$$

The average D_p is

$$\bar{D} = \frac{1}{N} \sum_{p} D_p. \tag{A4}$$

Using Equation (A3), Equation (A4) becomes

$$\bar{D} = \frac{1}{N} \sum_{p} \sum_{i,g} n_{ig} a_{pig}. \tag{A5}$$

The double summation in Equation (A5) is over all paths and airways. The order of the summation is irrelevant. Thus,

$$\bar{D} = \frac{1}{N} \sum_{i,g} n_{ig} \left[\sum_{p} a_{pig} \right]. \tag{A6}$$

The terms in the brackets are the estimates of the dose of the (i,g) real airway based on the APLLM simulations; the sum in the brackets equals to $N_{ig}\bar{a}_{ig}$ (Equation A1). Inserting this into Equation (A6) and using Equation (A2), we have

$$\bar{D} = \sum_{i,g} \bar{a}_{ig}, \qquad \text{(A7)}$$

proving Equation (4).

REFERENCES

Boorman, G. A., Schwartz, L. W., and Dungworth, D. L. (1980). Pulmonary effects of prolonged ozone insult in rats. *Lab. Invest.* **43,** 108–115.

Hoppin, F. G., Jr., Hughes, J. M. B., and Mead, J. (1977). Axial forces in the bronchial tree. *J. Appl. Physiol. Respir. Environ. Exercise Physiol.* **42,** 773–781.

Horsfield, K., and Cumming, G. (1968). Morphology of the bronchial tree in man. *J. Appl. Physiol.* **24,** 373–383.

Horsfield, K., and Thurlbeck, A. (1981). Relation between diameter and flow in branches of the bronchial tree. *Bull. Math. Biol.* **43,** 101–109.

Menzel, D. B. (1984). Ozone: An overview of its toxicity in man and animals. *In* "Fundamentals of Extrapolation Modeling of Inhaled Toxicants: Ozone and Nitrogen Dioxide" (F. J. Miller and D. B. Menzel, eds.), pp. 3–24. Hemisphere, Washington, D.C.

Mercer, R. R., and Crapo, J. D. (1987). 3-D reconstruction of the rat acinus. *J. Appl. Physiol.* **63,** 785–794.

Miller, F. J., Overton, J. H., Jr., Jaskot, R. H., and Menzel, D. B. (1985). A model of the regional uptake of gaseous pollutants in the lung. I. The sensitivity of the uptake of ozone in the human lung to lower respiratory tract secretions and to exercise. *Toxicol. Appl. Pharmacol.* **79,** 11–27.

Overton, J. H., Graham, R. C., and Miller, F. J. (1987). A model of the regional uptake of gaseous pollutants in the lung. II. The sensitivity of ozone uptake in laboratory animal lungs to anatomical and ventilatory parameters. *Toxicol. Appl. Pharmacol.* **88,** 418–432.

Paiva, M., and Engel, L. A. (1984). Model analysis of gas distribution within human lung acinus. *J. Appl. Physiol: Respir. Environ. Exercise Physiol.* **56,** 418–425.

Raabe, O. G., Yeh, H. C., Schum, G. M., and Phalen, R. F. (1976). Tracheobronchial geometry: Human, dog, rat, hamster. LF-53. Lovelace Foundation, Albuquerque, New Mexico.

Schwartz, L. W., Dungworth, D. L., Mustafa, M. G., Tarkington, B. K., and Tyler, W. S. (1976). Pulmonary responses of rats to ambient levels of ozone. *Lab. Invest.* **34,** 565–578.

Weibel, E. R. (1963). "Morphometry of the Human Lung," pp. 1–151. Academic Press, New York.

Yeh, H. C., Schum, G. M., and Duggan, M. T. (1979). Anatomic models of the tracheobronchial and pulmonary regions of the rat. *Anat. Rec.* **195,** 483–492.

Chapter 24

Predicting Respiratory Tract Clearance in Man

R. G. Cuddihy
H. C. Yeh
R. O. McClellan
Inhalation Toxicology Research Institute
Lovelace Biomedical and Environmental Research Institute
Albuquerque, New Mexico 87185

I. INTRODUCTION

Knowledge of the relationships between exposure air concentrations of toxic substances and the probability of inducing biological effects provides crucial input for conducting risk assessments and establishing occupational and environmental standards and guidance for limiting exposures, thereby minimizing human health impacts of airborne toxicants (Cuddihy and McClellan, 1988). It is increasingly recognized that an understanding of the dose of toxicants to critical biological units (from specific macromolecules to cells to tissues) of the respiratory tract is valuable in linking exposure and biological responses, i.e., establishing exposure → dose → response relationships. This approach has been used for many years in the field of radiation protection, due no doubt in part to the ease with which radiation doses can be measured and calculated.

Biomathematical models describing the deposition and clearance of inhaled radioactivity have played an important role in the development of radiation protection standards and practices for airborne radioactive materials. Such models have also provided a useful framework for contemplating our knowledge of mechanisms, pathways, and rates of clearance of inhaled materials from various regions of the respiratory tract. Such models not only help integrate and clarify what is known, but also aid in identifying gaps that may be filled through further research. The most notable past contribution of this type is the report of the Task Group on Lung Dynamics (1966), a group chaired by Morrow and convened by the International Commission on Radiation Protection (ICRP).

Currently, committees of the ICRP and the National Council on Radiation Protection and Measurements (NCRP) are developing updated reports on our knowledge of respiratory tract deposition and clearance of inhaled materials. Although primarily oriented toward radioactive materials, the reports will undoubtedly also be of major value in considering the biological disposition of inhaled chemical toxicants. In this abbreviated paper, we summarize some of the concepts and information on inhaled particulate material being considered for inclusion in the report of the NCRP committee. A more extensive review has been published by Cuddihy and Yeh (1988). In addition, more detailed information on inhaled particles is available in recent reviews by Schlesinger (1988a,b), Sun *et al.* (1988), and Bond (1988). Information on inhaled gases is included in the reviews of Ultman (1988), Overton and Miller (1988), and Miller *et al.* (1988).

II. BASIC CONCEPTS FOR MODELING

In developing biomathematical models to describe the deposition and clearance of inhaled materials, it is useful to consider the respiratory tract as consisting of four regions; the naso-oro-pharyngo-laryngeal region, tracheobronchial region, pulmonary region, and regional lymph nodes (Fig. 24-1). These subunits are definable both anatomically and functionally. The naso-oro-pharyngo-laryngeal airways extend from the nares to the larynx. The region serves to warm and humidify inhaled air, entrap large particles through inertial processes and very small particles by diffusion, and is the site of sensory input for olfaction. The tracheobronchial airways extend from the larynx to the distal end of the terminal bronchioles and act as a distribution and collecting conduit for inspired and expired air between the larynx and the pulmonary compartment. In the upper portion, large volumes of air move at high velocity. As the air is distributed to airways with increasing cross-sectional areas, the air velocity is markedly

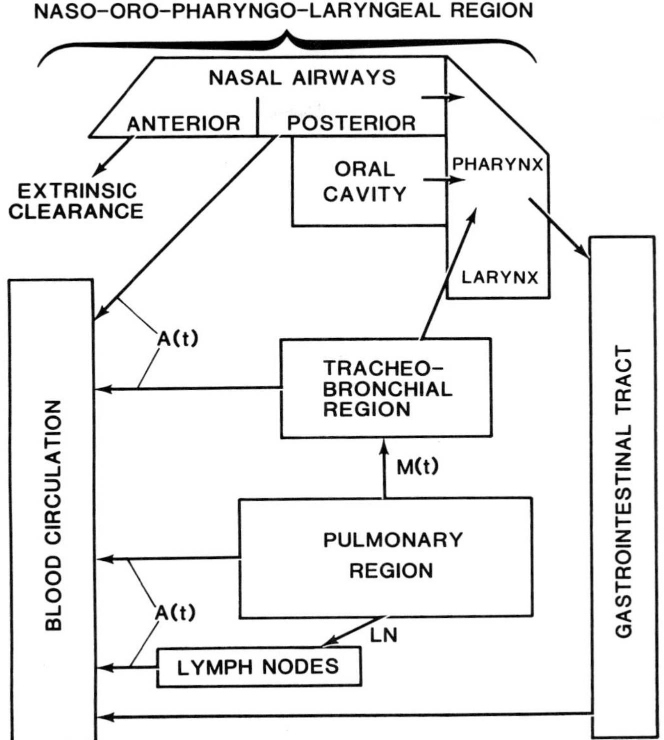

Fig. 24-1 Respiratory tract clearance model illustrating the role of mechanical, M(t), and absorptive, A(t), processes in clearing inhaled particles. (From Cuddihy and Yeh, 1988.)

reduced. In this section, additional particulate material is removed primarily by inertial, sedimentation, and diffusion processes. The pulmonary compartment consists of the structures distal to the terminal bronchioles. It is in this region that gas exchange between the alveoli and pulmonary capillaries occurs. Here particles are deposited primarily by sedimentation and diffusion. The regional lymph nodes are connected to lymphatic vessels draining the respiratory tract. The nodes filter and entrap materials carried by the lymphatics and have a role in the immune function in the respiratory tract.

The clearance of deposited materials from the first three compartments may be described in terms of competing mechanical and adsorptive processes (Fig. 24-1). Mechanical processes move deposited particles to the oropharynx, where they are swallowed, or to the interstitium. Particles may be removed from the interstitium to the regional lymph nodes. Particles smaller than 1 μm in diameter are found in virtually all types of cells and may pass into the interstitium and the endothelial cells of vessels without the aid of macrophages. Larger particles are mainly cleared to the airways, lymphatics, and the interstitium by macrophages. Macrophages, in concert with the mucous layer and the underlying ciliated epithelium, play a major role in the clearance of particles from the airways.

Absorptive processes transfer materials from the respiratory tract and lymph nodes to the blood circulation. Absorption occurs mainly for material that dissolves, elutes, or is metabolized from particles. However, some material removed from particles may interact with tissue constituents and be retained for extended times in the respiratory tract. Conversely, a small portion of deposited particles, especially those of the smallest size, may pass into the blood circulation and transfer to other organs or be excreted. In the model described here, absorptive processes represent the net transfer of material from the respiratory tract to blood, regardless of the mechanisms involved.

III. MECHANICAL CLEARANCE

In the anterior area of the nasal cavity, clearance is toward the nostrils. The material that moves forward may remain in the nasal vestibule for hours before it is cleared by extrinsic mechanisms. In the posterior portion of the nasal cavity the effective retention time is about 10 min, with material cleared to the oropharynx where it is swallowed. It is reasonable to assume, for most particle sizes, that half of the deposit is in the anterior area and half in the posterior area.

Clearance velocities for insoluble particles deposited in the trachea have been measured using a variety of techniques and range from about 5 to 20 mm/min. Particle clearance velocities for tracheobronchial airways have been calculated based on *in vivo* clearance measurement of inhaled material as well as on airway dimensions. The effective clearance velocities were estimated to decrease from a range of 5 to 20 mm/min in the trachea to about 1 mm/min for airway generation 6, down to 0.01 to 0.001 mm/min in airway generation 16, the terminal bronchioles.

Mechanical clearance is quite ineffective for the pulmonary region of humans, with deposited materials remaining for several hundred days. The rates are sufficiently slow that long-term observation can only be made with very insoluble materials. This is the case, because the fractional mechanical clearance of particles from the pulmonary region is only about 6×10^{-3}/day, corresponding to a half-time of 115 days, which decreases to about 1×10^{-3}/day, or a half-time of about 700 days, at 200 days after inhalation. It is apparent that if dissolution and absorption occur at this rate, or more rapidly, considerable uncertainty in the estimates of mechanical clearance will exist, especially if inadequate data are obtained to correct for dissolution and absorption. The daily fractional clearance rate of particles from the pulmonary region can be represented by the following mathematical function where time is equal to (t)

$$M(t) = 0.005e^{-0.02t} + 0.001$$

The rates at which insoluble particles clear from the pulmonary region is quite variable among different species. Clearance in mice and rats is faster than in humans, and in guinea pigs and dogs, it is somewhat slower (Fig. 24-2) (Cuddihy and Yeh, 1988; Snipes *et al.*, 1983; Snipes, 1988). The reasons for the species differences in clearance are not known. They may

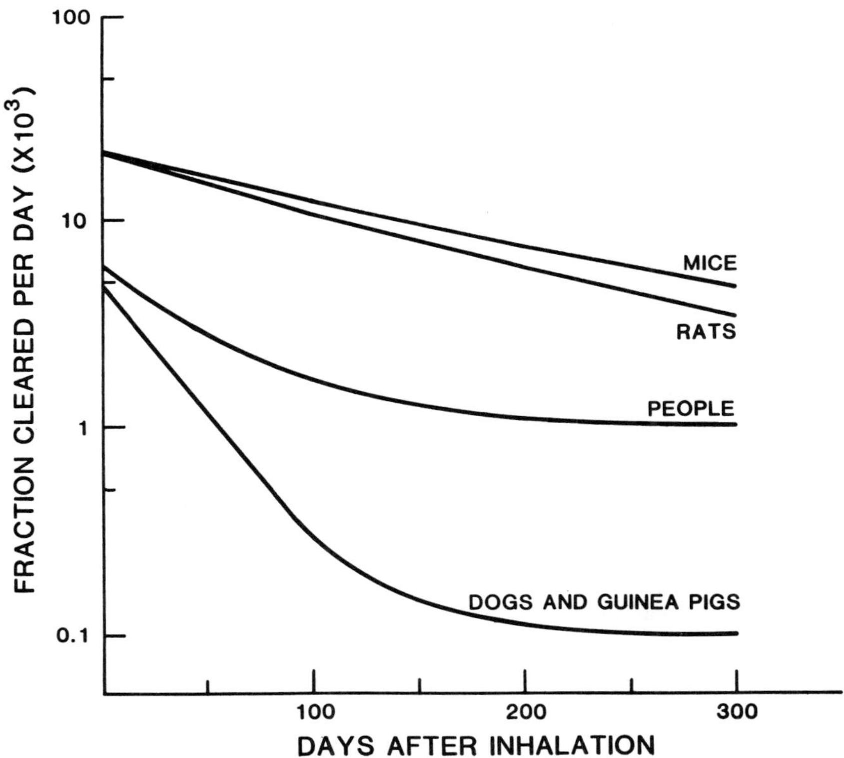

Fig. 24-2 Daily fractional clearance rate, M(t), from the pulmonary region of the tracheobronchial region, for insoluble particles inhaled by four species of laboratory animals or humans. (From Cuddihy and Yeh, 1988.)

relate to structural differences in the region of the terminal bronchioles and alveoli. Clearly, additional research is warranted to help understand the basis for the variations.

The rate of transport of insoluble particles from the pulmonary region to the lymph nodes has not been established quantitatively in humans. Thomas (1972) suggested that this rate is relatively independent of species and arrived at a value of 1×10^{-4} for the daily fractional transfer rate constant.

In considering the above transfer rates, it should be noted that most of the deposition and clearance studies have involved deposition of small quantities of material ranging from tens to hundreds of micrograms. The extent to which the kinetics may be altered by inhalation and deposition of larger quantities of material is open to question. Wolff *et al.* (1987) in a study with rats exposed to diesel exhaust for 7 hr/day, 5 days/week, found impairment of lung clearance at exposure concentrations of 3.5 and 7.0 mg/m^3, with lung burdens progressively becoming larger than predicted from the retention patterns observed in rats exposed to 0.35 mg/m^3. The impaired clearance and lung pathology, including lung tumor formation observed by Wolff *et al.* (1987) and Mauderly *et al.* (1987), may have been related in a cause–effect manner. This emphasizes the need for further research to clarify the influence of exposure concentration on the retention kinetics for inhaled particles.

IV. CLEARANCE BY ABSORPTION

A more detailed coverage of absorption from the respiratory tract to the blood circulation as a primary mode of clearance is given by Cuddihy and Yeh (1988). Many factors including chemical form and structure, surface area, previous temperature history, radionuclide specific activity, presence of other chemicals, and the metabolic activity of the respiratory tract influence the rate at which materials deposited in the respiratory tract are absorbed into the blood circulation. Mercer (1967) suggested that particle dissolution is the major determinant of the rate of absorption of inorganic cations into the blood and that the dissolution rate is proportional to particle surface area. This concept has had a major influence on work in this area and, if adequate information were available, it would provide an excellent basis for modeling pulmonary clearance. Unfortunately, the data base is not sufficient and, thus, a more pragmatic approach is being advocated which draws on the results of several hundred radionuclide inhalation studies reported in the literature.

The key measurements required for this approach are of radioactivity in the head airways, lungs (including trachea and bronchi), gastrointestinal tract, internal organs, feces, and urine as a function of time after inhalation exposure. There is also a need for knowledge of the rate of absorption from the gastrointestinal tract and endogenous fecal excretion for material reaching the blood. Obviously, few studies report such complete data sets, therefore some model parameters must be assumed or estimated from related studies.

A starting point is assessment of radioactivity in the thorax, including the pulmonary region, tracheobronchial airways, and lymph nodes. The decrease in thoracic radioactivity as a function of time, $\mathrm{TR}(t)$, results from radioactive decay, mechanical clearance via the airways, and absorption into the blood. After correcting for radioactive decay, the overall daily fractional clearance rate, $\mathrm{FC}(t)$, can be calculated as

$$\mathrm{FC}(t) = \frac{d\mathrm{TR}(t)}{dt} \Big/ \mathrm{TR}(t)$$

For modeling purposes, the total fractional clearance rate is assumed to consist of the mechanical clearance, $\mathrm{M}(t)$, via the airways and gastrointestinal tract, and absorption, $\mathrm{A}(t)$, to the blood:

$$\mathrm{FC}(t) = \mathrm{A}(t) - \mathrm{M}(t)$$

In using this equation to calculate absorption, it implies that the rate of absorption to the blood is similar for all materials retained in the thorax, including the pulmonary region, tracheobronchial region, and lymph nodes. In using data derived from studies with laboratory animals, it is assumed that the absorption rate is independent of species. This assumption appears valid from the limited information available, however, it is clear that more interspecific comparative studies with a range of materials are needed to validate the assumption.

As a check on the reasonableness of the computed value for $\mathrm{A}(t)$, an additional calculation can be made using information on radionuclide uptake by internal body organs and excretion rates. The amount of radionuclide absorbed from the respiratory tract into the blood circulation $(\mathrm{ABS_{RT}})_t$ between the exposure and time, t, is calculated from

$$(\mathrm{ABS_{RT}})_t = \sum (\text{Internal Organ Burdens})_t + \sum\nolimits_t (\text{Urine} + \text{Endo. Feces}) - \text{GI Absorption}$$

The amount of radioactivity absorbed from the respiratory tract should also be equal to the integrated product of the fractional absorption rate and the amount of radioactivity remaining in the respiratory tract, where $\mathrm{HR}(t)$ is the amount of radioactivity retained in the head airways.

$$(\mathrm{ABS_{RT}})_t = \int_0^t \mathrm{A}(t)[\mathrm{TR}(t) + \mathrm{HR}(t)]\,dt$$

V. ORGANIC

In contrast to the abundant literature on inorganic cations, much less information is available on the clearance of organic compounds. No doubt, this relates in part to the ease of studying various radioactive cations which have easily detected gamma and beta radiation emissions, and the much greater difficulty of working with ^{14}C- or ^{3}H-labeled organic compounds. In the case of the cations, the radiotracers are specific for the toxicant (the particular element under study). With organic compounds, the compounds inevitably undergo metabolism to one, or very frequently, a series of metabolites. Elucidation of those metabolic pathways is time consuming and analytically demanding.

Most of the information available on organic compounds has been obtained in studies with pure compounds. Examples of this type of information have been reviewed by Dahl (1988). In recent years, increasing attention has been directed to studies of organic compounds associated with particles, an interest driven by the recognition that organic compounds are rarely encountered in pure form in occupational or environmental exposure situations. Sun *et al.* (1988) and Bond (1988) have recently reviewed the literature in this area and emphasized the extent to which the association of organic compounds with particles can markedly influence the retention of the organic compounds in the respiratory tract. Additional research of this type with mixtures of materials typically found in occupational and environmental exposure settings is needed. Because of the association of specific organic compounds, or their metabolites, with critical macromolecules such as desoxyribonucleic acid, these studies will undoubtedly improve our understanding of the relationship between exposure concentrations and the most relevant biological dose.

VI. VALIDATION OF MODELS

A crucial need is to test the validity of biomathematical models using the limited human data that is available or by applying the models to data sets derived from independent studies with laboratory animals. Significant progress has already been made using data from humans accidentally exposed to various radionuclides, work that has been facilitated by the ease of radioactivity measurements. In the future, new techniques for tracing organic compounds using biomarkers such as adducts may help in expanding our base of knowledge on the disposition of organic compounds in humans. To achieve this end, it is important that we give added emphasis to building bridges between research using laboratory animals and opportunistic studies of exposed humans.

VII. SUMMARY

Biomathematical models describing the deposition and clearance of inhaled materials in laboratory animals and humans provide a useful tool for integrating our knowledge in this area, applying what is known to assessing and controlling inhalation hazards, and identifying information needs that can be filled by additional research. In this chapter, we have called attention to past contributions from modeling the disposition of inhaled radionuclides and described an updated model being considered by several advisory bodies concerned with the control of risks associated with radioactive materials. The model emphasizes the role of two competing processes, mechanical clearance and absorption of materials into the blood stream, in determining the overall clearance of deposited materials. The data on which the model has been developed is most substantial for early times after an inhalation exposure. Fewer data are available for longer time intervals. There is a particular demand for additional data on mechanical clearance and absorption in people at long time intervals following exposure. It is also necessary to extend the models developed primarily with data on inhaled radioactive inorganic cations to organic compounds. Finally, as with any model, it is essential to test the validity of the model and the assumptions on which it is based using other data sets. A special need exists to acquire and use human data whenever the opportunity presents itself.

ACKNOWLEDGMENTS

The authors gratefully acknowledge the valuable input received from their colleagues at the Inhalation Toxicology Research Institute and those serving on the NCRP and ICRP advisory committees. This activity was supported by the Office of Health and Environmental Research, Department of Energy under Contract Number DE-AC04-76EV01013.

REFERENCES

Bond, J. A. (1988). Factors modifying the disposition of inhaled organic compounds. *In* "Concepts in Inhalation Toxicology" (R. O. McClellan and R. F. Henderson, eds.), pp. 251–275. Hemisphere, New York.

Cuddihy, R. G., and McClellan, R. O. (1988). Risk assessment for inhaled toxicants. *In* "Concepts in Inhalation Toxicology" (R. O. McClellan and R. F. Henderson, eds.), pp. 521–546. Hemisphere, New York.

Cuddihy, R. G., and Yeh, H. C. (1988). Respiratory tract clearance of particles and substances dissociated from particles. *In* "Inhalation Toxicology. The Design and Interpretation of Inhalation Studies and Their Use in Risk Assessment," (U. Mohr, ed.) Chap. 12. Springer-Verlag, Berlin and New York.

Dahl, A. R. (1988). Metabolic characteristics of the respiratory tract. *In* "Concepts in Inhalation Toxicology" (R. O. McClellan and R. F. Henderson, eds.), pp. 143–165. Hemisphere, New York.

Mauderly, J. L., Jones, R. K., Griffith, W. C., Henderson, R. F., and McClellan, R. O. (1987). Diesel exhaust is a pulmonary carcinogen in rats exposed chronically by inhalation. *Fundam. Appl. Toxicol.* **9,** 208–221.

Mercer, T. T. (1967). On the role of particle size in the dissolution of lung burdens. *Health Phys.* **13,** 1211–1221.

Miller, F. J., Overton, J. H., and Graham, R. C. (1988). Regional deposition of inhaled reactive gases. *In* "Concepts in Inhalation Toxicology" (R. O. McClellan and R. F. Henderson, eds.), pp. 231–251. Hemisphere, New York.

Overton, J. H., and Miller, F. J. (1988). Dosimetry modeling of inhaled toxic reactive gases. *In* "Air Pollution, The Automobile, and Public Health" (A. Y. Watson, R. R. Bates, and D. Kennedy, eds.), pp. 367–389. National Academy Press, Washington, D.C.

Schlesinger, R. B. (1988a). Biological disposition of airborne particles: Basic principles and application to vehicular emission. *In* "Air Pollution, The Automobile, and Public Health" (A. Y. Watson, R. R. Bates, and D. Kennedy, eds.), pp. 239–299. National Academy Press, Washington, D.C.

Schlesinger, R. B. (1988b). Deposition and clearance of inhaled particles. *In* "Concepts in Inhalation Toxicology" (R. O. McClellan and R. F. Henderson, eds.), pp. 165–195. Hemisphere, New York.

Snipes, M. B. (1988). Species comparisons for pulmonary retention of inhaled particles. *In* "Concepts in Inhalation Toxicology" (R. O. McClellan and R. F. Henderson, eds.), pp. 195–231. Hemisphere, New York.

Snipes, M. B., Boecker, B. B., and McClellan, R. O. (1983). Retention of monodisperse or polydisperse aluminosilicate particles inhaled by dogs, rats and mice. *Toxicol. Appl. Pharmacol.* **69,** 345–362.

Sun, J. D., Bond, J. A., and Dahl, A. R. (1988). Biological disposition of vehicular emissions. *In* "Air Pollution, The Automobile, and Public Health" (A. Y. Watson, R. R. Bates, and D. Kennedy, eds.), pp. 299–323. National Academy Press, Washington, D.C.

Task Group on Lung Dynamics (1966). Deposition and retention models for internal dosimetry of the human respiratory tract. *Health Phys.* **12,** 173–201.

Thomas, R. G. (1972). An interspecies model for retention of inhaled particles. *In* "Assessment of Airborne Particulates" (C. C. Thomas, ed.), pp. 405–419. Thomas, Springfield, Illinois.

Ultman, J. S. (1988). Transport and uptake of inhaled gases. *In* "Air Pollution, The Automobile, and Public Health" (A. Y. Watson, R. R. Bates, and D. Kennedy, eds.), pp. 323–367. National Academy Press, Washington, D.C.

Wolff, R. K., Henderson, R. F., Snipes, M. B., Griffith, W. C., Mauderly, J. L., Cuddihy, R. G., and McClellan, R. O. (1987). Alterations in particle accumulation and clearance in lungs of rats chronically exposed to diesel exhaust. *Fundam. Appl. Toxicol.* **9,** 154–166.

Chapter 25

The Role of Particle Hygroscopicity in Aerosol Therapy and Inhalation Toxicology

T. B. Martonen
Health Effects Research Laboratory
U.S. Environmental Protection Agency
Research Triangle Park, NC 27711

W. Hofmann*
Center for Extrapolation Modeling
Duke University Medical Center
Durham, NC 27710

A. D. Eisner
M. G. Ménache
NSI Technology Services Corporation
Research Triangle Park, NC 27709

I. INTRODUCTION

The extrathoracic compartment of the human respiratory tract (i.e., the naso- and oropharyngeal regions) effectively conditions inhaled air by warming and humidifying processes even in extreme ambient environments (Ingelstedt, 1956). Additional air conditioning takes place within downstream, distal airways of the lung (Ingelstedt and Toremalm, 1960, 1961). The extent to which heating and humidifying may be continuous

* Permanent address Abteilung für Biophysik, Universität Salzburg, A-5020 Salzburg, Austria.

Disclaimer: The research described in this article has been reviewed by the Health Effects Research Laboratory, U.S. Environmental Protection Agency, and approved for publication. Approval does not signify that the contents necessarily reflect the views and policies of the Agency nor does mention of trade names or commercial products constitute endorsement or recommendation for use.

processes active throughout the lung has been studied by McFadden *et al.* (1982).

When airborne particles are inhaled, the effectiveness of interactive particle-gas thermodynamic processes will be affected by the nature of surrounding fluid dynamics patterns. There is a paucity of data characterizing internal heat, mass transfer, and gas kinetics parameters because of the obvious technical difficulties in making such measurements *in situ* with human subjects. Here, available data are organized and related to environments within the lung for oral and nasal breathing modes.

To conduct mathematical modeling of effects of hygroscopicity upon deposition, the growth characteristics of particles within defined environments must be known. Laboratory tests which yield data regarding the hygroscopic nature of some therapeutic aerosols and ambient pollutants have been performed and will be discussed. We will show how these *in vitro* data, which must be utilized in the restrictive formats presented in the literature, can be associated with the *in vivo* data, and incorporated into a validated deposition model to estimate and study effects of aerosol hygroscopicity within the lung.

II. AEROSOL DEPOSITION WITHIN LUNG AIRWAYS

Experimental studies have determined that the regional distribution of nonhygroscopic (i.e., water insoluble) particles among tracheobronchial (TB) and pulmonary (P) airways is primarily a function of their aerodynamic diameter(s), D_{ae}(s), when $D_{ae} > 0.5\ \mu m$ (e.g., Task Group on Lung Dynamics, 1966; Lippmann and Altshuler, 1976; Heyder *et al.,* 1986). By definition, for a spherical particle of geometric diameter D_g and density ρ, $D_{ae} = \sqrt{\rho} D_g$. The D_{ae} of a nonspherical particle is defined (Stöber, 1972) as the geometric diameter of a spherical particle of unit density, in the cgs system of units, that has the same Stokes terminal settling speed as the nonspherical particle considered. The deposition of smaller submicron particles is primarily dependent upon their geometric diameters.

The compartmental (i.e., TB and P) deposition fractions of larger particles are relatively unaffected by the range of sizes present in a polydisperse aerosol; as described for example, by the geometric standard deviation (σ_g) of an aerosol with a lognormal particle size distribution when $\sigma_g < 2$ (Task Group on Lung Dynamics, 1966). The mass median aerodynamic diameter (MMAD) of such an aerosol is usually sufficient to predict its deposition pattern within the lung for practical purposes.

Aerosol particles composed of hygroscopic (i.e., water soluble) material(s) will have an affinity for the water vapor present in the lung's warm and humid atmosphere. Therefore, identifying and estimating factors affecting hygroscopicity are necessary for accurate assessment of the behavior and fate of such particles, which will change in size and density subsequent to the uptake of water vapor while traveling through the lung. Hygroscopic growth effects have important implications to:

- Aerosol therapy, where knowledge of the influence of hygroscopicity upon medicinal aerosols may permit drugs to be selectively deposited so as to elicit optimum therapeutic effects; and,
- Inhalation toxicology, where the potential threat to human comfort and health following inhalation exposure to irritant or toxic particulate matter must be ascertained.

A deterministic mathematical model, validated by comparisons of theoretically predicted TB, P, and thoracic (i.e., TB + P) deposition patterns with data from inhalation exposure tests using nonhygroscopic aerosols with human subjects, has been presented elsewhere (Martonen, 1983a,b; Martonen and Graham, 1987a,b). In this study, we expand its application to include other aerosols, deposition locations, and methods of breathing. Moreover, we will use the predictive deposition computer code for extrapolation modeling purposes, that is, to predict effects of hygroscopicity upon the deposition of representative pollutant and medicinal aerosols. The primary factors to be addressed in our analyses are, relative to nonhygroscopic aerosols of identical preinspired geometric or aerodynamic sizes,

- Does hygroscopicity qualitatively influence total aerosol deposition within the human respiratory system and its spatial distribution among TB and P airways?
- What is the magnitude of the effect of hygroscopic growth?

III. RESPIRATORY TRACT MORPHOLOGY AND ENVIRONMENT

Computation of aerosol deposition probabilities within the human lung requires the selection of a suitable anatomy regarding individual airway geometries and their relative spatial orientations. Different descriptions of the lung have been proposed for deposition modeling purposes, varying from abstract conceptual airway arrangements (e.g., Findeisen, 1935) to schemes based upon laboratory measurements of cadaveric lungs (e.g., Weibel, 1963; Horsfield *et al.,* 1971). We will use airway dimensions of a statistically modified (Soong *et al.,* 1979) version of Weibel's Model A to describe the human lung. It is a symmetric, dichotomously branching network of 24 generations, i. The TB compartment consists of $i = 0$ (trachea) to 16 (terminal conducting bronchioles), and the P compartment of $i = 17$ (partially alveolated respiratory bronchioles) to 23 (alveolar sacs). There are 2^i number of identical airways within a given generation, but dimensions vary from generation to generation. From the data of Horsfield and Cumming (1967), we shall assume a mean bifurcation angle between branching airways of a generation to be 70°. This lung morphology has been shown to be acceptable for modeling applications, even when compared to what may have been considered, *a priori,* to be a more physiologically realistic asymmetric lung (Martonen, 1983a).

The behavior and fate of inhaled hygroscopic particulate matter are certainly affected by the environment within the human respiratory system. The two most significant components of the *in situ* surroundings are the:

- Dynamics of fluid motion within airways, which influences the degree of mixing between inhaled aerosols and resident air; and
- Spatial atmospheric (i.e., temperature and relative humidity) profiles to which particles are exposed, which will affect heat and mass transfer processes between the aerosols and surrounding air.

The subject of gas kinetics has been reviewed by Pedley *et al.* (1977) and will not be discussed in detail here. The primary, or longitudinal, flow patterns within the tracheobronchial and pulmonary passages of the lung are quite complex. Flow instabilities exist within the upper bronchi, initiated by the vibrating glottis and laryngeal jet, and air movement over protruding cartilagenous rings. Indeed, the very nature of a branching network of airways *per se* may introduce unstable flow conditions at bifurcation sites. Moreover, secondary, multiple vortex-like air currents within the cross-sections of airway tubes may be created at airway branching sites, and will vary in pattern and intensity during the respective inspiratory and expiratory phases of a breath. In the smaller, more peripheral passages of the lung, flow instabilities can even be induced by the beating action of the heart. Because of obvious difficulties in classifying air movement patterns *in vivo,* most attempts to study flow anomalies have been conducted using replica airway casts and models. The influences of selected airflow patterns upon deposition probabilities of inhaled particles have been formulated by various authors with specific applications and airway morphologies in mind (e.g., Findeisen, 1935; Landahl, 1950; Beeckmans, 1965). The system of equations proposed by Martonen (1982) shall be utilized in this work.

The subject of the lung's temperature (T) and relative humidity (RH) atmosphere as related to aerosol hygroscopicity has been reviewed by Morrow (1986). The transitory T and RH profiles within the human lung have not been simultaneously and categorically mapped (i.e., in an airway-by-airway fashion). The closest effort to a systematic protocol are the *in situ* works of McFadden *et al.* (1982, 1985) which were restricted to T measurements within the relatively large, upper bronchi. Most other published T and RH data are for single, selected airways and, in many instances, it is not possible to ascertain precise geometric locations for the sites of the measurements relative to the nares or lips. Frequently, the data were merely lower-order, or tangential, aspects of the respective scientific investigations. Nevertheless, we have attempted to pool the limited data to make an estimate of T and RH distributions within the human lung, and to put them in a format suitable for aerosol deposition modeling purposes. The data that we have culled from the literature to simulate the lung's atmosphere for nose-only and mouth-only breathing are presented in Table 25-1. Such breathing modes, while being realistic simulations of controlled laboratory tests, are obviously bounding limits of ambient breathing patterns. The influences of mixed, oral–nasal breathing modes on aerosol deposition within the human lung have been discussed elsewhere (Miller *et al.,* 1988).

The measurements tabulated were made under a variety of laboratory conditions (e.g., Were the subjects anaesthetized or not? Were the breathing patterns monitored? See the references for specific details). The listed data of McFadden *et al.* (1985) were the closest to the most common testing conditions, and they did conduct experiments ranging from minute ventilations of 15 to 100 liters/min. Therefore, with these considerations in mind, and based upon the data in Table 25-1, in our calculations we have tacitly assumed a RH of 99.5% in the trachea and all distal airways for inhalation via the nose. For the oral breathing mode, however, we considered the RH pattern to be 90% in the trachea and to increase monotonically by 1% increments for each peripheral bronchial airway until a saturation level of 99.5% was achieved in generation $i = 10$ airways. For either respiratory mode a temperature of 37°C was taken to be appropriate for the lung.

IV. SIMULATION RESULTS

The influences of hygroscopic growth upon the deposition patterns of selected aerosols due to the uptake of water vapor present in the warm, moist atmosphere of the lung will now be investigated. This will be assessed by using appropriate computer subroutines simulating particle growth in a validated predictive aerosol deposition model, in conjunction with published hygroscopic growth data characterizing aerosols of therapeutic applications and those posing health hazards. To emphasize the ubiquitous

Table 25-1 Description of Atmospheres within the Human Lung

Anatomical Location	Oral[a]				Nasal[a]				References
	T_{in}	T_{ex}	RH_{in}	RH_{ex}	T_{in}	T_{ex}	RH_{in}	RH_{ex}	
Laryngeal cavity	30.6 ± .8	36.2 ± .3	90.	99.	32.3 ± 8.	36.4 ± .2	98.–99.	98.–99.	Ingelstedt (1956)
Airway generation, i									
$i = 0$ (trachea)	33.–34.	36.–37.	—	—	34.–35.	36.–37.	—	—	Cole (1954)
0	—	—	—	—	35.3	35.7	98.	99.9	Perwitzschky (1928)
0	34.5	35.8	—	—	35.4	36.2	—	—	Verzar *et al.* (1953)
0	32.9	34.4	—	—	32.6	35.3	—	—	Herlitzka (1921)
0	26.7	—	82.7	—	—	—	—	—	Déry *et al.* (1967)
0	31.4–31.9	33.2–33.4	73.6–80.4	—	—	—	—	—	Déry (1973)
0	31.2	32.6	—	—	—	—	—	—	McFadden *et al.* (1985)
0	32.	33.	—	—	—	—	—	—	McFadden *et al.* (1985)
0	32.2	33.4	—	—	—	—	—	—	Déry *et al.* (1967)
$i = 1$ (main)	30.6	—	85.8	—	—	—	—	—	Déry (1973)
1	32.2	33.7	87.	—	—	—	—	—	McFadden *et al.* (1985)
2 (lobar)	33.	34.	—	—	—	—	—	—	McFadden *et al.* (1985)
3 (segmental)	33.1	—	91.3	—	—	—	—	—	Déry *et al.* (1967)
4 (subsegmental)	33.9	—	94.6	—	—	—	—	—	McFadden *et al.* (1985)
4–5	33.9	35.	—	—	—	—	—	—	Déry (1973)
10–11[b]	34.6	36.	—	—	—	—	—	—	McFadden *et al.* (1985)

[a] The subscripts in and ex refer to the inspiratory and expiratory components of a breathing cycle, respectively.

[b] This site is based upon a thermistor's location in a Weibel (1963) morphology. McFadden *et al.* (1985) stipulated its distance to be 5.8 cm beyond the orifice of a subsegmental bronchus.

nature of hygroscopic materials, we will consider implications based upon saline aerosols, chosen to be representative of medicinal drugs, and sulfuric acid droplets, selected to portray ambient acid aerosols posing a threat to human health.

The kinetics of hygroscopic growth within lung airways may be simplistically conceived in the following manner. A particle enters a generation i airway with $D_{ae_i} = D_{g_i}\sqrt{\rho_i}$, absorbs water vapor while traversing it (and thereby changes in size and density), and exits it with $D_{ae_{i+1}} = D_{g_{i+1}}\sqrt{\rho_{i+1}}$. The density of the particle entering an airway of generation $i + 1$ can be written as

$$\rho_{i+1} = \left[\frac{D_{g_i}}{D_{g_{i+1}}}\right]^3 (\rho_i - \rho^*) + \rho^*$$

where ρ^* is the density of water. In this formulation, it has been assumed that changes in particle size are due only to the uptake of water vapor; that is, spontaneous molecular transformations or variations in the physical state not related to condensation and absorption of the vapor are not considered. In most instances, $\rho^* < \rho_i$, so that $\rho_{i+1} < \rho_i$. Therefore, $D_{ae_{i+1}} > D_{a_i}$ only if the particle's growth rate is such that $D_{g_{i+1}}$ is sufficiently greater than D_{g_i}. An example of a decrease in particle D_{ae} with hygroscopic growth has been presented by Martonen and Wilson (1983).

Information describing the hygroscopic growth potential of certain medicinal and airborne acidic pollutants is available in two rather disparate formats. Bell and Ho (1981), Martonen *et al.* (1982), and Martonen and Lowe (1984) have reported on the growth rates of saline and two bronchodilator aerosols within a simulated lung chamber held at physiologically realistic, but constant, T and RH atmospheres. The RH values in the surrogate lung varied between 90 and 95%, although they were constant in any given test. The T was controlled at 37°C, the body's core temperature. Their data were presented as changes in initial particle sizes as functions of residence times within the chamber. The appropriate time-dependent size and density to use in computing an inhaled particle's deposition probability within the lung are determined by the amount of time it takes to reach a given airway location. For a stated inspiratory flow rate and specified dimensions of an assumed morphology, the calculation is straightforward. This computed time value may then be used as the abscissa in graphs contained in the aforementioned references to determine relevant particle physical characteristics to be used in deposition computations.

Regarding pollutants, Tang (1976), Tang *et al.* (1977), and Tang and Munkelwitz (1977) have determined, theoretically and experimentally, the final equilibrium sizes attained by numerous sulfate aerosols within a "continuous flow" laboratory apparatus. The T value was monitored and fixed at 25°C, and the RH was varied from about 40% to a saturation value expected within the lung. To simulate effects of growth upon deposition, it is necessary to assume T and RH atmospheric patterns among lung airways, as in Table 25-1, and to use corresponding particle parameters from the cited manuscripts to represent changes resulting from the assimilation of water vapor in the computational manner of Martonen *et al.* (1985).

A. Medicinal Drugs

Saline solution droplets are frequently employed for therapeutic procedures. Our simulation program has been used to investigate the effects of hygroscopic growth on thoracic deposition of NaCl aerosols within the lung relative to like-sized nonhygroscopic particles, namely Fe_2O_3 aerosols, used in experimental inhalation exposures with human test subjects (Martonen and Graham, 1987a). Briefly, those results indicated that hygroscopic growth is associated with decreased total deposition for particles originally smaller than about 0.6 μm, but with increased deposition probabilities for particles originally larger than about 0.8 μm.

Because the TB and P regions of the lung are two relatively distinct anatomical compartments and have different clearance mechanisms that vary in effectiveness and rate of operation, we have segmented the previously reported total deposition values for NaCl into predicted compartmental, TB and P, deposition patterns. Comparisons are made using NaCl particles and a control, like-sized nonhygroscopic particle as shown in Fig. 25-1 and 25-2 for the TB and P compartments, respectively. A realistic oral breathing regimen simulating aerosol therapy practices and laboratory aerosol deposition test conditions was used in these calculations. From Fig. 25-1 it is apparent that the effect of hygroscopic growth is to reduce deposition of the smaller particles (i.e., with an initial D_{ae} of less than about 0.9 μm). Conversely, the percentage change in the deposition fraction is greater for the larger hygroscopic particles, by as much as 40 to 65% for particles in the range of about 1.3 to 3.0 μm. That is, for a particle with an entry D_{ae} of 2 μm, the predicted deposition fraction, not considering hygroscopic growth, would be 0.14, but the effect of hygroscopic growth would be to increase the deposition fraction to 0.22 which is a 65% difference in predicted fractions. As shown in Fig. 25-2, pulmonary deposition can also be substantially over- or underestimated if the hygroscopic quality of an aerosol is not considered. The effect of hygroscopicity is to reduce the predicted deposition fraction for submicron particles. Forty to fifty percent enhancement of the deposition fraction, relative to nonhygroscopic aerosols of similar size, is found for particles in the range of 1.3 to 2.5 μm. These observations, together with findings of previous works conducted with bronchodilator drugs (Martonen *et al.*, 1982; Martonen and

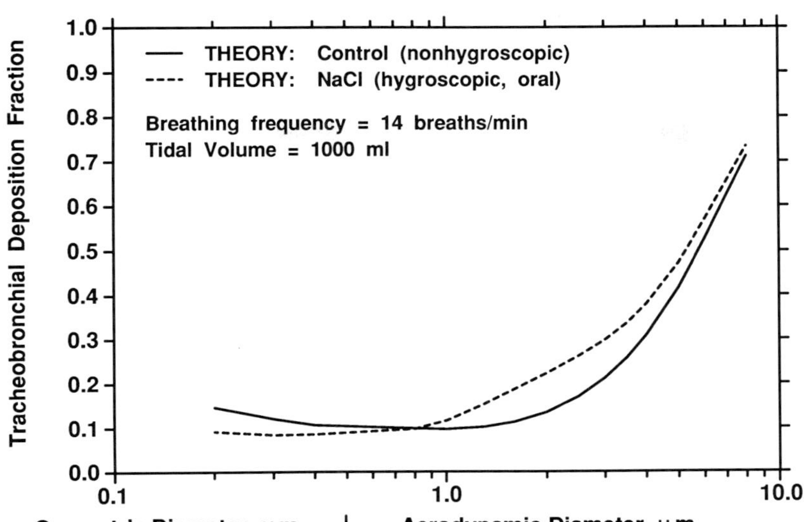

Fig. 25-1 Deposition of therapeutic saline aerosols within conducting airways of the human lung.

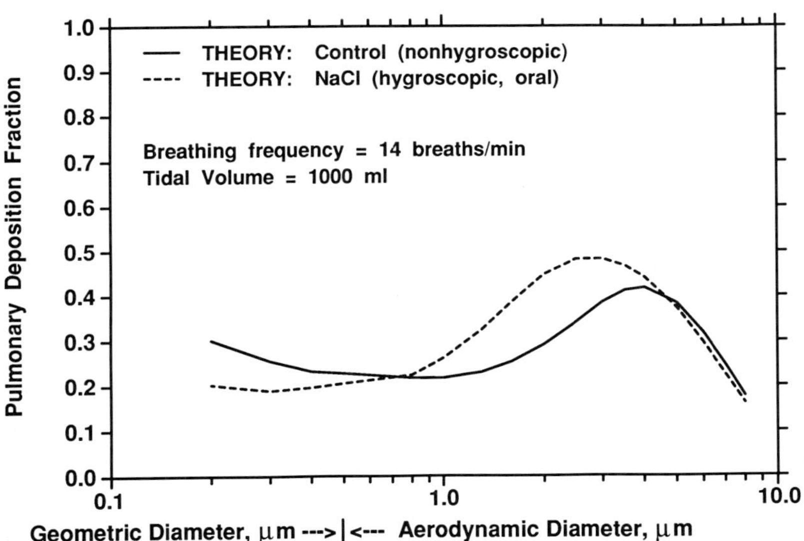

Fig. 25-2 Deposition of therapeutic saline aerosols among alveolated airways of the human lung.

Wilson, 1983), indicate that hygroscopic growth is an important factor affecting aerosols used in therapy protocols. By properly accounting for hygroscopicity in the administration of airborne medicinal agents, it may be possible, in concert with mathematical modeling efforts, to deposit drugs selectively at prescribed sites within the lung.

B. Ambient Pollutant Aerosols

In Fig. 25-3 through 25-5, the effect of the hygroscopic growth of H_2SO_4 particles is compared to predicted deposition probabilities for like-sized nonhygroscopic particles for the thoracic, TB, and P compartments of the lung, respectively. Martonen and Graham (1987b) have used the validated model to study effects of hygroscopicity upon the thoracic dispersion of NH_4HSO_4 aerosols. Under ambient exposure conditions, individuals might be expected to be breathing with air flow division in some nose–mouth combination. Because hygroscopic growth rates will be affected by breathing patterns, the effects of two extreme modes (i.e., exclusively nasal and exclusively oral) were examined to provide an envelope of potential hygroscopic growth effects. The deposition fractions shown on the figures are adjusted, or normalized, to the amount of particulate matter entering the trachea. The differences between nose and mouth breathing, therefore, are due strictly to the various rates at which the temperature and relative humidity profiles change for these two breathing modes. Our calculated

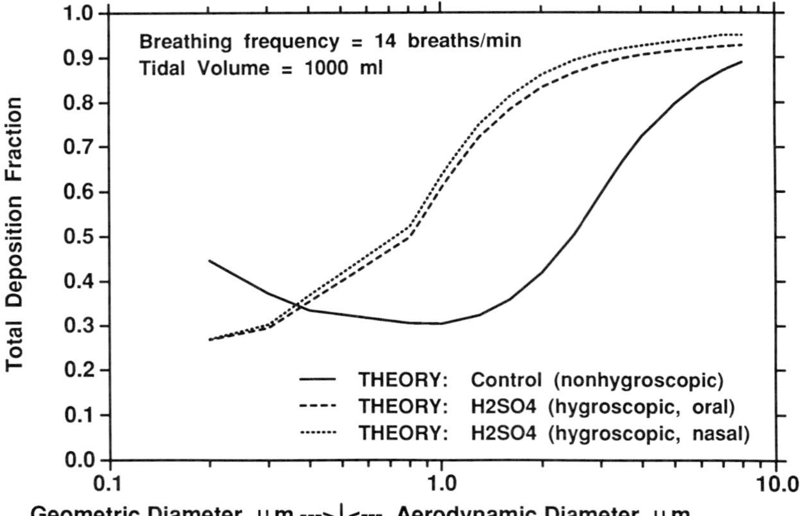

Fig. 25-3 The influence of type of breathing upon thoracic deposition of pollutant sulfuric acid aerosols in humans.

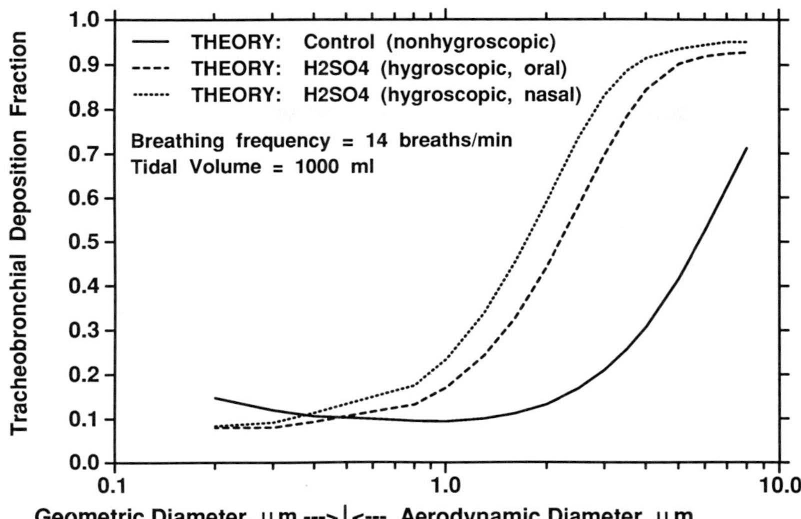

Fig. 25-4 Deposition fraction of ambient acid aerosols in the tracheobronchial compartment of the human lung as a function of breathing mode.

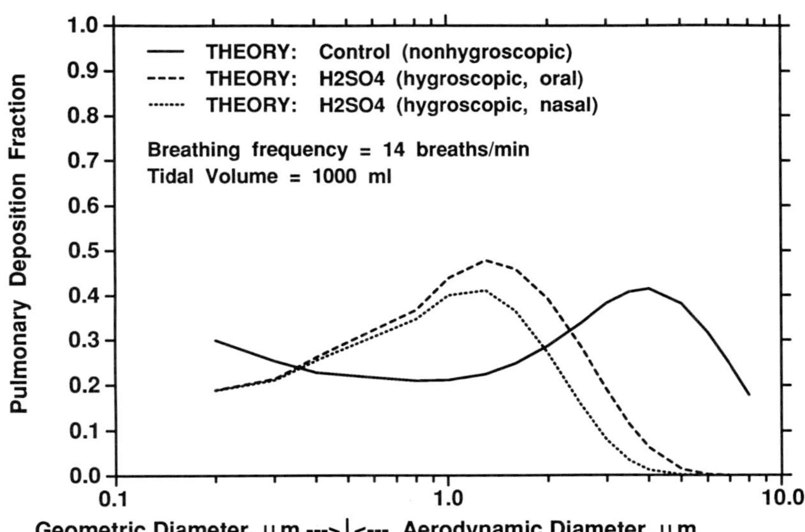

Fig. 25-5 Deposition fraction of atmospheric acid aerosols in the pulmonary region of the human lung as related to mode of breathing.

deposition curves are compared to a theoretical, control aerosol with identical properties, except hygroscopicity.

Thoracic deposition is shown in Fig. 25-3. Qualitatively similar deposition fractions are predicted for ambient acid aerosols under the two extreme breathing modes, with a slightly higher deposition predicted for nose-breathing conditions. This difference may be attributed to the relatively greater humidification occurring in the nasal region which encourages more rapid growth of the inhaled particles. If hygroscopic growth of H_2SO_4 particles were to be ignored, severe underestimation of deposition would occur, as illustrated by the comparison with the control aerosol. For particles with an inhaled D_{ae} of about 1 to 4 μm, the assumption of a nonhygroscopic H_2SO_4 aerosol underestimates actual deposition by a factor of nearly 2. For submicron particles, the differences between nose and mouth breathing are minimal, but not to consider the hygroscopicity *per se* of H_2SO_4 can result in an overestimation of the deposition fraction by nearly 50%.

Investigation of the split of thoracic deposition between the TB and P regions indicates the importance of including hygroscopic growth when studying atmospheric acid aerosol behavior. For both regions, the predicted deposition curves shift to the left of the corresponding curves for nonhygroscopic particles. In the TB region (Fig. 25-4), about a 10% difference in deposition can be found due to mode of breathing when D_{ae} is greater than about 1.0 μm. A similar difference is seen in the P region (Fig. 25-5). Inspection of Fig. 25-4 and 25-5 shows that deposition in the TB region is greater for breathing via the nose, but deposition is greater in the P region for breathing via the mouth. This offsetting difference in compartmental deposition patterns accounts for the small variations observed in total deposition. Thus, if thoracic deposition, adjusted to the aerosol mass entering the trachea, is of interest, then the differences between nose and mouth breathing will be of minimal importance. If the regional distribution of the particulate matter within the lung is of interest, particularly regarding the relative efficiencies of the TB mucociliary and P macrophage clearance mechanisms, the breathing mode will have a significant impact on deposition predictions for hygroscopic particles of health effects concern.

V. CONCLUSIONS

Data concerning the environment of the human respiratory tract are rare. By incorporating available T, RH, and hygroscopic growth information into a validated predictive aerosol deposition code, computations suggest that hygroscopicity is an important determinant regarding inhaled particle behavior. Therefore, particle growth must be accounted for in aerosol therapy protocols and risk assessment analyses of ambient acidic aerosols. More detailed *in situ* measurements of the respiratory tract's environment are necessary for improved simulation of hygroscopic growth effects.

REFERENCES

Beeckmans, J. M. (1965). The deposition of aerosols in the respiratory tract. I. Mathematical analysis and comparison with experimental data. *Can. J. Physiol. Pharmacol.* **43,** 157–172.

Bell, K. A., and Ho, A. T. (1981). Growth rate measurements of hygroscopic aerosols under conditions simulating the respiratory tract. *J. Aerosol Sci.* **12,** 247–254.

Cole, P. (1954). Recordings of respiratory air temperature. *J. Laryngol. Otol.* **68,** 295–307.

Déry, R. (1973). The evolution of heat and moisture in the respiratory tract during anaesthesia with a non-rebreathing system. *Can. Anaesth. Soc. J.* **20,** 296–309.

Déry, R., Pelletier, J., Jacques, H., Clavet, M. and Houde, J. J. (1967). Humidity in anaesthesiology. III. Heat and moisture patterns in the respiratory tract during anaesthesia with the semi-closed system. *Can. Anaesth. Soc. J.* **14,** 287–298.

Findeisen, W. (1935). Über das Absetzen kleiner, in der Luft suspendieter Teilchen in der menschlichen Lunge bei der Atmung. *Pflügers Arch. Ges. Physiol.* **236,** 367–379.

Herlitzka, A. (1921). Sur la température trachéale de l'air inspiré et expiré. *Arch. Int. Physiol.* **18,** 587–600.

Heyder, J., Gebhart, J., Rudolf, G., Schiller, C. F., and Stahlhofen, W. (1986). Deposition of particles in the human respiratory tract in the size range 0.005–15μm. *J. Aerosol Sci.* **17,** 811–825.

Horsfield, K., and Cumming, G. (1967). Angles of branching and diameters of branches in the human bronchial tree. *Bull. Math. Biophys.* **29,** 245–259.

Horsfield, K., Dart, G., Olson, D. E., Filley, G. F., and Cumming, G. (1971). Models of the bronchial tree. *J. Appl. Physiol.* **31,** 207–217.

Ingelstedt, S. (1956). Studies on the conditioning of air in the respiratory tract. *Acta Otolaryngol. (Suppl.)* **131,** 1–80.

Ingelstedt, S., and Toremalm, N. G. (1960). Aerodynamics within the larynx and trachea. *Acta Otolaryngol. (Suppl.)* **158,** 81–92.

Ingelstedt, S., and Toremalm, N. G. (1961). Air flow patterns and heat transfer within the respiratory tract. *Acta Physiol. Scand.* **51,** 204–217.

Landahl, H. D. (1950). On the removal of air-borne droplets by the human respiratory tract. I. The lung. *Bull. Math. Biophys.* **12,** 43–56.

Lippmann, M., and Altshuler, B. (1976). Regional deposition of aerosols. *In* "Air Pollution and the Lung" (E. F. Aharanson, A. Ben-David, and M. A. Klingberg, eds.), pp. 25–48. Wiley, New York.

McFadden, E. R., Jr., Denison, D. M., Waller, T. F., Assoufi, B., and Peacock, A. (1982). Direct recordings of the temperatures in the tracheobronchial tree in normal man. *J. Clin. Invest.* **69,** 700–705.

McFadden, E. R., Jr., Pichurko, B. M., Bowman, F. H., Ingenito, E., Burns, S., Dowling, N., and Solway, J. (1985). Thermal mapping of the airways in humans. *J. Appl. Physiol.* **58,** 564–570.

Martonen, T. B. (1982). Analytical model of hygroscopic particle behavior in human airways. *Bull. Math. Biol.* **44,** 425–442.

Martonen, T. B. (1983a). On the fate of inhaled particles in the human: A comparison of experimental data with theoretical computations based on a symmetric and asymmetric lung. *Bull. Math. Biol.* **45,** 409–424.

Martonen, T. B. (1983b). Deposition of inhaled particulate matter in the upper respiratory tract, larynx, and bronchial airways: A mathematical description. *J. Toxicol. Environ. Health* **12,** 787–800.

Martonen, T. B., and Graham, R. C. (1987a). Hygroscopic growth: Its effect on aerosol therapy and inhalation toxicology. *In* "Deposition and Clearance of Aerosols in the Human Respiratory Tract" (W. Hofmann, ed.), pp. 200–206. Facultas, Vienna.

Martonen, T. B., and Graham, R. C. (1987b). The effect of hygroscopic growth upon the regional dispersion of particulates in human airways. *In* "The Sixth International Congress on Aerosols in Medicine" (J. M. Aiache and C. Molina, eds.), pp. 221–234. Librairie Lavoisier, Paris.

Martonen, T. B., and Lowe, J. E. (1984). Measurements of hygroscopic growth rates of medicinal aerosols. *In* "Aerosols" (B. Y. H. Liu, D. Y. H. Pui, and H. J. Fissan, eds.), pp. 1003–1006. Elsevier, New York.

Martonen, T. B., and Wilson, A. F. (1983). The influence of hygroscopic growth upon the deposition of bronchodilator aerosols in upper human airways. *J. Aerosol Sci.* **14,** 208–211.

Martonen, T. B., Bell, K. A., Phalen, R. F., Ho, A., and Wilson, A. F. (1982). Growth rate measurements and deposition modelling of hygroscopic aerosols in human tracheobronchial models. *In* "Inhaled Particles V" (W. Walton, ed.), pp. 93–108. Pergamon, Oxford.

Martonen, T. B., Barnett, A. E., and Miller, F. J. (1985). Ambient sulfate aerosol deposition in man: Modeling the influence of hygroscopicity. *Environ. Health Perspect.* **63,** 11–24.

Miller, F. J., Martonen, T. B., Ménache, M. G., Graham, R. C., Spektor, D. M., and Lippmann, M. (1988). Influence of breathing mode and activity level on the regional deposition of inhaled particles and implications for regulatory standards. *In* "Inhaled Particles VI" (J. Dodgson, R. I. McCallum, M. R. Bailey, and D. R. Fisher, eds.), pp. 3–10. Pergamon, Oxford.

Morrow, P. (1986). Factors determining hygroscopic aerosol deposition in airways. *Physiol. Rev.* **66,** 330–376.

Pedley, T. J., Schroter, R. C., and Sudlow, M. F. (1977). Gas flow and mixing in the airways. *In* "Lung Biology in Health and Disease, Vol. 3, Bioengineering Aspects of the Lung" (J. B. West, ed.), pp. 163–265. Dekker, New York.

Perwitzschky, R. (1928). Die Temperatur and Feuchtigkeitsverhältnisse der Atemluft in den Luftwegen. 1. *Mitt. Arch. Ohren Nasen Kehlkopfh.* **117,** 1–36.

Soong, T. T., Nicholaides, P., Yu, C. P., and Soong, S. C. (1979). A statistical description of the human tracheobronchial tree geometry. *Respir. Physiol.* **37,** 161–172.

Stöber, W. (1972). Dynamic shape factors of nonspherical aerosol particles. *In* "Assessment of Airborne Particles" (T. T. Mercer, P. E. Morrow, and W. Stöber, eds.), pp. 249–289. Thomas, Springfield, Illinois.

Tang, I. N. (1976). Phase transformation and growth of aerosol particles composed of mixed salts. *J. Aerosol Sci.* **7,** 361–371.

Tang, I. N., and Munkelwitz, H. R. (1977). Aerosol growth studies - III. Ammonium bisulfate aerosols in a moist atmosphere. *J. Aerosol Sci.* **8,** 321–330.

Tang, I. N., Munkelwitz, H. R., and Davis, J. G. (1977). Aerosol growth studies - II. Preparation and growth measurements of monodisperse salt aerosols. *J. Aerosol Sci.* **8,** 149–159.

Task Group on Lung Dynamics (1966). Deposition and retention models for internal dosimetry of the human respiratory tract. *Health Phys.* **12,** 173–207.

Verzar, F., Keith, T., and Parchet, V. (1953). Temperatur and Feuchtigkeit der Luft in den Atemwegen. *Pflügers Arch. Ges. Physiol.* **257,** 400–416.

Weibel, E. R. (1963). "Morphometry of the Human Lung." Springer-Verlag, Berlin.

Chapter 26

Age-Dependent Lung Dosimetry of Radon Progeny

W. Hofmann*
Center for Extrapolation Modeling
Duke University Medical Center
Durham, North Carolina 27710

T. B. Martonen
Toxicology Branch
Health Effects Research Laboratory
U.S. Environmental Protection Agency
Research Triangle Park, North Carolina 27711

M. G. Ménache
NSI Technology Services Corporation
Environmental Sciences
Research Triangle Park, North Carolina 27709

I. INTRODUCTION

Inhalation exposure of the general population to ambient radon progeny constitutes the most significant health hazard of the natural radiation environment (Pohl *et al.*, 1976). It is generally accepted by the scientific community that a considerable fraction of all naturally occurring lung tumors is caused by inhalation of the short-lived radon progeny (NCRP,

* Permanent address: Abteilung für Biophysik, Universität Salzburg, A-5020 Salzburg, Austria.

Disclaimer: The research described in this article has been reviewed by the Health Effects Research Laboratory, U.S. Environmental Protection Agency, and approved for publication. Approval does not signify that the contents necessarily reflect the views and policies of the Agency nor does mention of trade names or commercial products constitute endorsement or recommendation for use.

1984). This risk is exacerbated for certain identifiable subpopulations, such as tobacco smokers and uranium miners (American Cancer Society, 1981; Ochsner *et al.,* 1960; Saccomanno *et al.,* 1964; Sevc *et al.,* 1976; Wagoner *et al.,* 1965; Whittemore and McMillan, 1983). While most radiation protection standards are concerned with occupational exposure, affecting only a small, well-defined fraction of the adult population, radon progeny are radiological carcinogens which affect the whole population, including children. Children are, therefore, a critical subpopulation which warrants special consideration in risk assessment analyses following inhalation exposure to ambient pollutants, particularly radon progeny.

The clearest evidence of a relatively higher risk for individuals irradiated at very young ages has been found in the incidence of leukemia in atomic bomb survivors (ABS) of Hiroshima and Nagasaki (Beebe *et al.,* 1978; BEIR, 1972). While a similar relationship could also be observed for all cancers (except leukemia) pooled together, BEIR (1972) and Beebe *et al.* (1978) concluded that there were too few lung cancer cases to yield statistically reliable, age-specific risk estimates. A recent analysis of the ABS cancer mortality data (Preston *et al.,* 1987) has demonstrated, however, a statistically significant increase in relative lung cancer risk in the age group 0 – 19 years old at the time of exposure compared to all ages greater than 20 years. Does this epidemiological evidence then imply that lung cancer risk in children exposed to radon progeny is also higher than in adults? There are significant differences between the two exposure situations. While the atomic bomb survivors were exposed to an acute, external neutron and gamma irradiation, radon progeny inhalation represents a chronic, internal alpha irradiation. For the latter irradiation conditions, Crawford-Brown (1983) has developed a predictive model, in which he has combined age-dependent bronchial basal cell doses with a theory relating dose, the number of irradiated cells, and the rate of turnover of those cells to the risk of inducing lung cancers. The model results indicate that the relative risk is, indeed, greatly increased in young children when compared with calculated tumor incidence in adults, if the age-dependent pattern of risk factors discerned at Hiroshima and Nagasaki holds true for all radiations acting on individual cells.

To study lung cancer risk in children following inhalation of radon progeny, we take the radiation dose in lung tissue as an approximate index of lung cancer risk, thus neglecting potential differences in cellular radiosensitivity, cell turnover rates, and efficiency of repair. Previous calculations of the radiation dose to tracheobronchial and pulmonary regions (Hofmann *et al.,* 1979), and later to bronchial basal cells (Hofmann *et al.,* 1980; Hofmann, 1982a) in the maturing human lung have indicated that doses in children may be significantly higher than those in adults, under identical exposure conditions. For defined breathing parameters, e.g., sedentary activity, radiation doses decreased monotonically with increasing age from a maximum at birth to a minimum at age 30 years, with distinct differences among the various bronchial generations. When a typical age-dependent physical activity pattern (Hofmann *et al.,* 1979) was factored into the dose calculations, a pronounced dose maximum could be observed at an age of about 6 years, which was approximately a factor of 3 greater than the corresponding adult value. Increased bronchial basal cell doses in children following exposure to radon progeny have also been predicted by Harley and Pasternack (1981, 1982), Crawford-Brown (1982), and James (1987).

Here, we utilize a new morphometric model of the growing human lung (Martonen *et al.,* 1989) in radon progeny lung dosimetry calculations and compare the results with previously published dose calculations (Hofmann *et al.,* 1980; Hofmann, 1982a) which used a different morphological model (Hofmann, 1982b). Radiation doses are computed for two different target cells in the bronchial epithelium, for deep lying basal cells, which have been regarded for many years as the critical target cells, and for more uniformly distributed secretory cells (SMGC), which are currently believed to be the main progenitor cells for bronchial carcinomas (McDowell *et al.,* 1978; McDowell and Trump, 1983; Masse, 1984).

Subject age affects radon progeny lung dosimetry in all stages of dose calculations, and requires information on:

1. Morphological changes of the airway system occurring during postnatal growth.
2. Deposition patterns of inspired aerosols dependent upon the aforementioned morphometric changes and differences in breathing parameters.
3. Mucociliary clearance rates in bronchial airways as a function of age.
4. Location of sensitive target cells in the bronchial epithelium of the developing human lung.

Because of the rather limited, and sometimes even contradictory, experimental data available, different modeling assumptions will be compared at each stage of age-dependent dose calculations. To familiarize the reader with the objectives of this simulation effort, the different modeling assumptions at the various stages of lung dosimetry are summarized in Table 26-1 and these will be described in detail in the following sections. For the sake of brevity, only a limited number of possible modeling parameter combinations will be presented in this chapter to illustrate the effect of human subject age on the lung dosimetry of inhaled radon progeny.

Table 26-1 Summary of Modeling Assumptions Used to Investigate the Effect of Human Subject Age on the Lung Dosimetry of Inhaled Ambient Radon Progeny

Lung morphology	Model 1 (age-dependent TB morphology based on Hofmann, 1982b)
	Model 2 (age-dependent TB morphology based on Phalen *et al.*, 1985)
Respiration parameters	Sedentary activity breathing pattern
	Maximal activity breathing pattern
Mucociliary clearance	Clearance velocities are independent of age, i.e., clearance rates are proportional to airway length at a given age.
	Clearance velocities are proportional to airway length at a given age, i.e., clearance rates are independent of age
	Clearance velocities are exponential functions of age (Whaley *et al.*, 1987)
Target cells	Basal cells of bronchial epithelium
	Secretory (SMGC) cells (epithelial tissue)

II. AGE-DEPENDENT MORPHOLOGY AND PHYSIOLOGY

The newborn human lung is not simply the adult lung in miniature. After birth, each of the lung's structural components has a different pattern of maturation regarding increase in number as well as in size of airways. The postnatal development of each structure has been summarized in the three laws of lung development (Reid, 1977). Morphometric measurements suggest that the number of conductive airways down to the level of terminal bronchioles is actually complete at birth (Dunnill, 1962; Hislop *et al.*, 1972; Reid, 1977). With age, each individual airway grows in length and diameter in a constant relation to the whole organ (Hislop *et al.*, 1972). Respiratory airways and alveoli grow rapidly after birth, increasing in number until the age of about 8 years, and increasing in size until the growth of the chest wall is completed at adulthood (Dunnill, 1962; Reid, 1977).

The growth of the tracheobronchial (TB) and pulmonary (P) compartments of the lung have not been systematically measured and characterized in a single research effort. Data from different authors, therefore, must be assembled and incorporated into an anatomical description of the growing lung. An updated discussion of postnatal lung growth has been presented by Ménache and Graham (Chapter 22, this volume).

Most measurements of age-dependent parameters of human airways have been made in the TB region. The majority of data have been obtained for the relatively large, upper airways; only a few measurements of the smaller, more peripheral bronchioles have been reported. The morphometric data currently available suggest two viable, alternative methods of constructing an age-dependent geometry of the human TB region: (1) a model based upon the relatively many measurements of the upper airways, though from different authors, with systematic features identified there extrapolated to lower airways; or (2) a model based upon the relatively few complete TB pathway measurements made in a single lobe, and the findings extrapolated to all other lobes. In this study we compare two different models of the TB branching system. Model 1 will be based upon the work of Hofmann (1982b) who followed the former method, and model 2 will be based upon Phalen *et al.* (1985) who followed the latter. The model of the developing alveolated, pulmonary airway structure will be the same for both models, and will be based on Dunnill (1962). The actual airway dimensions of the two developing lung morphologies for five different ages are in the Appendix (Tables 26-A1–26-A3). Other theoretical models of age-dependent changes of the complete human lung have been proposed by Crawford-Brown (1982), Hofmann (1982b), and Xu and Yu (1986).

A. Tracheobronchial Airways

The age-dependent lung model developed by Hofmann (1982b) is based on Weibel's Model A (Weibel, 1963) which describes the tracheobronchial tree by 17 symmetrically dividing generations, where the trachea is airway generation $i = 0$ and the last of the terminal bronchioles is generation $i = 16$. Morphometric data for diameters and lengths of the trachea and main bronchi ($i = 1$), gleaned from the published literature, were fitted to exponentional functions of age and normalized to the Weibel adult dimensions. Available experimental evidence suggests a nearly threefold increase of linear dimensions from birth to adulthood, and similar growth patterns for diameters and lengths. However, no experimental information on linear airway dimensions for other distal bronchial airways was available when this model was developed. Since limited data for the terminal bronchioles (Hislop *et al.,* 1972) suggested a functional relationship with age for peripheral bronchial air passages similar to that for the main bronchi, all bronchial generations from $i = 2$ through $i = 16$ use the same, scaled-down equations suitable for the main bronchi.

Recently, diameters and lengths for all maturing TB airways in the right upper lobe of 20 individuals age 11 days to 21 years were measured by Phalen *et al.* (1985). These measurements were applied to the ordering scheme of Yeh and Schum (1980) in which the TB region is described by 16 airway generations (generations $i = 0$ to 15). Lengths and diameters of each generation's airways were expressed as linear functions of height, which were converted here to functions of age by interpolation from tables presented by Altman and Dittmer (1972). It is interesting to note that the growth of bronchial airway dimensions in the Phalen *et al.* (1985) geometry decreases monotonically with increasing generation number, resulting in almost constant diameters and lengths in terminal bronchioles regardless of age. Several inherent inconsistencies encountered in the compilation of an age-dependent human lung morphology, incorporating the latest Phalen *et al.* (1985) data with findings from other authors, have been discussed in more detail by Martonen *et al.* (1989). Inspection of Appendix Tables 26-A1 and 26-A2 shows that there are not only differences in the relative growth of bronchial airways between the two TB models, but also significant differences in the adult lung dimensions.

Age-dependent changes in gravity angles of conducting airways have been reported by Walker *et al.* (1987), illustrating a consistent pattern for all types of airways. Gravity angles affect deposition by sedimentation, but not by diffusion. Since deposition by diffusion is characteristic of submicron radon progeny, these observations have not been incorporated into either of the proposed lung morphologies. No information about age-dependent changes of branching angles could be found in the literature.

B. Pulmonary Airways

While our morphometric knowledge of the tracheobronchial region needs some improvement, it is completely inadequate for the pulmonary region. Dimensions of pulmonary airways as functions of age were derived from Dunnill's (1962) data for total respiratory and alveolar duct volumes, and number of respiratory airways in a given generation, assuming Weibel's adult pulmonary structure of 7 dichotomously branching generations. To determine the number of pulmonary airway generations at a specific age, it is assumed that there are 2^{17} airways in the first pulmonary generation, and that the number of airways per generation is constant with age. The total number of pulmonary airway generations was chosen so that the total number of airways agrees with that proposed by Dunnill (1962). The number of pulmonary generations obtained in this manner ranges from five at age 7 months to seven at the age 20 years (i.e., the adult). The diameters and lengths in each generation were then calculated by assuming that the length-to-diameter ratio in any given generation was independent of age, and that alveolar and alveolar duct volumes were apportioned among individual generations in the same relative proportion as the first pulmonary generations in Weibel's Model A. In this manner, airway models for the pulmonary region were developed for ages 7 months, 22 months, 48 months, 98 months, and the adult. At these ages the assumption of 2^i total pulmonary airways was consistent with Dunnill's data for i being an integer number. This age-dependent pulmonary airway model is used in this manuscript with both tracheobronchial models for aerosol deposition simulations.

C. Respiration Parameters

In addition to morphological changes in the developing lung, tidal volume, V_T, and respiratory frequency, f, also vary with age. While f decreases with rising age, V_T increases during lung development, though at a much higher rate. This leads to a higher respiratory minute volume, $\dot{V}_E (\dot{V}_E = V_T \times f)$, in the adult than in the child. While a number of studies have related breathing parameters to different physical activities in the adult (ICRP, 1975), there is little information relating the effects of activity on lung functions of children. For infants, almost all published measurements seem to relate to the sedentary state.

Two sets of breathing parameters, representing extreme breathing conditions, sedentary and maximal activity (Table 26-2), were used for our computer simulations to explore the effect of breathing patterns on aerosol particle deposition in the maturing lung. For sedentary activity or rest, tidal volumes and respiratory frequencies were calculated from equations proposed by Hofmann (1982b). For breathing during maximal activity, tidal volumes and frequencies were taken from ICRP Report No. 23 (1975) for ages 48 months, 98 months, and adult. For ages 7 and 22 months, tidal volumes were derived from Dunnill's total alveolar volumes, and respiratory frequencies were assumed to approach the limit of 70 breaths/min for infants (Martonen *et al.*, 1989).

III. AEROSOL DEPOSITION IN THE DEVELOPING HUMAN LUNG

The deposition of inhaled aerosol particles of specified size within the human respiratory tract depends on airway dimensions and breathing pattern, both being functions of age (see Table 26-2, and Appendix Tables

Table 26-2 Age-Dependent Respiratory Parameters for Sedentary and Maximal Breathing

Age		Sedentary Activity[a]		Maximal Activity[b]	
(yr)	(mo)	Frequency (breaths/min)	Tidal volume (ml)	Frequency (breaths/min)	Tidal volume (ml)
0.58	7	35	42	70	165
1.83	22	28	84	70	328
4.00	48	22	152	68	570
8.17	98	18	266	60	1190
20.00	240	14	500	40	3050

[a] The values for sedentary activity are calculated from equations proposed by Hofmann (1982b).

[b] The values for maximal activity are based upon values reported in ICRP Report Number 23 (ICRP, 1975).

26-A1–26-A3). The removal of particles flowing through straight tubes is mainly due to three physical mechanisms: Brownian diffusion, inertial impaction, and gravitational settling. While deposition by diffusion is computed according to Ingham (1975), equations proposed by Martonen (1982) and Martonen *et al.* (1985) have been used for impaction and sedimentation deposition.

Deposition within the human airway network is calculated for a complete breathing cycle, consisting of equal inspiration and expiration times with no pause, and assuming constant flow rates. For small particle sizes, i.e., below 0.5 μm, deposition fractions are plotted versus the geometric diameter, D_g, while for larger particles deposition is expressed as a function of the aerodynamic diameter, D_{ae}.

Since the majority of lung tumors in uranium miners exposed to high levels of radon progeny have been observed in upper bronchial airways, which is consistent with a predicted dose maximum at such locations, all deposition calculations performed here shall refer exclusively to the tracheobronchial region. In Fig. 26-1 to 26-5, therefore, tracheobronchial deposition is shown as a function of age for the two breathing patterns specified in Table 26-2. Corresponding data for pulmonary and total deposition are presented in Martonen *et al.* (1989). The theoretical deposition model used in the present calculations has been validated by comparisons of predicted total and regional deposition patterns with experimental data from inhalation exposure tests with adult human test subjects, using aerosol particle diameters ranging over almost two orders of magnitude, i.e., from 0.2 to 9 μm (Martonen, 1983a,b; Martonen and Graham, 1987a,b). Deposition fractions represent aerosol mass fractions deposited within the tracheobronchial compartment during a complete breathing cycle, normalized to the mass penetrating the extrathoracic region and entering the trachea.

The effect of age on tracheobronchial deposition for sedentary and maximal activity as a function of particle size is illustrated in Fig. 26-1 and 26-2, using the Hofmann (1982b) tracheobronchial morphology, denoted as Model 1 (Hofmann). For both breathing conditions, computations show an age dependency effect on TB deposition within the maturing lung. As the lung develops, the aerosol mass collected within tracheobronchial airways decreases for any particular particle size. This difference between infant and adult is even more significant if deposition is expressed per unit surface area, which is the relevant measure of deposition for radon progeny dosimetry. The total TB surface area of a 7-month-old is by a factor 6.8 smaller than that of an adult. For inhaled radon progeny attached to submicron carrier aerosols, diffusion is the most effective collection mechanism. TB deposition for the smallest, 0.2-μm, particles tested is increased for the 7-month-old by about a factor of 2 compared to the adult. This is in

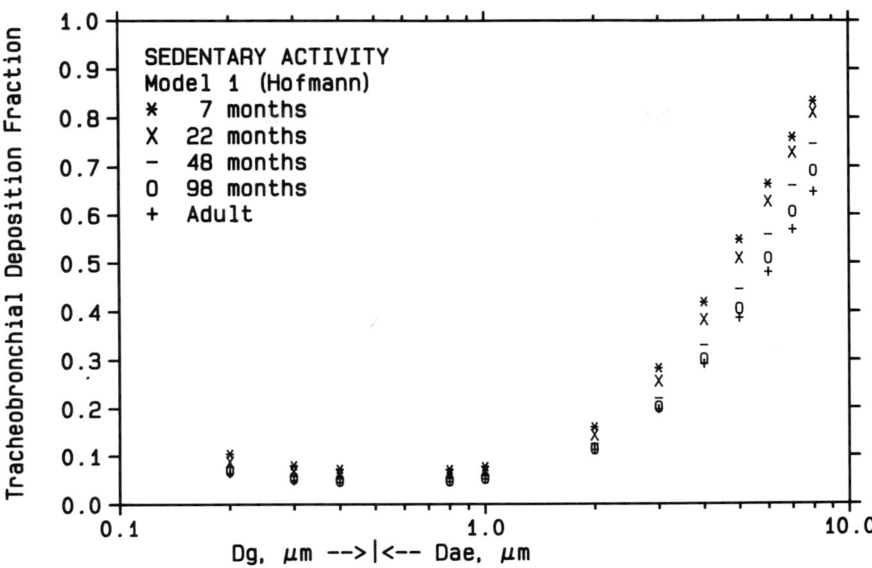

Fig. 26-1 Tracheobronchial deposition within the developing human lung for the sedentary breathing state, using the Model 1 (Hofmann) morphology.

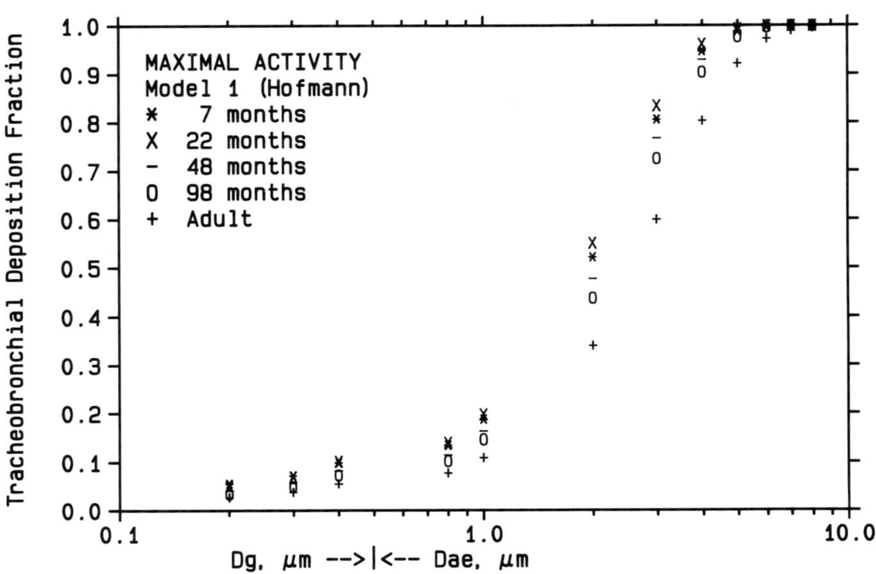

Fig. 26-2 Tracheobronchial deposition within the developing human lung for the maximal breathing state, using the Model 1 (Hofmann) morphology.

accordance with increased probabilities of deposition by diffusion as a result of reduced linear airway dimensions which offset the effect of slightly reduced particle residence times within airways. The effect of respiratory activity upon inhaled submicron particle deposition can be observed by comparing Fig. 26-1 and 26-2. TB deposition for 0.2-μm particles for maximal activity is about half of that for sedentary activity. This is the result of reduced particle residence times in bronchial air passages at higher inspiratory flow rates, and subsequently reduced probabilities of deposition by diffusion. The total number of particles, however, inhaled per minute under maximal breathing conditions is 8 to 17 times higher than that for sedentary activity, depending on the age considered (see Table 26-1).

Comparisons of tracheobronchial deposition as a function of age using the age-dependent tracheobronchial morphology proposed by Phalen *et al.* (1985), which is denoted as Model 2 (Phalen *et al.*), were made and showed similar results. The effect of the two tracheobronchial morphology models on particle deposition is illustrated in Fig. 26-3 to 26-5 for three selected ages. Deposition for sedentary activity is consistently slightly less in Model 2 than in Model 1 for all selected ages. Deposition for maximal breathing, however, shows an interesting behavior: for the 7-month-old (Fig. 26-3), Model 1 predicts higher deposition fractions than Model 2 for all particle sizes; for the 48-month-old (Fig. 26-4) the differences between the two

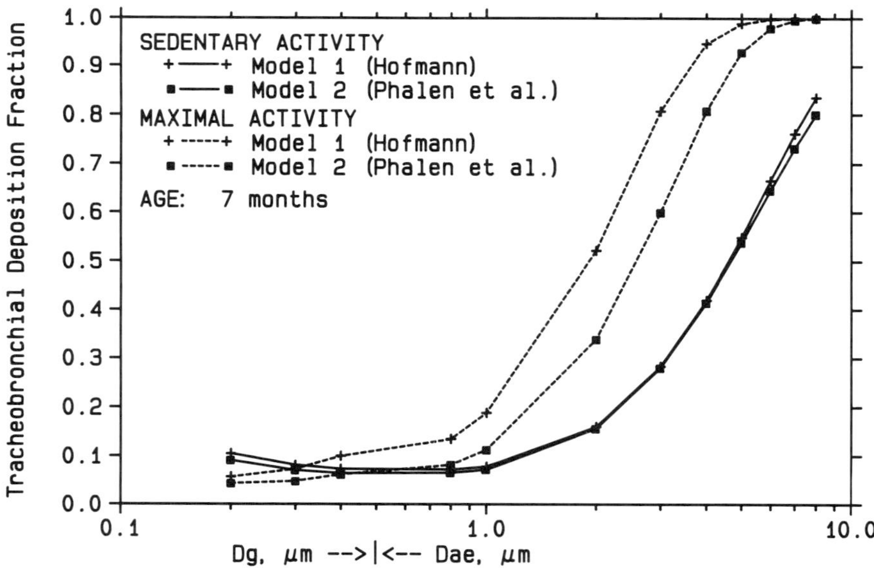

Fig. 26-3 Comparison of tracheobronchial deposition within two morphological descriptions of a 7-month-old human lung for conditions of sedentary and maximal breathing.

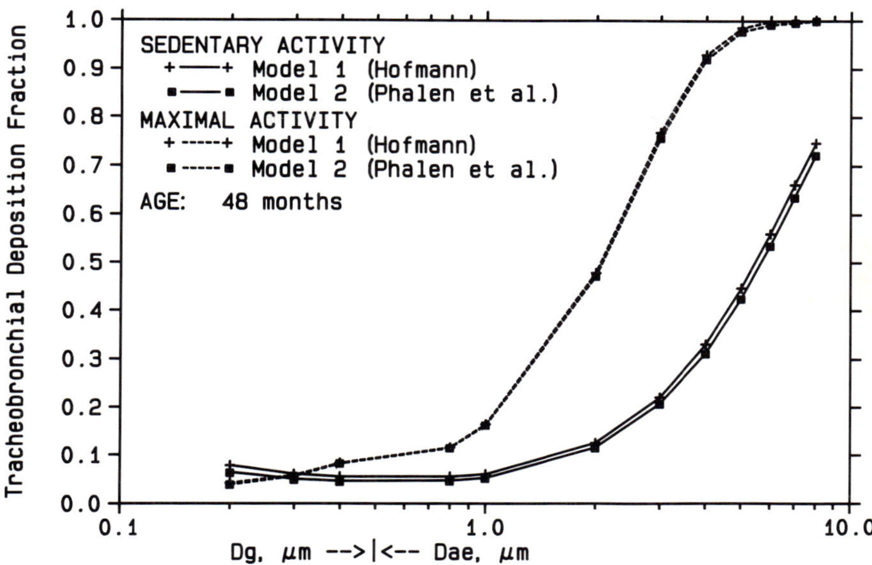

Fig. 26-4 Comparison of tracheobronchial deposition within two morphological descriptions of a 48-month-old human lung for conditions of sedentary and maximal breathing.

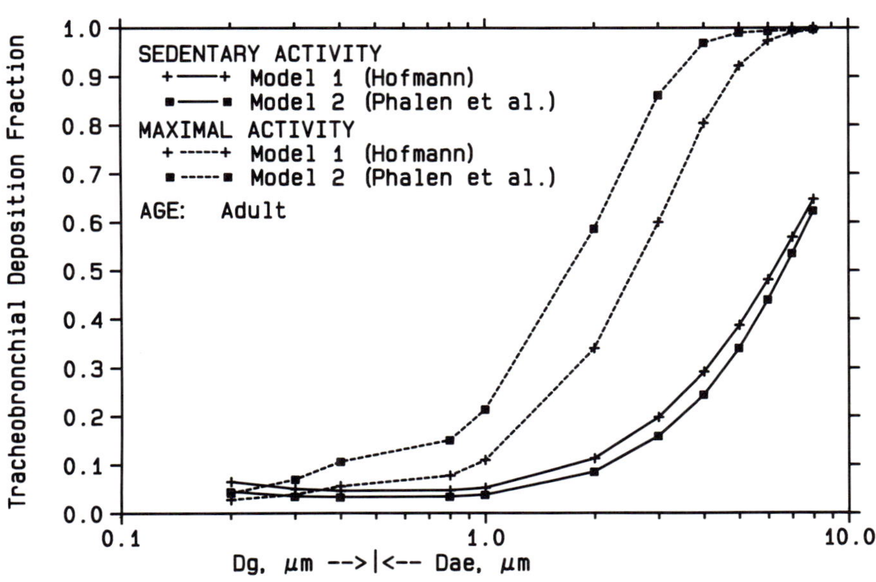

Fig. 26-5 Comparison of tracheobronchial deposition within two morphological descriptions of an adult human lung for conditions of sedentary and maximal breathing.

models are negligible; and for the adult (Fig. 26-5) deposition is consistently greater in Model 2. Computations were also made for 22- and 98-month-old subjects, and the results were intermediate with Fig. 26-3 and 26-4, and 26-4 and 26-5, respectively. In conclusion, tracheobronchial deposition under sedentary breathing conditions is always slightly higher using the Model 1 morphology compared to Model 2. Under maximal breathing, Model 2 systematically predicts increasing deposition as a function of age relative to the deposition predictions using the Model 1 morphology. It should be noted that the observed differences in total tracheobronchial deposition between both morphologies are less significant in upper bronchial airway generations because differences in the assumed airway growth pattern are greatest in the terminal bronchioles.

Similar results have been obtained for pulmonary and total deposition (Martonen *et al.,* 1989). Inspection of Fig. 26-3 to 26-5 also shows that the effects of the change in respiratory breathing patterns from sedentary to maximal activity are qualitatively similar in both morphological models.

IV. RADON PROGENY LUNG DOSIMETRY

For the dosimetry of inhaled short-lived radon progeny in bronchial airways, two additional factors must be considered subsequent to initial deposition. First, mucociliary clearance competes with radioactive decay, since radioactive half-times of the short-lived radon progeny ^{218}Po (3.05 min), ^{214}Pb (26.8 min), and ^{214}Bi (19.7 min; with ^{214}Po in radioactive equilibrium) are comparable with mucus clearance half-times in a given airway generation. Thus, the surface activity distribution of alpha-emitting radionuclides within the bronchial compartment will not be identical to the initial deposition pattern. Second, the radiation dose deposited by alpha particles (^{218}Po: 6.00 MeV; ^{214}Po: 7.69 MeV) in the surrounding tissue decreases with depth in bronchial epithelium. Consequently, the dose in sensitive target cells depends on the location of these cells within the depth-dose distribution. Because of a decrease of the bronchial epithelial thickness with depth in the bronchial tree, the cellular dose distribution within the various generations of the bronchial region is again different from the surface activity distribution.

A. Mucociliary Clearance

The few reported measurements of mucus clearance velocities in human bronchial air passages have been conducted exclusively in the trachea and the main bronchi of adult test subjects (Foster *et al.,* 1982; Yeates *et al.,* 1975, 1982). Extrapolation to peripheral airways in this manuscript is based on analyses of particle deposition and clearance studies reported by Stahlhofen *et al.* (1980). The resulting estimates of effective particle clearance velocities, V_k, are in Table 26-3. Clearance rates, λ_k, for both TB morphologies are obtained from the clearance velocities, V_k, and the airway lengths, L_k, in the respective morphometric models by $\lambda_k = \ln2 \times V_k/L_k$.

Based on theoretical considerations (Altshuler *et al.*, 1964) of mucus transit times in bronchial airways, Hofmann (1982a) derived two scenarios for potential alterations of mucus clearance times with progressing age: (1) clearance rates remain constant throughout infancy and childhood (columns 3 and 4 in Table 26-3); and (2) clearance rates are higher in the immature lung because of a thinner mucus layer and increased mucus production per unit surface area. The latter criterion is consistent with the assumption that clearance velocities are independent of age (column 2 in Table 26-3). Crawford-Brown (1982) arrived at the same conclusion, though his judgment was based on different assumptions about mucus production and absorption.

Recent measurements of tracheal mucociliary clearance velocities in beagle dogs of different ages suggest, however, reduced clearance rates at

Table 26-3 Effective Particle Clearance Velocities and Clearance Rates for All Tracheobronchial Airway Generations of the Adult Human Lung

Generation k	Clearance Velocities, V_k (mm/min)	Clearance Rates, λ_k (hr^{-1})	
		Weibel (1963)	Phalen *et al.* (1985)
0	5.5	1.906	2.574
1	4.0	3.495	4.355
2	2.0	4.378	5.658
3	1.3	7.109	5.147
4	1.0	3.275	5.154
5	0.9	3.498	5.422
6	0.7	3.234	5.866
7	0.6	3.282	6.931
8	0.4	2.599	5.697
9	0.3	2.311	3.713
10	0.2	1.808	2.829
11	0.1	1.066	1.535
12	0.04	0.504	0.671
13	0.024 (0.02)[a]	0.370	0.368
14	0.009 (0.005)[a]	0.163	0.102
15	0.0037 (0.001)[a]	0.077	0.023
16	0.001 (—)[a]	0.025	—

[a] Numbers in parentheses refer to the Yeh and Schum (1980) morphology.

very young ages (Whaley *et al.*, 1987). This pattern of age-related changes seems to be similar to age-related changes reported for certain pulmonary function measurements. Since Felicetti *et al.* (1981) have shown that canine tracheal mucociliary clearance patterns are very similar to those of humans, their exponential growth function may be used to describe age-dependent changes in the human trachea, normalized to adult tracheal velocities. In the absence of any information on age-dependent changes in bronchial generations other than the trachea, the same functional relationship is assumed for all conductive airways. In this paper, we will consistently use the data of Whaley *et al.* (1987) to simulate age-dependent changes of mucociliary clearance rates. Only in Fig. 26-13 will we compare these three different assumptions and how they affect the dosimetry of radon progeny in bronchial epithelium.

Radon progeny (^{218}Po: $n = 1$, ^{214}Pb: $n = 2$, ^{214}Bi/^{214}Po: $n = 3$) deposited on bronchial airway generations k (with deposition fractions D_k^n) may be cleared from the initial deposition site by mucociliary action (with clearance rate λ_k) or diffusion through epithelial tissue (with clearance rate λ_b), or may decay on-site into their respective daughter products (with radioactive decay constants λ_r^n). The interplay between the various processes involved in a given bronchial airway generation k, which determine the retention pattern of the inhaled activity, is schematically illustrated in Fig. 26-6. The dynamic equilibrium conditions occurring in any bronchial generation can be described mathematically by a system of first-order differential equations (Hofmann, 1982a) or, in the case of chronic exposure (as in this chapter), approximated by their steady state solutions (Hofmann and Daschil, 1986). The resulting surface activities, i.e., the activities per unit surface area for a given generation, of the alpha-emitting progeny ^{218}Po and ^{214}Po finally determine the number of alpha particles emitted per unit time and, consequently, the dose in bronchial tissue. For subsequent dosimetry calculations, radon progeny are assumed to be uniformly distributed within the upper, viscous part of the mucus layer.

B. Alpha Particle Dosimetry in Bronchial Epithelium

The dose rate per unit surface activity for ^{218}Po and ^{214}Po alpha particles decreases in an almost linear fashion with depth in the bronchial epithelium (Fig. 26-7), dependent on alpha particle ranges in soft tissue. Harley and Pasternack (1972) have shown that the depth-dose distribution is virtually independent of airway diameter, the difference between trachea and terminal bronchioles being less than 10%. We, therefore, assume the same depth-dose distribution for all bronchial generations. In the absence of further information, the same distribution is also used for all ages, since

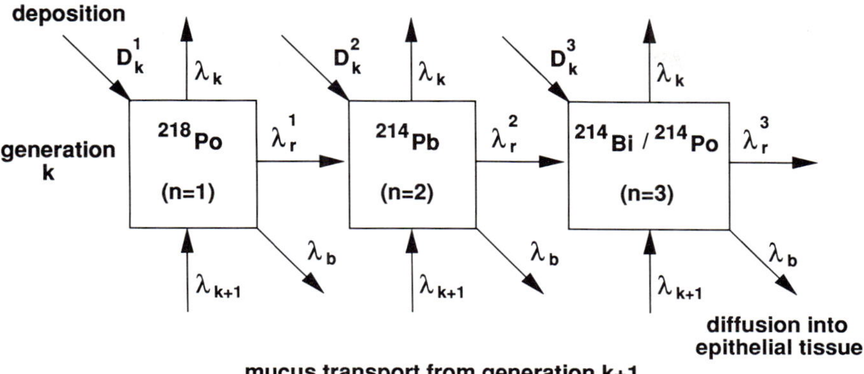

Fig. 26-6 Compartment model for deposition, clearance, and radioactive decay of inhaled radon progeny in bronchial airway generations. D_k^n is the deposition fraction of nuclide n in generation k, λ_r^n is the radioactive decay constant of nuclide n, λ_k is the mucociliary clearance rate in generation k, and λ_b is the clearance rate for diffusion into epithelial tissue.

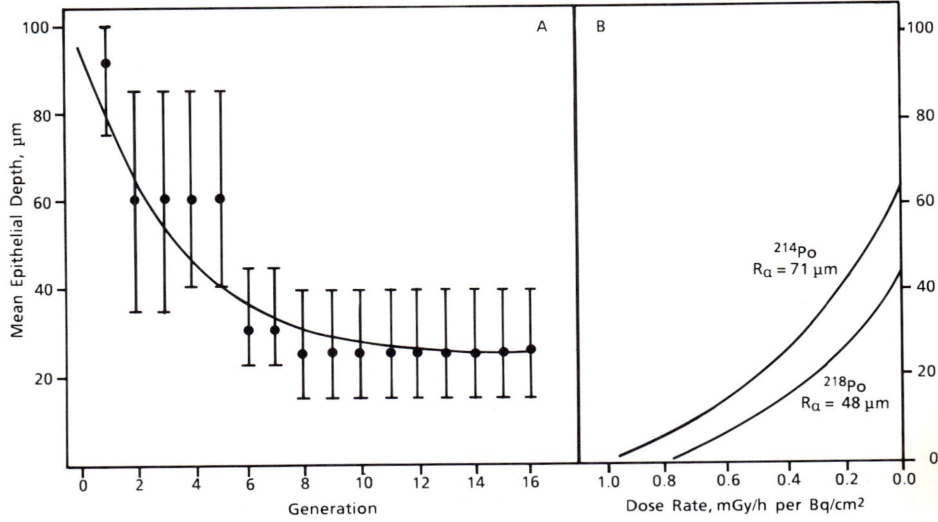

Fig. 26-7 Mean epithelial depth, i.e., thickness of epithelium and lining mucus layer, within the adult tracheobronchial region (A) in relation to the depth-dose rate distribution per unit surface activity of ^{218}Po and ^{214}Po nuclides (B).

age-dependent changes of diameters within a given generation are small compared to differences between various generations.

Though the thickness of the bronchial epithelium in different bronchial generations has been identified as the most sensitive parameter in dose calculations (Hofmann and Daschil, 1986), the only direct measurements have been made by Gastineau *et al.* (1972) from a rather small sample of surgical specimens of human bronchi, presumably from adult lungs. These authors found that bronchial epithelial thickness is highly variable within a given class of bronchioles and that it decreases significantly in peripheral airways. We have used this limited set of data to define mean basal cell depths for all bronchial generations (Hofmann, 1982a), and also mean epithelial depths, including the thickness of the covering mucus layer (Fig. 26-7). Crawford-Brown (1982) has suggested that the thickness of the bronchial epithelium is proportional (roughly a factor of 0.01) to the diameter of the corresponding airways. Since the thickness of the bronchial epithelium as a function of age is unknown, we assume the same relationship for age-dependent changes within a given generation, constrained, however, by a minimum average thickness of 15 μm, which is the median value for terminal bronchioles reported by Gastineau *et al.* (1972).

The basal cells of the bronchial epithelium have long been regarded as the sole critical target cells for lung cancer induction following exposure to radon progeny (Altshuler *et al.*, 1964; Jacobi, 1964; ICRP, 1980). Consequently, nearly all previously published dose calculations refer to basal cell doses (Crawford-Brown, 1982; Harley and Pasternack, 1972, 1982; Hofmann, 1982a; Hofmann and Daschil, 1986; ICRP, 1981; Jacobi and Eisfeld, 1980; James, 1987; OECD, 1983; NCRP, 1984). Recently, however, another cell type of the bronchial epithelium, namely the secretory (McDowell and Trump, 1983) or small mucus granule cells (SMGC) (Trump *et al.*, 1978) (denoted also as serous cells in Jeffery and Reid, 1977), have been identified as the main progenitor cells for bronchial carcinomas (McDowell *et al.*, 1978; McDowell and Trump, 1983; Masse, 1984). Since McDowell *et al.* (1978) have shown that 88% of all lung carcinomas in humans arise from basal and/or secretory cells, without any reliable, quantitative information on their relative contributions, we computed alpha radiation doses to both target cells.

Because of the limited range of alpha particles in soft tissue (^{218}Po: 48 μm; ^{214}Po: 71 μm), the location of cells relative to the depth-dose distribution determines the amount of energy they receive. Basal cells line the basal membrane of the bronchial epithelium (Jeffery and Reid, 1977), and thus are situated most distantly to the radiation sources on the mucus layer. It is interesting to note that average basal cell depths in the trachea and main bronchi are larger than alpha particle ranges (see Fig. 26-6). Doses to basal cell nuclei are calculated by applying generation-specific average basal cell depths (Hofmann, 1982a; Hofmann and Daschil, 1986). Nuclei of secretory cells (SMGC), however, are likely to be uniformly distributed throughout the depth of the epithelium (Jeffery and Reid, 1977). Identification of secretory cells as sensitive cells also supports the argument by James (1987) that the average bronchial epithelial dose, absorbed by all cells of the bronchial epithelium, is the relevant quantity for lung cancer risk assessment. Doses to secretory cell nuclei, or average epithelial cell doses, in a given generation are calculated by integrating the depth-dose distribution over the entire thickness of the bronchial epithelium in that generation.

C. Age-Dependent Cellular Doses

Ambient radon progeny attached to environmental aerosols have characteristic lognormal distributions with activity median diameters (AMDs) between 0.1 and 0.3 μm, and geometric standard deviations (GSDs) between 2 and 3 (George *et al.,* 1975; George and Breslin, 1980). In the present calculations for a normally ventilated indoor atmosphere (OECD, 1983; Hofmann and Daschil, 1986), a size distribution characterized by an AMD of 0.15 μm and a GSD of 2.5 is assumed with unattached fractions of 10% for ^{218}Po atoms, 1% for ^{214}Pb atoms, and a decay product ratio of $^{222}\text{Rn}:^{218}\text{Po}:^{214}\text{Pb}:^{214}\text{Bi} = 1:0.9:0.6:0.4$ (Hofmann *et al.,* 1979). This reference atmosphere is used for all calculations in this manuscript.

The effect of the subject's age on bronchial basal and secretory cell doses under sedentary and maximal breathing conditions is illustrated in Fig. 26-8–26-11. The basal and secretory cell doses plotted in these figures represent "mean" doses for the whole tracheobronchial region defined by averaging all cellular doses in singular generations with equal weight, regardless of the number of cells in each generation. This procedure is conceptually equivalent with a protocol that cells in all bronchial generations contribute to lung cancer risk (ICRP, 1981).

Some interesting features can be observed in these figures:

1. In general, cellular radiation doses decrease with progressing age, regardless of lung morphology, breathing mode, and target cell. Only in the case of Model 2 (Phalen *et al.*) and maximal activity breathing, bronchial cellular doses in the adult are approximately the same as those in the 7-month-old, exhibiting a maximum at the age of 98 months. The main reason for this deviation from the generally observed pattern is the finding that TB deposition of inhaled radon progeny aerosols under the above conditions is virtually independent of subject age.

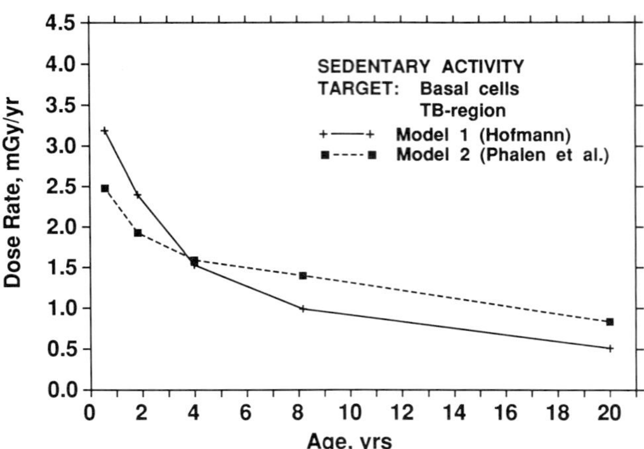

Fig. 26-8 Comparison of mean basal cell doses for the TB region of the two morphometric models under conditions of sedentary breathing as a function of age.

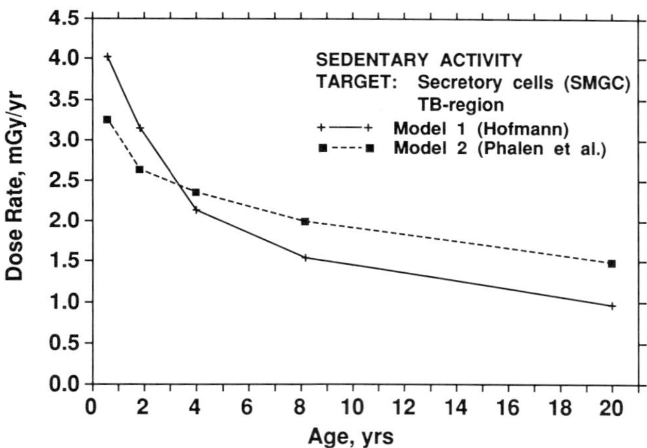

Fig. 26-9 Comparison of mean secretory cell doses, or mean epithelial cell doses, for the TB region of the two morphometric models under conditions of sedentary breathing as a function of age.

2. Age dependency is less significant for secretory cells than for basal cells, for both respiratory patterns and morphology descriptions simulated. This illustrates the effect of age-dependent variations of target cell locations relative to the depth-dose profile (Fig. 26-7) as a result of corresponding variations of the bronchial epithelium thickness.

3. The effect of human subject age is greater for sedentary breathing than for maximal activity, for both target cells and morphology models. This

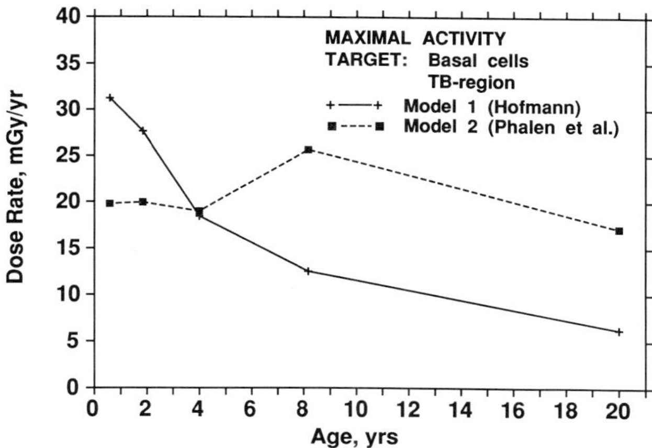

Fig. 26-10 Comparison of mean basal cell doses for the TB region of the two morphometric models under conditions of maximal breathing as a function of age.

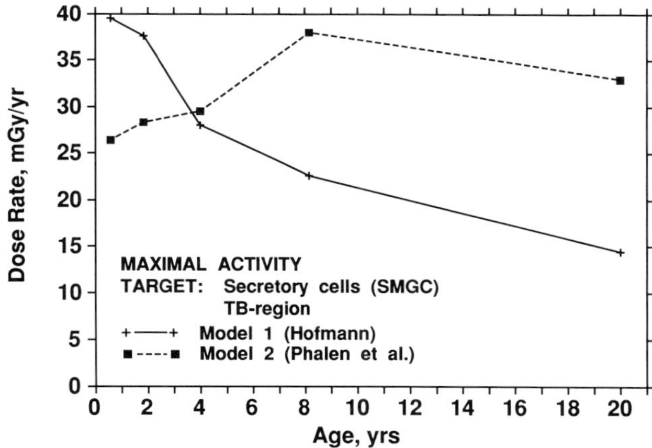

Fig. 26-11 Comparison of mean secretory cell doses, or mean epithelial cell doses, for the TB region of the two morphometric models under conditions of maximal breathing as a function of age.

is caused by differences in the deposition pattern of lognormally distributed submicron aerosols in the developing lung.

The comparison between the two morphology descriptions used reveals that Model 1 consistently predicts higher doses than Model 2 below the age of about 4 to 5 years, but smaller doses at higher ages, regardless of breathing parameters and target cells. This results from differences in the assumed growth patterns of bronchial airways in both tracheobronchial morphologies. The TB volume in the Model 1 morphology is smaller at birth, but larger in the adult (Ménache and Graham, Chapter 22, this volume), which affects deposition fractions, surface areas, clearance rates, and location of target cells. While the shape of the dose-versus-age curve is only moderately sensitive to the choice of the target cell, it is strongly affected by the breathing pattern considered. For a sedentary breathing mode, differences between the two model predictions are not very significant. For maximal activity, however, the effect of age is substantially reduced in Model 2 relative to Model 1, where doses even increase at intermediate ages.

While Figure 26-8–26-11 refer to mean bronchial doses, Fig. 26-12 presents corresponding information on the age dependency of secretory cell doses in airway generation 4, selected to represent bronchial generations with preferential tumor occurrence (Hofmann, 1982; Hofmann and Daschil, 1986). The choice of generation 4 is compatible with an observation that the preference of lung tumors in upper bronchial airways may be correlated with the predicted dose maximum in these airways and that local dose to a generation, not average dose throughout the TB region, is the relevant dosimetric index of lung cancer risk. A comparison with the data of Fig. 26-9 shows that there is virtually no difference between the shapes of the two curves. Qualitatively similar results have also been found for the modeling parameters used for Fig. 26-8, 26-10, and 26-11, suggesting that the effect of age dependency is similar for local and average bronchial doses.

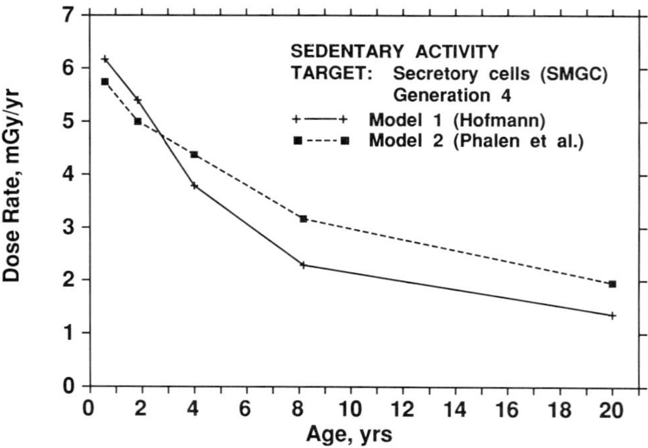

Fig. 26-12 Comparison of secretory cell doses, or epithelial cell doses, in generation 4 of the two morphometric models under conditions of sedentary breathing as a function of age.

Model 1 is used in Fig. 26-13 to treat the effect of different assumptions regarding clearance velocities in children:

1. Clearance velocities increase exponentially with age (Whaley *et al.*, 1987);
2. Clearance velocities are proportional to airway lengths as functions of age, which is equivalent to constant clearance rates; and
3. Clearance velocities are independent of age.

While all computed results presented in Fig. 26-8–26-12 are based on clearance assumption 1, assumptions 2 and 3 have been used and discussed in previous dose calculations (Crawford-Brown, 1982; Harley and Pasternack, 1982; Hofmann, 1982a; James, 1987). The data of Fig. 26-12 demonstrate that assumptions on mucociliary clearance velocities in infants and children significantly affect lung dose calculations, particularly during the early years of life. In view of the scarcity of available information, this comparison further emphasizes the need for more experimental data on clearance velocities within the maturing human lung. Qualitatively similar results have been obtained using the Model 2 morphology, and for other breathing patterns and target cells, but for the sake of brevity these will not be given here.

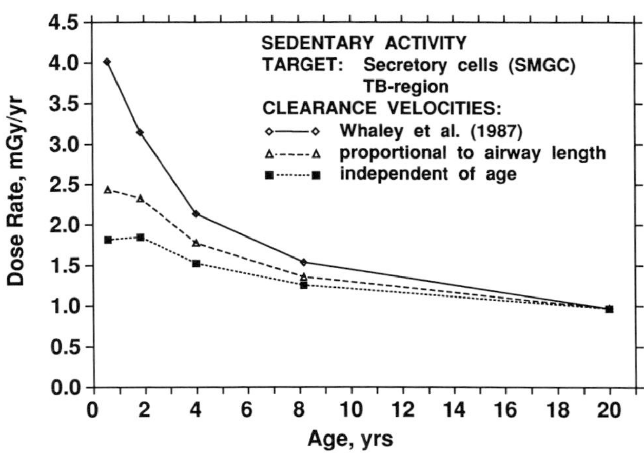

Fig. 26-13 Mean secretory cell doses, or mean epithelial cell doses, for the TB region of the developing lung described by Model 1 (Hofmann) for sedentary activity, comparing three different assumptions of mucociliary clearance changes with age: clearance velocities are exponential functions of age (Whaley *et al.*, 1987); clearance velocities are proportional to airway length at a given age; and clearance velocities are independent of age.

V. CONCLUSIONS

The age-dependent lung dosimetry models for inhaled radon progeny presented in this paper predict that:

1. Deposition fractions of inhaled radon progeny are greater in children than in adults, particularly if normalized to unit surface area.
2. Mucociliary clearance in bronchial airways of children may be less effective or more efficient than those of adults, dependent on modeling assumptions.
3. Target cells in the bronchial epithelium of children are closer to alpha-emitting radiation sources on inner airway surfaces than those in adults.

As a result of these model predictions, radiation doses in the bronchial epithelium of children are generally believed to be higher than those for adults; the extent of this increase depends on assumptions regarding lung morphology, breathing parameters, clearance velocities, and salient target cells for cancer induction. Secretory cell doses in Model 2 under maximal breathing conditions are a noteworthy exception. The effect of age dependency is greatest for a Model 1 morphology, sedentary breathing conditions, the clearance velocities reported by Whaley *et al.* (1987), and basal cells.

The comparisons between both tracheobronchial morphologies employed in these simulations show that both models produce qualitatively similar dose-versus-age curves. Model 1, however, consistently predicts higher doses in the newborn, but smaller doses in the adult, with the transition occurring at an age of about 4 to 5 years. It is interesting to note that differences between both models are greatest in the adult. Particularly for sedentary breathing, which is representative for domestic radon exposure, the differences between the two descriptions of maturing lung morphology are rather insignificant. Thus, despite the apparent variations in the growth pattern assumed in both models, lung morphology of the developing lung does not seem to be a very sensitive parameter in radon progeny lung dosimetry, particularly if biological intersubject variability is taken into account.

APPENDIX

Table 26-A1 Diameters and Lengths of TB Airways in the Model 1 (Hofmann, 1982) Morphology for Different Ages[a]

	7 Months		22 Months		48 Months		98 Months		Adult	
Generation	Diameter	Length	Diameter	Length	Diameter	Length	Diameter	Length	Diameter	Length
0	0.647	4.388	0.757	5.329	0.927	6.713	1.189	8.659	1.799	12.000
1	0.466	1.851	0.575	2.211	0.729	2.740	0.934	3.484	1.227	4.760
2	0.310	0.739	0.380	0.882	0.486	1.094	0.620	1.391	0.816	1.900
3	0.214	0.296	0.260	0.353	0.336	0.437	0.428	0.556	0.564	0.760
4	0.177	0.494	0.217	0.590	0.276	0.731	0.355	0.930	0.466	1.270
5	0.133	0.416	0.165	0.497	0.208	0.616	0.267	0.783	0.350	1.070
6	0.106	0.350	0.131	0.418	0.166	0.518	0.213	0.659	0.279	0.900
7	0.085	0.296	0.105	0.353	0.133	0.437	0.171	0.556	0.225	0.760
8	0.071	0.249	0.087	0.297	0.111	0.368	0.142	0.468	0.186	0.640
9	0.059	0.210	0.073	0.251	0.092	0.311	0.117	0.395	0.154	0.540
10	0.049	0.179	0.061	0.214	0.077	0.265	0.099	0.337	0.130	0.460
11	0.042	0.152	0.051	0.181	0.065	0.224	0.084	0.285	0.110	0.390
12	0.037	0.128	0.045	0.153	0.057	0.190	0.073	0.242	0.096	0.330
13	0.032	0.105	0.040	0.125	0.050	0.155	0.064	0.198	0.085	0.270
14	0.028	0.089	0.035	0.107	0.044	0.132	0.057	0.168	0.074	0.230
15	0.025	0.078	0.030	0.093	0.039	0.115	0.050	0.146	0.065	0.200
16	0.022	0.064	0.028	0.077	0.035	0.095	0.045	0.121	0.059	0.165

[a] All measurements in centimeters.

Table 26-A2 Diameters and Lengths of TB Airways in the Model 2 (Phalen *et al.*, 1985) Morphology for Different Ages[a]

	7 Months		22 Months		48 Months		98 Months		Adult	
Generation	Diameter	Length	Diameter	Length	Diameter	Length	Diameter	Length	Diameter	Length
0	0.715	3.390	0.901	4.310	1.105	5.330	1.351	6.560	1.815	8.890
1	0.535	1.840	0.663	2.170	0.804	2.540	0.972	2.990	1.294	3.820
2	0.396	0.699	0.481	0.828	0.575	0.971	0.689	1.140	0.901	1.470
3	0.307	0.467	0.368	0.565	0.436	0.673	0.517	0.803	0.670	1.050
4	0.210	0.422	0.241	0.487	0.276	0.559	0.318	0.645	0.397	0.807
5	0.161	0.327	0.187	0.388	0.216	0.455	0.251	0.536	0.316	0.690
6	0.132	0.254	0.154	0.295	0.178	0.340	0.208	0.394	0.264	0.496
7	0.095	0.217	0.108	0.241	0.122	0.268	0.139	0.300	0.172	0.360
8	0.076	0.204	0.083	0.219	0.091	0.235	0.101	0.255	0.120	0.292
9	0.070	0.220	0.076	0.240	0.082	0.261	0.090	0.287	0.104	0.336
10	0.063	0.204	0.067	0.219	0.071	0.236	0.076	0.256	0.085	0.294
11	0.054	0.194	0.056	0.207	0.058	0.221	0.061	0.239	0.066	0.271
12	0.053	0.184	0.054	0.195	0.056	0.207	0.059	0.221	0.063	0.248
13	0.049	0.174	0.051	0.183	0.052	0.192	0.053	0.204	0.056	0.226
14	0.047	0.164	0.048	0.171	0.049	0.178	0.049	0.187	0.051	0.203
15	0.044	0.156	0.048	0.160	0.045	0.165	0.046	0.170	0.048	0.181

[a] All measurements in centimeters.

Table 26-A3 Diameters and Lengths of P Airways for Different Ages[a,b]

Generation	7 Months		22 Months		48 Months		98 Months		Adult	
	Diameter	Length	Diameter	Length	Diameter	Length	Diameter	Length	Diameter	Length
17[c] (16)[d]	0.033	0.086	0.032	0.085	0.032	0.084	0.040	0.104	0.054	0.141
18 (17)	0.031	0.071	0.031	0.071	0.030	0.070	0.037	0.086	0.050	0.117
19 (18)	0.029	0.060	0.029	0.060	0.028	0.059	0.035	0.073	0.048	0.099
20 (19)	0.028	0.051	0.027	0.050	0.027	0.049	0.033	0.061	0.045	0.083
21 (20)	0.027	0.043	0.027	0.042	0.026	0.042	0.032	0.052	0.044	0.070
22 (21)	—	—	0.026	0.036	0.025	0.035	0.031	0.044	0.042	0.059
23 (22)	—	—	—	—	—	—	0.031	0.037	0.042	0.050

[a] Data used in connection with both TB models.
[b] All measurements in centimeters.
[c] Generation numbers refer to the Weibel (1963) terminology.
[d] Generation numbers refer to the Yeh and Schum (1980) terminology.

REFERENCES

Altman, P. L., and Dittmer, D. S. (1972). "Biology Data Book." Fed. of Am. Soc. for Exp. Biol., Bethesda, Maryland.

Altshuler, B., Nelson, N., and Kuschner, M. (1964). Estimation of lung tissue dose from the inhalation of radon and daughters. *Health Phys.* **10,** 1137–1161.

American Cancer Society (1981). "Cancer Facts and Figures." ACS, New York.

Beebe, G. W., Kato, H., and Land, C. E. (1978). Studies of the mortality of A-bomb survivors. 6. Mortality and radiation dose, 1950–1974. *Radiat. Res.* **75,** 138–201.

(BEIR) Advisory Committee on the Biological Effects of Ionizing Radiations (1972). "The Effects on Population of Exposure to Low Levels of Ionizing Radiation." National Academy of Sciences, Washington, D.C.

Crawford-Brown, D. J. (1982). Identifying critical human subpopulations by age groups: Radioactivity and the lung. *Phys. Med. Biol.* **27,** 539–552.

Crawford-Brown, D. J. (1983). On a theory of age dependence in the incidence of lung carcinomas following inhalation of a radioactive atmosphere. *In* "Current Concepts in Lung Dosimetry" (D. R. Fisher, ed.), pp. 178–188. Battelle Pacific Northwest Laboratories, Richland, Washington.

Dunnill, M. S. (1962). Postnatal growth of the lung. *Thorax* **17,** 329–333.

Felicetti, S. A., Wolff, R. K., and Muggenburg, B. A. (1981). Comparison of tracheal mucous transport in rats, guinea pigs, rabbits, and dogs. *J. Appl. Physiol.* **51,** 1612–1617.

Foster, W. M., Langenback, E. G., and Bergofsky, E. H. (1982). Lung mucociliary function in man: Interdependence of bronchial and tracheal mucus transport velocities with lung clearance in bronchial asthma and healthy subjects. *In* "Inhaled Particles V" (W. H. Walton, ed.), pp. 227–244. Pergamon, Oxford.

Gastineau, R. M., Walsh, P. J., and Underwood, N. (1972). Thickness of bronchial epithelium with relation to exposure to radon. *Health Phys.* **23,** 857–860.

George, A. C., and Breslin, A. J. (1980). The distribution of ambient radon and radon daughters in residential buildings in the New Jersey-New York area. *In* "The Natural Radiation Environment III" (T. F. Gesell and W. M. Lowder, eds.), pp. 1272–1292. National Technical Information Service, Springfield, Virginia.

George, A. C., Hinchliffe, L., and Sladowski, R. (1975). Size distribution of radon daughter particles in uranium mine atmospheres. *Am. Ind. Hyg. Assoc. J.* **34,** 484–490.

Harley, N. H., and Pasternack, B. S. (1972). Alpha absorption measurements applied to lung dose from radon daughters. *Health Phys.* **23,** 771–782.

Harley, N. H., and Pasternack, B. S. (1981). A model for predicting lung cancer risks induced by environmental levels of radon daughters. *Health Phys.* **40,** 307–316.

Harley, N. H., and Pasternack, B. S. (1982). Environmental radon daughter alpha dose factors in a five-lobed human lung. *Health Phys.* **42,** 789–799.

Hislop, A., Muir, D. C. F., Jacobson, M., Simon, G., and Reid, L. (1972). Postnatal growth and functions of pre-acinar airways. *Thorax* **27,** 265–274.

Hofmann, W. (1982a). Dose calculations for the respiratory tract from inhaled natural radioactive nuclides as a function of age—II. Basal cell dose distributions and associated lung cancer risk. *Health Phys.* **43,** 31–44.

Hofmann, W. (1982b). Mathematical model for the postnatal growth of the human lung. *Respir. Physiol.* **49,** 115–129.

Hofmann, W., and Daschil, F. (1986). Biological variability influencing lung dosimetry for inhaled ^{222}Rn and ^{220}Rn decay products. *Health Phys.* **50,** 345–367.

Hofmann, W., Steinhäusler, F., and Pohl, E. (1979). Dose calculations for the respiratory tract from inhaled natural radioactive nuclides as a function of age—I. Compartmental deposition, retention and resulting dose. *Health Phys.* **37,** 517–532.

Hofmann, W., Steinhäusler, F., and Pohl, E. (1980). Age-, sex- and weight-dependent dose distribution patterns for human organs and tissues due to inhalation of natural radioactive nuclides. *In* "The Natural Radiation Environment III" (T. F. Gesell and W. M. Lowder, eds.), pp. 1116–1143. National Technical Information Service, Springfield, Virginia.

Ingham, D. B. (1975). Diffusion of aerosols from a stream flowing through a cylindrical tube. *J. Aerosol Sci.* **6,** 125–132.

(ICRP) International Commission on Radiological Protection (1975). Report of the Task Group on Reference Man. ICRP Publication No. 23. Pergamon, Oxford.

(ICRP) International Commission on Radiological Protection (1980). Report of the Task Group on Biological Effects of Inhaled Radionuclides. *Ann. ICRP* **4** (1/2).

(ICRP) International Commission on Radiological Protection (1981). Limits for Inhalation of Radon Daughters by Workers. *Ann. ICRP* **6** (1).

Jacobi, W. (1964). The dose to the human respiratory tract by inhalation of short-lived ^{222}Rn- and ^{222}Rn-decay products. *Health Phys.* **10,** 1163–1174.

Jacobi, W., and Eisfeld, K. (1980). Dose to Tissue and Effective Dose Equivalent by Inhalation of Radon-222, Radon-220 and Their Short-Lived Daughters. GSF Report S-626. Gesellschaft für Strahlen und Umweltforschung, Munich.

James, A. C. (1987). A reconsideration of cells at risk and other key factors in radon daughter dosimetry. *In* "Radon and Its Decay Products: Occurrence, Properties, and Health Effects" (P. K. Hopke, ed.), pp. 400–418. American Chemical Society, Washington, D.C.

Jeffery, P. K., and Reid, L. M. (1977). The respiratory mucous membrane. *In* "Respiratory Defense Mechanisms" (J. D. Brain, D. F. Proctor, and L. M. Reid, eds.), Part 1, pp. 193–245. Dekker, New York.

McDowell, E. M., and Trump, B. F. (1983). Histogenesis of preneoplastic and neoplastic lesions in tracheobronchial epithelium. *Surv. Synth. Pathol. Res.* **2,** 235–279.

McDowell, E. M., McLaughlin, J. S., Kerenyi, D. K., Kieffer, R. F., Harris, C. C., and Trump, B. F. (1978). The respiratory epithelium. V. Histogenesis of lung carcinomas in the human. *J. Natl. Cancer Inst.* **61,** 587–606.

Martonen, T. B. (1982). Analytical model of hygroscopic particle behavior in human airways. *Bull. Math. Biol.* **44,** 425–442.

Martonen, T. B. (1983a). On the fate of inhaled particles in the human: A comparison of experimental data with theoretical computations based on a symmetric and asymmetric lung. *Bull. Math. Biol.* **45,** 409–424.

Martonen, T. B. (1983b). Deposition of inhaled particulate matter in the upper respiratory tract, larynx, and bronchial airways: A mathematical description. *J. Toxicol. Environ. Health* **12,** 787–800.

Martonen, T. B., and Graham, R. C. (1987a). The effect of hygroscopic growth upon the regional dispersion of particulates in human airways. *Int. Soc. Aerosols Med. Proc. Congr., 6th* pp. 221–234.

Martonen, T. B., and Graham, R. C. (1987b). Hygroscopic growth: Its effect on aerosol therapy and inhalation toxicology. *In* "Deposition and Clearance of Aerosols in the Human Respiratory Tract" (W. Hofmann, ed.), pp. 200–206. Facultas, Vienna.

Martonen, T. B., Barnett, A. E., and Miller, F. J. (1985). Ambient sulfate aerosol deposition in man: Modeling the influence of hygroscopicity. *Environ. Health Perspect.* **63,** 11–24.

Martonen, T. B., Graham, R. C., and Hofmann, W. (1989). Human subject age and activity level: Factors addressed in a biomathematical deposition program for extrapolation modeling. *Health Phys.* (in press).

Masse, R. (1984). Cells at risk. *In* "Lung Modelling for Inhalation of Radioactive Materials," (H. Smith and G. Gerber, eds.) pp. 227–243. Report EUR 9384 EN. Commission of the European Communities, Luxembourg.

(NCRP) National Council on Radiation Protection and Measurements (1984). "Evaluation of Occupational and Environmental Exposures to Radon and Radon Daughters in the United States." NCRP Report No. 78 National Council on Radiation Protection and Measurements, Bethesda, Maryland.

Ochsner, A., Ochsner, A., Jr., H'Doubler, C., and Blalock, J. (1960). Bronchogenic carcinoma. *Dis. Chest* **37,** 1–12.

(OECD) Organization for Economic Cooperation and Development, Nuclear Energy Agency (1983). "Dosimetry Aspects of Exposure to Radon and Thoron Daughter Products." OECD, Paris.

Phalen, R. F., Oldham, M. J., Beaucage, C. B., Crocker, T. T., and Mortensen, JD (1985). Postnatal enlargement of human tracheobronchial airways and implications for particle deposition. *Anat. Rec.* **212,** 368–380.

Pohl, E., Steinhäusler, F., Hofmann, W., and Pohl-Rüling, J. (1976). Methodology of measurements and statistical evaluation of radiation burden to various population groups from all internal and external natural sources. *In* "Biological and Environmental Effects of Low-Level Radiation," pp. 305–315. IAEA, Vienna.

Preston, D. L., Kato, H., Kopecky, K. J., and Fujita, S. (1987). Studies of the mortality of A-bomb survivors. 8. Cancer mortality, 1950–1982. *Radiat. Res.* **111,** 151–178.

Reid, L. (1977). The lung: Its growth and remodeling in health and disease. *Am. J. Roentgenol.* **129,** 777–788.

Saccomanno, G., Archer, V. E., Saunders, R. P., James, L. A., and Beckler, P. A. (1964). Lung cancer of uranium miners on the Colorado plateau. *Health Phys.* **10,** 1195–1201.

Sevc, J., Kunz, E., and Placek, V. (1976). Lung cancer in uranium miners and long term exposure to radon daughter products. *Health Phys.* **30,** 433–437.

Stahlhofen, W., Gebhart, J., and Heyder, J. (1980). Experimental determination of regional deposition of aerosol particles in the human respiratory tract. *Am. Ind. Hyg. Assoc. J.* **41,** 385–398.

Trump, B. F., McDowell, E. M., Glavin, F., Barrett, L. A., Becci, P. J., Schürch, W., Kaiser, H. E., and Harris, C. C. (1978). The respiratory epithelium. III. Histogenesis of epidermoid metaplasia and carcinoma *in situ* in the human. *J. Natl. Cancer Inst.* **61,** 563–575.

Wagoner, J. K., Archer, V. E., Lundin, F. E., Jr., Holaday, D. A., and Lloyd, J. W. (1965). Radiation as the cause of lung cancer among uranium miners. *N. Engl. J. Med.* **278,** 181–188.

Walker, J. T., Nelson, N. S., and Mortensen, JD (1987). Age dependent changes in gravity angles of conducting airways in the human lung. *In* "Deposition and Clearance of Aerosols in the Human Respiratory Tract" (W. Hofmann, ed.), pp. 129–134. Facultas, Vienna.

Weibel, E. R. (1963). "Morphometry of the Human Lung." Academic Press, New York.

Whaley, S. L., Muggenburg, B. A., Seiler, F. A., and Wolff, R. K. (1987). Effect of aging on tracheal mucociliary clearance in beagle dogs. *J. Appl. Physiol.* **62,** 1331–1334.

Whittemore, A. S., and McMillan, A. (1983). Lung cancer mortality among U.S. uranium miners: A reappraisal. *J. Natl. Cancer Inst.* **71,** 489–499.

Xu, G. B., and Yu, C. P. (1986). Effects of age on deposition of inhaled aerosols in the human lung. *Aerosol Sci. Technol.* **5,** 349–357.

Yeates, D. B., Aspin, N., Levison, H., Jones, M. T., and Bryan, A. C. (1975). Mucociliary tracheal transport rates in man. *J. Appl. Physiol.* **39,** 487–495.

Yeates, D. B., Gerrity, T. R., and Garrard, C. S. (1982). Characteristics of tracheobronchial deposition and clearance in man. *In* "Inhaled Particles V" (W. H. Walton, ed.), pp. 245–257. Pergamon, Oxford.

Yeh, H. C., and Schum, G. M. (1980). Models of human lung airways and their application to inhaled particle deposition. *Bull. Math. Biol.* **42,** 461–480.

Chapter 27

Deposition and Retention Modeling of Inhaled Cadmium in Rat and Human Lung: An Example for Extrapolation of Effects and Risk Estimation

G. Oberdörster
Environmental Health Sciences Center and Department of Biophysics
University of Rochester
Rochester, New York 14642

I. INTRODUCTION

A. Lung Dosimetry and Pulmonary Effects

Applying results of experimental inhalation studies to risk assessment is one goal of extrapolation modeling of effects of inhaled substances. This modeling involves both extrapolation from animal to man and from high (experimental) to low (e.g., environmental) exposures or to low doses resulting from these exposures. Respiratory tract dosimetry defines such doses and represents an important aspect in extrapolation modeling of effects. Unfortunately, there is no generally accepted definition of a dose to the lung. For the purpose of this paper, lung dosimetry doses are defined in Fig. 27-1, which outlines a simple scheme of the interrelation of doses and effects for inhaled substances and also depicts the principle of extrapolating from rat to humans. The exposure to an airborne particulate substance—often mislabeled as dose—results in an inhaled dose (e.g., mg of the

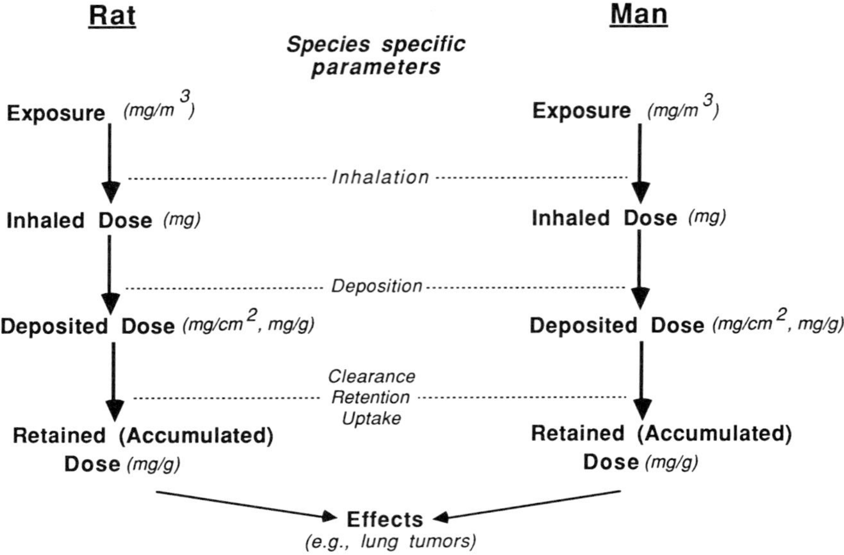

Fig. 27-1 Interrelation of doses and effects in inhalation studies and principle of extrapolation modeling of results from rat to humans. Exposure concentrations (mg/m^3) which lead to the same effects in rat and humans are different for the two species. An equivalent human exposure concentration can be derived from a known rat exposure concentration by using species-specific inhalation deposition and retention data for the inhaled compound in question.

substance inhaled) or an inhaled dose rate (mg inhaled per unit time). The inhaled dose (or dose rate) depends only on respiratory parameters (breathing minute volume) and exposure concentration, and is independent of particle characteristics [as long as all of the particles are inspirable, as defined by the Air Sampling Procedures Committee of the American Conference of Governmental and Industrial Hygienists (Phalen, 1984)], and independent of the anatomy of the airways. A fraction of the inhaled dose is deposited in different regions of the respiratory tract, and this depends on respiratory parameters (tidal volume, respiratory rate, respiratory pause), physicochemical particle characteristics (size, shape, density, hygroscopicity, exposure concentration), and airway geometry (length, diameter, branching angle, angle of gravity). The deposited dose can be expressed as mg per square cm epithelial surface area, mg per g lung tissue, or as dose rate. Due to physical and biochemical processes in the respiratory tract after deposition (mechanical clearance, solubilization, uptake into epithelial cells, interstitium, or systemic circulation), a retained or accumulated dose at a local or systemic target site will result from the deposited dose which eventually may cause effects. The retained or accumulated dose, which can be expressed as milligram per g target tissue, will change with time as a function of the retention half-time of the compound which depends on its metabolism in the respiratory tract. However, the accumulated dose will reach an equilibrium value under constant exposure conditions, provided the daily deposited dose and the kinetics of the compound are not altered during chronic exposure.

It appears that the length of exposure, i.e., the time factor, is important for the onset of effects, especially for chronic effects like lung cancer. By using the equilibrium value of the accumulated dose for describing a "dose–effect" relationship, the time factor becomes an integral part of this relationship since the value of the accumulated dose is determined by the pulmonary retention half-time of a compound: A tenfold longer retention half-time of a compound in one species requires a tenfold longer exposure time to reach the equilibrium of the accumulated dose compared to another species.

Dose–effect relationships of inhaled substances are often described as exposure–effect or inhaled dose–effect relationships. Such indirect correlations (Fig. 27-1) do not consider respiratory tract deposition and retention. However, especially during chronic exposure, clearance and retention processes become important, and the initially deposited dose may have been cleared by the time the effects occur. Thus, chronic effects should correlate well with the retained or accumulated dose. On the other hand, it could be argued that chronic pulmonary effects may well be described as being correlated to the deposited dose or deposited dose rate, since the accumulated dose correlates with the deposition rate [see Section IIB, Equation (1)]. However, for extrapolation from one species to another, species specific pulmonary retention must be taken into account which, when considering a dose to the lung, could not be achieved by using the deposition rate only. Even acute effects involve uptake mechanisms, i.e., retention and accumulation are important. Thus, in both acute and chronic inhalation exposures, a dose–response relationship can probably best be described with the accumulated, rather than the deposited dose. This accumulated (retained) dose–effect relationship appears to be best suited to extrapolate results from animal inhalation studies to humans, since it takes into account differences in both deposition and retention between the two species (Fig. 27-1).

In the following sections, these dosimetric principles will be applied to a practical example, the induction of lung tumors in rats and humans after chronic inhalation exposure to cadmium. Two objectives will be addressed to demonstrate the usefulness of lung dosimetry for extrapolation modeling: to partially explain differences in localization of cadmium induced lung cancer between rat and human; and to extrapolate the carcinogenic effects observed in a rat inhalation study to predict human risk from exposure to airborne environmental cadmium. This will include lung model calculations on respiratory tract deposition and retention of inhaled cadmium in rat and human and a calculation of an exposure concentration in both species which leads to the same lung equilibrium concentration of cadmium after chronic inhalation exposure. For a better understanding of the discussed data, a brief background on environmental airborne cadmium and on the carcinogenic effects of inhaled cadmium will be presented first.

B. Background: Lung Cancer and Airborne Cadmium

1. Carcinogenic Effects of Inhaled Cadmium

Nonpulmonary carcinogenic effects of injected cadmium compounds in animals were first described in the early 1960s (Haddow *et al.,* 1961, 1964; Kazantzis, 1963; Gunn *et al.,* 1963, 1964), whereas pulmonary carcinogenic effects in animals were first reported by Heering *et al.* (1979) and in a more detailed study by Takenaka *et al.* (1983). In this latter study, about 40 rats each were exposed to aerosols of $CdCl_2$ at concentrations of 12.5, 25, and 50 μg Cd/m^3 23 hr/day for 18 months. The particle size was 0.55 μm (mass median aerodynamic diameter, MMAD) with a geometric standard deviation (GSD) of 1.8. After an additional 18 month observation period, a classical exposure concentration–lung carcinoma incidence relationship was found (Fig. 27-2). Primary lung carcinomas developed in 15.4, 52.6, and 71.4% of the rats of the respective groups plus one adenoma each in the low- and high-exposure group. No lung tumors were found in the sham exposed control group. Most of the carcinomas were adenocarcinomas which originated in the bronchiolar–alveolar region, based on histologic typing and the appearance of numerous type II-like alveolar cells in the tumors. The experimentally found lung tumor incidences, *P* (excluding the two adenomas), as a function of exposure concentration are described in Fig. 27-2 by a logit function which does not reflect a dose–response relationship. Long-term studies with other Cd compounds (CdO, $CdSO_4$, CdS), designed after the study of Takenaka *et al.* (1983), also resulted in the induction of lung carcinomas (Oldiges *et al.,* 1989) which was not significantly different from the results of the $CdCl_2$ study. Thus, it appears that inhalation of many different cadmium compounds with different *in vitro* and *in vivo* solubilities have the same potency to induce significant numbers of lung tumors in the rat.

Based on the results of these animal studies it can be speculated that the

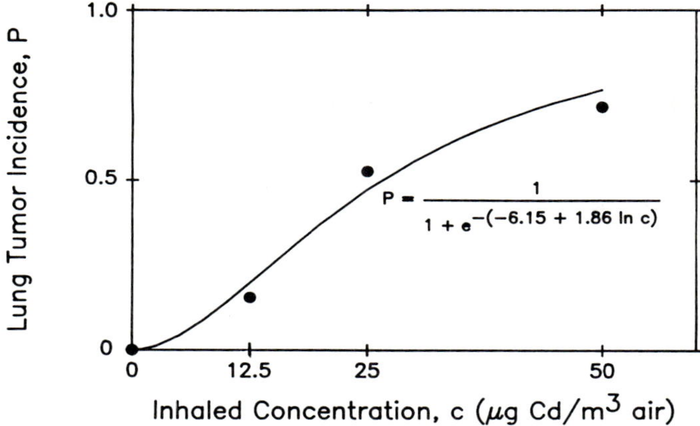

Fig. 27-2 Exposure–concentration response curve for lung tumor induction in rats after chronic exposure to $CdCl_2$ aerosols (constructed after results of Takenaka *et al.*, 1983). A logit function was used to describe the data, however, other functions might be used as well. One lung adenoma each in the 12.5-$\mu g/m^3$ group and in the 50-$\mu g/m^3$ group are not included in this analysis, i.e., only malignant tumors were considered.

different cadmium compounds have to be solubilized in the lung before they become effective. A rapid dissolution of CdO in the lung has indeed been reported (Hadley *et al.*, 1980), probably occurring after phagocytosis in the alveolar macrophages, so that pulmonary clearance kinetics of CdO are essentially the same as those for the water soluble $CdCl_2$ (Oberdörster *et al.*, 1979). This means that the particulate nature of CdO is lost shortly after its deposition in the lung and that pulmonary clearance mechanisms for cadmium are different from insoluble particles which are cleared by alveolar macrophages. Both $CdCl_2$ and CdO do not act like a hydrophilic solute which would be quickly eliminated from the lung, but they are slowly cleared by binding to and reacting with proteins (Hart, 1986; Oberdörster, 1988). A low dissolution rate in the lung has also been reported for CdS (Glaser *et al.*, 1986). However, since it is likely that pulmonary clearance and accumulation kinetics for CdS are different from the other cadmium compounds, the extrapolation modeling described in this paper may not apply to CdS. Pulmonary CdS kinetics need to be determined in order to modify the extrapolation model for inclusion of cadmium compounds of low *in vivo* solubility.

Epidemiological studies in cadmium-exposed workers have been summarized (Oberdörster, 1986). A significant pulmonary carcinogenic effect of inhaled cadmium in exposed workers was first described by Lemen *et al.* (1976) and was recently confirmed by Thun *et al.* (1985). This latest study represents the best evidence that inhaled cadmium, in this case mostly CdO dust and fume, may lead to an excess lung tumor risk. The particle size of the Cd containing particles to which the workers had been exposed must have varied over a wide range, judging from the description of the plant operation by Smith *et al.* (1980), with a median particle size for CdO dust presumably around 2 to 3 μm and for CdO fume in the submicron size range. The lung tumors were, as far as could be determined, mostly of bronchogenic origin. Lung cancer mortality rose with increasing cumulative exposure. Although the authors corrected for smoking and possible simultaneous arsenic exposure, the Carcinogen Assessment Group of the U.S. Environmental Protection Agency (EPA) felt that a possible exposure to arsenic and cigarette smoking may have confounded the results (EPA, 1985). Consequently, they concluded that the evidence that inhaled cadmium is a human lung carcinogen is limited. Nevertheless, based on the study by Thun *et al.* (1985), the Carcinogen Assessment Group of EPA estimated that the risk for lung cancer from lifetime exposure to 1 μg Cd/m^3 air is 1.87×10^{-3} (EPA, 1985).

In summary, there is sufficient evidence that inhaled cadmium is a lung carcinogen in rats, whereas the evidence is only limited for lung carcinogenicity in humans. It appears that rats are more susceptible than humans. However, results from the rat study cannot be extrapolated directly and quantitatively to humans. The local target sites of tumor induction seem to differ between the two species: more peripheral (bronchiolar–alveolar) in the rat, more central (bronchial) in humans. One goal of this chapter is to determine whether lung dosimetry can contribute to an explanation of this difference.

2. Environmental Airborne Cadmium

Cadmium is released into the atmosphere from mining, primary smelting, and other industrial processes [manufacture of batteries, pigments, stabilizers in plastics, nuclear reactor rods, in metallurgy, in the semiconductor industry, catalysts (NIOSH, 1984)]. In addition, natural sources contribute to atmospheric cadmium. Estimates of the source contributions of atmospheric cadmium show that cadmium from natural sources represents only about 10% of the total airborne cadmium (Nriagu, 1979). Waste incineration can also add significantly to cadmium in the ambient air. Although little is known about the chemical form of cadmium in ambient aerosols (Nriagu, 1979), it is conceivable that cadmium oxides prevail in those particles generated by smelting and other thermal processes. Elemental cadmium and cadmium sulfides may also be present. Measured concentrations of cadmium in air range from 0–6 ng/m^3 for rural areas,

5–500 ng/m^3 for urban areas, and 20–700 ng/m^3 for industrial areas (Nriagu, 1979; CEC, 1978). Workplace concentrations can be in the μg/m^3 range (Threshold Limit Value in USA: 50 μg/m^3).

With regard to particle size of the cadmium containing ambient particles, results of several measurements showed that more than 50 to 90% of the cadmium was associated with particles of less than 5 μm in diameter (CEC, 1978). The most complete measurements of particle size distributions using an 8-stage Anderson impactor were reported by Dorn *et al.* (1976). Fig. 27-3 and 27-4, showing particle size distributions in a rural area and near a cadmium emitting smelter, were constructed from their data. The MMAD of the particles with respect to cadmium mass collected in a rural area was 2.6 μm, with a GSD of 3.6 and with a concentration of 3.7 ng Cd/m^3 (Fig. 27-3). In an industrial area 100 m from a smelter, the Cd concentration was 24.8 ng/m^3 with a MMAD of the particles of 1.3 μm and a GSD of 2.6 (Fig. 27-4). The reasons for the association of Cd with larger particles in the rural area are not known (windblown dust from Cd contaminated soil?); however, these reported particle size distributions for a rural and an industrial site will be used for the discussion of the second goal of this chapter, describing a method of risk assessment of an inhaled environmental pollutant for humans based on the results of a rat inhalation study.

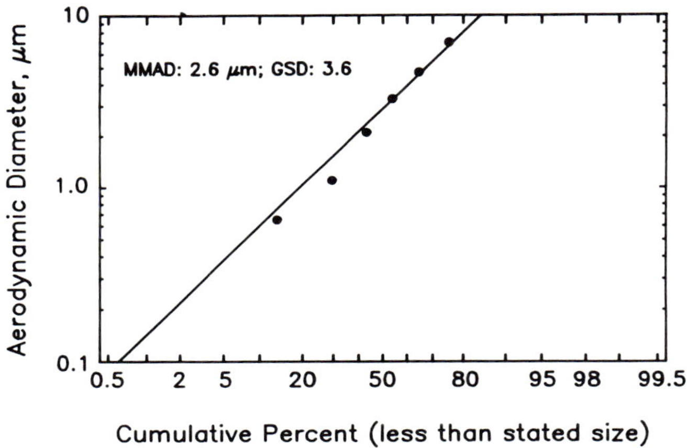

Fig. 27-3 Cadmium particle size distribution in rural area. Air concentration: 3.7 ng/m^3 (based on the data of Dorn *et al.*, 1976).

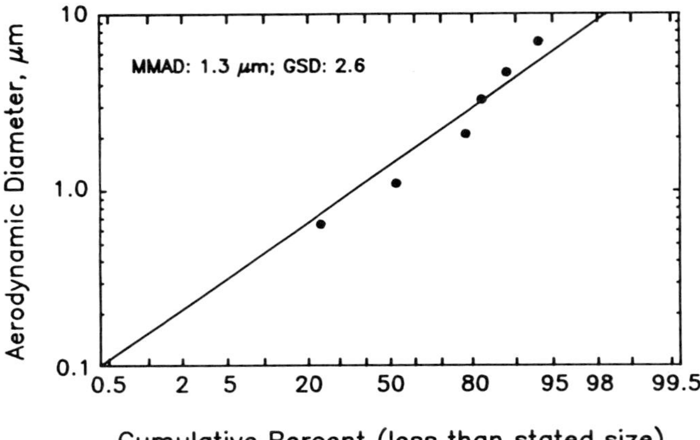

Fig. 27-4 Cadmium particle size distribution in industrial area. Air concentration: 24.8 ng/m^3 (based on the data of Dorn *et al.*, 1976).

II. LUNG MODEL CALCULATIONS

A. Deposition Modeling of Inhaled Cadmium in Rat and Human

The sites of deposition of an inhaled compound could be indicative of the sites of action, depending on the local concentration in the respiratory tract and on the sensitivity of the respective cells to injury. Whether the observed differences in tumor localization in rat and human after chronic inhalation of cadmium can be partly explained by differences in respiratory tract deposition will be discussed in this section. The deposition characteristics of inhaled particles in the nasopharyngeal, tracheobronchial, and pulmonary compartments for humans were described by the Task Group on Lung Dynamics (1966) and since then the findings have been refined by results from newer studies (Heyder, 1982). Similar deposition curves have been found for rats (Raabe, 1980), although fractional deposition in their pulmonary and tracheobronchial regions are lower than in humans, due to the more efficient filtering capacity of the rat's nose.

Based on anatomical data of the human and rat respiratory tract including measurement of length, diameter, branching angle, and angle of gravity of each generation (trachea = generation 1, terminal bronchioles = generation 16, respiratory bronchioles through alveoli = generation 17–25), Yeh and Schum (1980) and Schum and Yeh (1980) have described a mathematical model to predict deposition of inhaled particles in rats and humans. Their calculations are based on a symmetrical typical path whole lung model of rat and human airways. Although this is a simplification compared to the *in vivo* conditions, the authors could show that the regional depositions predicted by their model agree quite well with results from experimental studies in rats (Raabe *et al.*, 1977) and humans (Lippman, 1977). However, the predicted deposition in certain individual generations, in particular in the more monopodial airway structure of the rat, could be different from those derived by a more realistic stochastic model of particle deposition as described by Hofmann *et al.* (1989). Therefore, deposition calculations for an individual generation in the rat based on a symmetrical model may have to be used with caution.

Particle size, density, and heterodispersity and respiratory patterns (tidal volume, breathing rate) can be adjusted independently in the mathematical model of Yeh and Schum (1980) to reflect relevant values. Deposition of inhaled particles in generations 1–25 is assumed to occur by diffusion, sedimentation, and impaction during inhalation and by diffusion and sedimentation during exhalation, whereas modeling of nasopharyngeal deposition is based on empirical data. For details on the deposition equations, see Schum and Yeh (1980) and Yeh and Schum (1980). Modifications of the lung model calculations, including effects of particle size heterodispersity and mixed oro-nasal breathing on deposition (Wojciak *et al.*, 1987) are included in the following results. The microcomputer-based modified mathematical model is described in detail by Wojciak (1988). The deposited dose will be expressed as deposited surface area dose per 8 hr of exposure (ng Cd/cm^2 epithelial surface area). The surface area available for deposition at functional residual capacity (FRC) is used in these calculations, as determined from the anatomical data and corrected for overlap at the bifurcations (Wojciak, 1988).

Deposition of inhaled cadmium particles was determined according to the modified model. Respiratory parameters for the rat (tidal volume and respiratory rate) in the chronic inhalation study by Takenaka *et al.* (1983) are predicted by the formulas of Stahl (1967). A mean body weight over the 18 months of exposure of 400 g was assumed, which reflects very well the actual body weights during this time. For human workers exposed occupationally to airborne cadmium two conditions were simulated: nose breathing at rest and breathing under light activity, i.e., increasing the minute ventilation to 20 liters/min (ICRP, 1975). The respective respiratory values given by ICRP (1975) for Reference Man were used. Respiratory parameters for rat and man are listed in Table 27-1. For breathing under light activity, 50% oral and 50% nasal breathing was assumed. Although this may overemphasize the oral portion of respiration[1] this partitioning of

[1] Niinimaa *et al.* (1980, 1981) reported that switching to oro-nasal breathing in two thirds of their subjects occurred at a minute ventilation > 35 liters/min, and that even during heavy exercise about 40% of the respired air is still drawn through the nose.

Table 27-1 Respiratory Parameters for Rat and Man[a]

	Rat	Man (at rest)	Man (light activity)
Breathing mode	nasal	nasal	oro-nasal
Tidal volume, ml	2.97	500	1250
Resp. rate, min^{-1}	68	15	16

[a] For rats exposed under the conditions of the study by Takenaka *et al.* (1983), values were predicted according to Stahl (1967) assuming a mean body weight of 400 g. For humans, data for ICRP Reference Man (1975) with a body weight of 70 kg were used.

airflow was chosen to show the influence of oral breathing for increasing particle deposition in the respiratory tract.

Predicted deposition of the inhaled $CdCl_2$ particles (MMAD = 0.55 μm, GSD = 1.8) in the respiratory tract of the exposed rats of the study by Takenaka *et al.* (1983) is shown in Fig. 27-5. The figure shows the surface area dose per airway generation during 8 hr of exposure to 25 μg Cd/m^3. For comparison, deposition in the human respiratory tract during an 8-hr exposure (resting conditions) to the same aerosol is also shown. Two regions of high surface area deposition occur in the rat respiratory tract: one is in the central bronchial region, generation 4–7, and the other in the transitional zone, generation 16–18. The latter could correspond to the region from which the primary lung carcinomas in the rat inhalation study had originated. The fact that this peripheral pulmonary region has much less efficient clearance mechanisms than the central bronchial region,

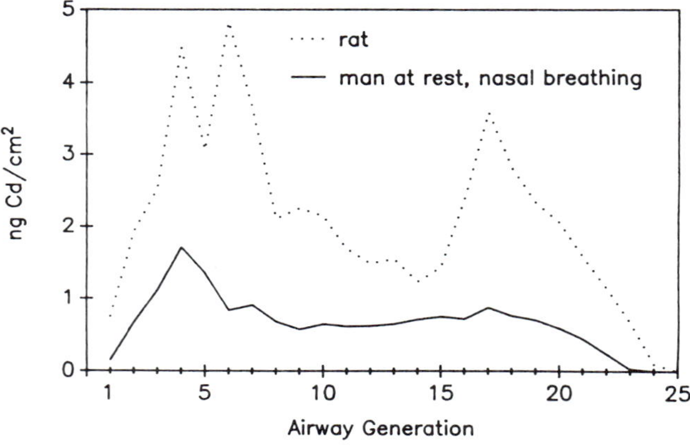

Fig. 27-5 Lung model prediction of cadmium deposition per surface area of airways in rat and man during 8-hr exposure to 25 μg Cd/m^3. Particles: MMAD = 0.55 μm; GSD = 1.8.

which also receives a high deposited cadmium surface area dose, may have contributed to the higher incidence of the tumors in the peripheral lung. However, variations in susceptibility of cells in different regions of the respiratory tract may be additional factors in the pathogenesis of carcinogenic effects which still need further investigation.

In humans exposed to the same aerosol, predicted deposited surface area doses are lower throughout the respiratory tract (Fig. 27-5). Within the human respiratory tract, a higher surface area dose is reached in the upper conducting airways. Due to the efficiency of bronchial clearance mechanisms, this may not be of consequence. However, as described in Section IB1, occupationally exposed human workers were often exposed to CdO dust where they encountered larger particle sizes. Predicted depositions of inhaled cadmium particles with MMAD of 2.5 μm and GSD of 1.8 are shown in Fig. 27-6, assuming again an 8-hr exposure to 25 $\mu g/m^3$. For a worker at rest, bronchial deposition in the upper conducting airway is much more pronounced under these conditions, exceeding the values reached in the rat respiratory tract.

If this same person increases his tidal volume and respiratory rate and switches to oro-nasal breathing (conditions of light activity), a further increase in deposition by a factor of up to eight for the bronchial region is predicted (Fig. 27-6). Although, as previously noted, bronchial clearance mechanisms are very effective, it is also known that cadmium causes an impairment of ciliary function (Adalis *et al.,* 1977) which conceivably occurs at higher concentrations and increases cadmium retention in the bronchial region. Indeed, a significant increase in the incidence of chronic bronchitis of cadmium-exposed workers confirms this assumption (Arm-

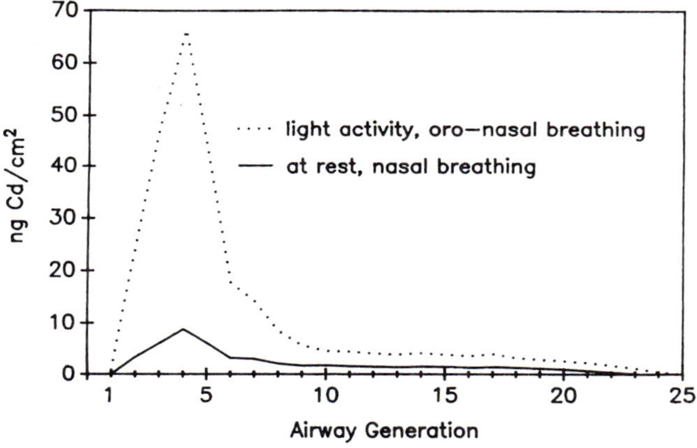

Fig. 27-6 Lung model prediction of cadmium deposition per surface area of airways in man during 8-hr exposure to 25 μg Cd/m^3. Particles: MMAD = 2.5 μm; GSD = 1.8.

strong and Kazantzis, 1983). Thus, the occurrence of mostly bronchogenic tumors in occupationally cadmium-exposed workers may be a consequence of a combination of high local deposited doses, chronic impairment of bronchial clearance, bronchitis, and increased local retention of cadmium.

Therefore, the different sites of origins of lung tumors found in the rat inhalation study and the epidemiological study may partly be due to differences in local deposition in the respiratory tract of rat and human. The total deposited amounts of cadmium for the tracheobronchial region (generations 1 – 16) and for the pulmonary region during the 8-hr exposure to 25 μg Cd/m^3 during the different exposure conditions of rat and human are shown in Table 27-2.

B. Retention Modeling of Inhaled Cadmium in Rat and Human

The accumulation of cadmium in the lung during chronic exposure is dependent on its initial deposition and its retention half-time. Knowing the biological half-time of cadmium in the lungs of both rat and man and assuming first order clearance kinetics, one can calculate the accumulated dose under constant exposure conditions, as described by the Task Group on Metal Accumulation (1973):

$$A_t = \frac{a}{b}(1 - e^{-bt}) \tag{1}$$

where A_t is the amount of cadmium per g lung at time t, a is the amount deposited daily per g lung, and b is the fraction eliminated from the lung per day. The biological half-time, $t_{1/2}$, is correlated to b according to $t_{1/2} = \ln 2/b$. After an exposure period of about 5 half-times has elapsed,

Table 27-2 Lung Model Prediction of Deposited Amount of Cadmium (ng)[a]

	0.55 μm[b]		2.5 μm[b]	
Particle Size (MMAD)	TB	P	TB	P
Rat	53	240	96	270
Man				
At rest	2350	7800	5500	12,500
Light activity	4650	34,100	18,100	65,000

[a] During 8 hr of exposure to 25 μg Cd/m^3.
[b] TB, tracheobronchial region; P, pulmonary region.

97% of an equilibrium value A_E is reached in the lung and reduces the equation to:

$$A_E = \frac{a}{b} \tag{2}$$

The values for a (deposited dose for the pulmonary region) can be deduced from Table 27-2 by considering the respective lung weights, which are assumed to be 2 g for a 400-g rat and 950 g for humans. These values per g of lung are shown in Table 27-3.

The assumption that pulmonary elimination of Cd follows first-order clearance processes is justified by earlier and recent results of our laboratory: the pulmonary retention of inhaled $CdCl_2$ and CdO aerosols, which are both very soluble in the lung, determined over periods of up to 240 days for rats and up to 650 days for monkeys could be best expressed by a monoexponential fit with half-times of 66–85 days for 2 different rat strains and of about 780 days for the monkeys (Oberdörster *et al.*, 1979, 1987). Chronic inhalation exposure to cadmium does not change the pulmonary clearance rate for cadmium (Oberdörster and Hochrainer, 1980). This is in contrast to pulmonary clearance of inert test particles, which is slowed during chronic cadmium inhalation exposure (Oberdörster and Hochrainer, 1980), implying that pulmonary cadmium is cleared via interstitial pathways after binding to proteins and not by alveolar macrophages and the mucociliary escalator (see Section IB).

Assuming that humans retain pulmonary cadmium similarly to monkeys, a retention half-time of 780 days for humans and of 75 days for rats will be used for calculating Cd accumulation during chronic exposure. Applying the species specific pulmonary deposition and retention data, the accumulation of cadmium under the different exposure conditions can be calculated according to Equation (1). If the exposure is for 5 days per week (simulation of occupational exposure), A_t has to be corrected by a factor of 5/7 by adjusting a correspondingly. Figure 27-7 shows the resulting pulmonary accumulated doses per g of lung for rat and humans for the exposure

Table 27-3 Deposited Pulmonary Doses of Cadmium (ng Cd/g lung)[a]

Particle Size (MMAD)	0.55 μm	2.5 μm
Rat	120	135
Man		
At rest	8.2	13.2
Light activity	35.9	68.4

[a] During 8 hr of exposure to 25 μg Cd/m^3.

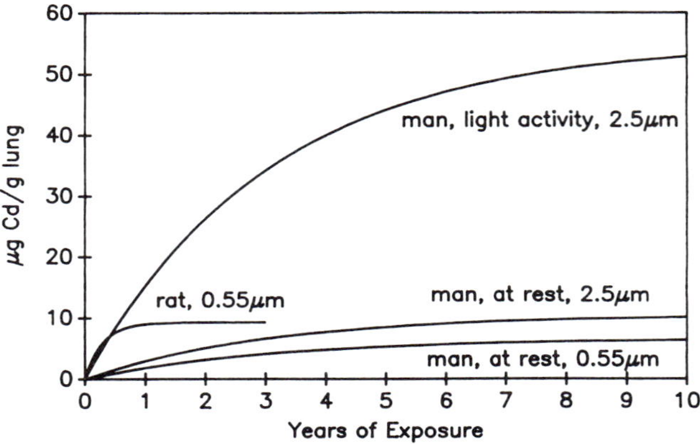

Fig. 27-7 Accumulation of cadmium in the lungs of rat and man during chronic inhalation exposure: 25 $\mu g/m^3$; 8 hr/day; 5 days/week. Particle MMAD = 0.55 μm and 2.5 μm; GSD = 1.8.

conditions described in the previous section IIA. Equilibrium concentrations reached in the lung are in the same range for rats and humans under resting conditions with nasal breathing (~6–10 μg Cd/g lung), irrespective of the particle size. Light activity with switching to oro–nasal breathing increases the accumulated pulmonary dose in humans considerably to 55 μg Cd/g lung.[2] However, as illustrated in Table 27-3, the deposited pulmonary dose rate is about a factor of 9–15 higher for the rat than for humans breathing at rest, and even under conditions of light activity the deposited pulmonary dose rate is still higher in the rat than in humans by a factor of 2. If the deposited dose rate of cadmium in the pulmonary region is important for its carcinogenic effect, this result may partly provide an explanation why rats are seemingly more susceptible to the carcinogenic effects of inhaled cadmium than humans (EPA, 1985).

As soon as the accumulated dose has reached an equilibrium, the deposited dose rate is equivalent to the elimination rate. Therefore, the result of Table 27-3 could also be discussed in terms of the higher pulmonary elimination rate of cadmium in rats, i.e., its higher pulmonary metabolism which very likely involves induction of and binding to metallothionein (Oberdörster, 1986). These differences in pulmonary cadmium turnover rates, i.e., deposited dose rates (Table 27-3) and/or elimination rates,

[2] Although nobody is breathing under such conditions continuously, they are included here to simulate higher exposures caused by partial mouth breathing. Breathing continuously under resting conditions will not occur either and, therefore, represents an unlikely lowest possible exposure.

become conceivably more important for human exposure to environmental airborne cadmium. In contrast to higher level occupational exposures, ambient air concentrations of cadmium are so low that no significant accumulation or effects in the tracheobronchial region, such as impairment of bronchial clearance mechanisms, are expected to occur and thus carcinogenic effects, if any, might occur predominantly in the pulmonary region. This presumption obviously needs to be verified. Under such environmental conditions, results of the rat inhalation study, where mostly adenocarcinomas were found, could be used to arrive at a risk estimate for human exposure. The importance of comparative lung dosimetry for such extrapolation modeling will be considered in the next section.

III. EXTRAPOLATION MODELING FROM RAT STUDY TO LOW INHALED CONCENTRATIONS IN HUMANS

Figure 27-2 shows the relationship between $CdCl_2$ concentrations in the air and experimentally induced lung tumors in rats. Several steps, including deposition and retention modeling of inhaled cadmium-containing particles, are necessary to arrive at a similar relationship for humans exposed to low environmental airborne concentrations of cadmium (see principles outlined in Fig. 27-1). The extrapolation model discussed considers differences in deposited doses to the pulmonary region of rat and human as well as differences in pulmonary metabolism of inhaled cadmium, which is reflected in the different pulmonary retention half-times for cadmium in rat and human. Basically, the assumption is made that equal accumulated doses of cadmium per g lung tissue at a steady state have the same carcinogenic potential in the peripheral lung in rat and human (Fig. 27-1). This assumption may be rather simplistic since it does not take into account the possibility of distinct sensitivities of the cells of the lung of rat and human, nor does it consider variations in the accumulation of cadmium in different cells, e.g., epithelial cells versus interstitial cells. Data to include such refinements are not presently available. However, since steady state conditions are reached after different time periods in rat and human, the use of the accumulated dose at steady state takes into account the time factor which may be important for cancer development, as mentioned previously (Section IA). Thus, as a first step in the extrapolation process, a relationship will be established between the pulmonary accumulated doses [equilibrium value A_E, Equation (2)] and tumor incidences derived from the rat study. In the second step, an interpolation to low accumulated doses is performed, and finally, human equivalent exposure concentrations of cadmium in air, which correspond to a given lung tumor risk, are calculated.

A. Pulmonary Deposition and Retention of Environmental Cadmium

The typical path whole lung model of Yeh and Schum (1980) as modified by Wojciak *et al.* (1987) was applied to predict pulmonary deposition of environmental airborne cadmium particles, and the pulmonary accumulated dose was calculated according to Equations (1) and (2). Particle sizes and concentrations for rural areas (MMAD = 2.6 μm; GSD = 3.6; conc. = 4 ng/m^3) and industrial areas (MMAD = 1.3 μm; GSD = 2.6; conc. = 25 ng/m^3) were used as described by Dorn *et al.* (1976; Fig. 27-3 and 27-4). Two activity states for humans living under those conditions were considered, i.e., nasal breathing at rest and oro-nasal breathing with light activity (Table 27-1). Table 27-4 shows the results of lung model predictions of the relative deposition (deposited fraction of inhaled mass) of the environmental cadmium-containing particles in a person living in a rural or industrial area. Predicted relative deposition efficiencies in nasopharyngeal and tracheobronchial regions at rest and during light activity turned out to be very similar for both particle sizes with MMADs of 2.6 μm and 1.3 μm and respective GSDs of 3.6 and 2.6. In contrast, pulmonary deposition, which is of interest for our model, increases from 10% (at rest) to 20% under light activity of inhaled 2.6-μm particles (rural areas) and from 11% to 20% for 1.3-μm particles (industrial area). By coincidence, the two particle size distributions give almost identical relative pulmonary depositions under conditions of rest (10% and 11%) and of light activity (20%). Because of this similarity, and since the accumulation of pulmonary cadmium is of main interest for this extrapolation modeling, there is no need to consider the different particle sizes of rural and industrial areas separately so that only the activity state (resting versus light activity) needs to be taken into account. Obviously, the absolute amount of cadmium being deposited and accumulating in the lung will be higher in the industrial area, as a function of the higher exposure concentration. Respective

Table 27-4 Lung Model Predictions of Relative Regional Deposition of Inhaled Environmental Cadmium Particles[c]

	Rural Area[a,d]		Industrial Area[b,d]	
	At Rest	Light Activity	At Rest	Light Activity
NP	44	35	23	18
TB	5	7	4	5
P	10	20	11	20

[a] MMAD = 2.6 μm; GSD = 3.6.
[b] MMAD = 1.3 μm; GSD = 2.6.
[c] NP, nasopharyngeal; TB, tracheobronchial; P, pulmonary region.
[d] Measurements in percentage deposition of inhaled mass.

Table 27-5 Accumulated Pulmonary Doses of Cadmium During Constant Exposure to Environmental Airborne Cadmium[a]

	At Rest	Light Activity
Rural area (4 ng Cd/m^3 air)[b]	5.1	27.1
Industrial area (25 ng Cd/m^3 air)[c]	35.8	174.8

[a] Equilibrium concentration of dose: ng Cd/g lung.
[b] MMAD = 2.6 μm; GSD = 3.6.
[c] MMAD = 1.3 μm; GSD = 2.6.

Table 27-6 Predicted Accumulated Cadmium in Lungs of Rats and Incidence of Primary Lung Tumors

Exposure Concentration (μg Cd/m^3 air)	Pulmonary Accumulated Dose (μg Cd/g lung) *d*	Lung Tumor Incidence (fraction of exposed rats)
0	0	0
12.5	37	0.15
25	74	0.53
50	149	0.71

[a] After 18 months of exposure and under the conditions of the study by Takenaka *et al.* (1983).

values of the pulmonary accumulated doses (ng Cd/g lung) are shown in Table 27-5. The values in this table reflect the equilibrium concentration (A_E) after continuous exposure (24 hr/day, 7 days/week). Compared to values reached in the experimental study in the lungs of rats (Table 27-6) and in occupationally exposed workers after an 8-hr daily exposure (derivable from Table 27-3), these values are lower by 2 to 4 orders of magnitude.

B. Linear Interpolation to Low Inhaled Concentrations

For extrapolation of the results of the rat inhalation study, the accumulated doses of pulmonary cadmium (μg Cd/g lung) of the rats at the end of the 18-month exposure period were calculated according to Equation (1). The values, shown in Table 27-6, are practically identical to the pulmonary equilibrium concentration (99%). These lung cadmium concentrations were plotted against the respective lung tumor incidences in Fig. 27-2 and Table 27-6 and the result is shown in Fig. 27-8. The method of linear interpolation after Gaylor and Kodell (1980) was adapted to calculate the upper limits of risk for low accumulated doses of lung cadmium which are predicted to result from exposure to low environmental cadmium. Basically, the interpolation method of Gaylor and Kodell (1980) involves four steps:

1. Fit a mathematical model to the experimental data.
2. Find the upper confidence limit (95 or 99%) in experimental dosage range.
3. Calculate a straight line from the upper confidence limit at lowest experimental dose to zero (interpolation line).
4. Obtain upper limits of risk for low doses from interpolation line.

As pointed out by Gaylor and Kodell (1980), the choice of the mathematical model, whether logit, probit, multistage, or any other model, is unimportant as long as the dose–response curve is convex at low doses. Furthermore, background cancer rates must also be considered in this approach when constructing the interpolation line. However, since spontaneous lung cancers are rare in rats—no lung cancers were found in the control group of the study by Takenaka *et al.* (1983), and from large cohorts of other studies spontaneous lung cancer rates appeared to be less than 0.5% (Altman and Goodman, 1979)—background cancer rates can probably be neglected in this case. Gaylor and Kodell also showed that this approach leads to conservative estimates of risks at low doses which are comparable to those obtained from the multistage model or using safety factors.

Both pulmonary deposition and pulmonary retention of cadmium—the latter reflecting its metabolism—are considered in the adaptation of the method of linear interpolation to the inhalation study shown in Fig. 27-8. As in Fig. 27-2, a logit analysis of the data points was performed and,

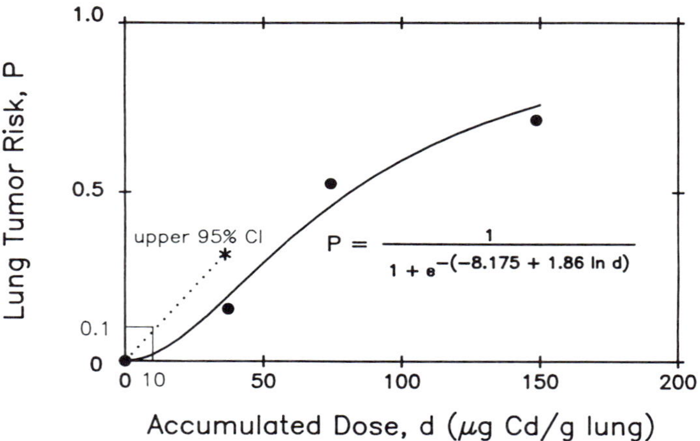

Fig. 27-8 Dose–response curve for lung tumor induction after cadmium inhalation derived from rat inhalation study.

unlike in Fig. 27-2, this yielded a true dose–response curve in the sense that the response is a function of the lung dose and not of the exposure concentration. The upper 95% confidence limit ($P = 0.32$) of the lowest experimental dose is used to construct the interpolation line, and this line is used to extrapolate to low dose risks. The dose is expressed as accumulated dose of pulmonary cadmium. These low dose risk estimates apply both to rats and humans, according to our assumption that at steady state, equal accumulated doses of cadmium per gram of lung have the same carcinogenic potential in rats and humans.

C. Human Equivalent Exposure Concentration and Lung Cancer Risk Estimate

Cadmium exposure in the rat study extended over 18 months, after which about 99% of the equilibrium lung dose should have been reached, equivalent to a human exposure of about 16 years. It is conceivable that additional continuous exposure to cadmium may further affect lung cancer rates. However, the impact of additional exposures beyond this period cannot be predicted at this time. We need to know more about the differences between relatively short and very long times of exposures on cadmium-induced lung cancer and about cellular–molecular mechanisms of cadmium carcinogenicity in order to overcome the present limitations of this extrapolation effort, which is based on a simple model and may have to be refined in the future. These limitations have to be kept in mind when reading the following section.

From the data in Fig. 27-8, a human equivalent exposure concentration for cadmium in the ambient air can be derived which, at a given particle size and breathing condition, will lead to the same accumulated pulmonary dose at steady state (*d* in Fig. 27-8) in the human lung as in the rat lung. The exposure duration should be long enough for lung cadmium to reach an equilibrium (i.e., >11 yr). This approach takes into account differences both in deposition and retention kinetics (reflecting metabolism) of cadmium in the lungs of rats and humans, as discussed in Section IIIB. A constant chronic exposure to low environmental cadmium concentrations is assumed. The low lung cadmium concentrations pointed out in the left lower corner of Fig. 27-8 will result from such exposures to low environmental airborne cadmium. Fig. 27-9 shows an expanded view of the interpolation line in this range of lower lung equilibrium concentration representing the upper 95% tumor risk for low lung cadmium levels. This approach assumes, in a conservative way, that there is no threshold dose for cadmium-induced lung cancer and that the risk relates to excess risk above background (Gaylor and Kodell, 1980).

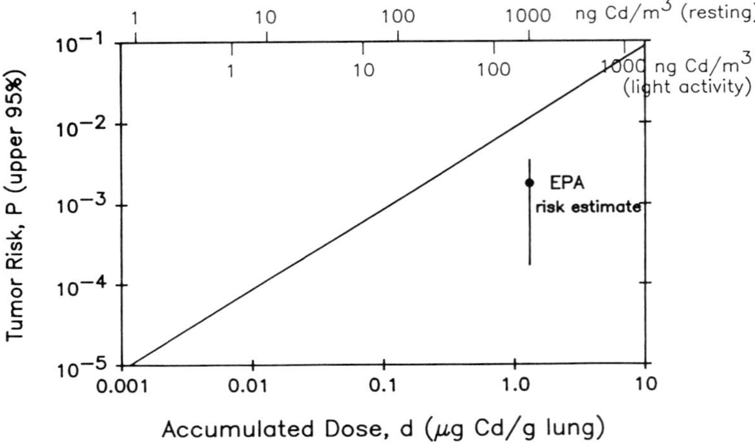

Fig. 27-9 Upper 95% confidence limit of lung cancer risk from environmental airborne cadmium (rural and industrial area) extrapolated from rat inhalation study. Also shown are the mean and upper and lower 95% confidence limits of the EPA unit risk estimate (for lifetime exposure to 1 $\mu g/m^3$) derived from an epidemiological study. The mean of the risk estimate for 1 $\mu g/m^3$ exposure extrapolated from the rat inhalation study is about 3×10^{-4} and lies in the range of the EPA risk estimate.

The upper x axis of Fig. 27-9 gives air concentrations of inhaled cadmium for humans calculated from the cadmium equilibrium concentration in the lung, d, shown on the lower x axis. The cadmium air concentrations are shown for two activity states, resting and light activity. Both apply to populations living in a rural as well as in an industrial area as discussed in section IIIA (Table 27-4), assuming that the inhaled particle sizes are the same as measured by Dorn *et al.* (1976) for these areas. This graph shows the 95% upper-bound lung cancer risk from chronic exposure to cadmium at a given concentration in the air. Increasing the breathing minute volume from resting level to light activity would increase the risk estimate by almost one order of magnitude (upper x axis in Fig. 27-9). Also shown in Fig. 27-9 is the EPA unit risk estimate (risk from exposure to 1 $\mu g/m^3$) with upper and lower 95% confidence limits as it was derived from the epidemiological study (EPA, 1985). This risk is less than one order of magnitude lower than extrapolation from the rat study for a person chronically exposed under resting conditions. This indicates that extrapolation from the rat study resulted in a slightly more conservative risk estimate. However, the mean of the risk estimate extrapolated from the rat study for human exposure to 1 $\mu g/m^3$ is about 3×10^{-4} (not shown in Fig. 27-9) which falls right in the range of the EPA risk estimate. Although this good agreement cannot be taken as a validation of the present extrapolation model, it indicates nevertheless that extrapolation using lung dosimetry leads to reasonable results.

Table 27-7 compares results of the risk estimates derived from this study with the lower risk estimate calculated by EPA. Since these are upper-bound risk estimates only, inclusion of both a lower and upper bound would show that estimates from both approaches overlap. Considering the completely different methodologies for arriving at these risk estimates, the results in Table 27-7 represent a good agreement. It should be remembered, however, that the EPA risk estimate is derived from rather high occupational exposures and that possibly the sites of tumor origin within the respiratory tract may be different for such exposures as compared to tumor sites originating from low environmental cadmium exposures.

In general, the numbers of Table 27-7 show that lung cancer risk from inhaled environmental cadmium is relatively low compared to smoking a pack of cigarettes per day (U.S.-DHHS, 1982). However, as pointed out above, a potential increase of this low risk may have to be considered due to continuous exposure to airborne cadmium after steady state conditions in the lung have been reached. In addition, the impact of possible synergistic interactions between inhaled cadmium and other carcinogens as described by Harrison and Heath (1986) may have to be considered. This aspect and further implications of this risk estimate, e.g., cadmium and cigarette smoking, are beyond the scope of this presentation and will not be discussed.

This risk extrapolation method based on principles of lung dosimetry has the advantage of being adaptable to different environmental conditions, such as different particle sizes and human activity conditions, which can change the dose to the lung and consequently considerably alter the risk. Using cadmium as an example for describing this approach should not be viewed as an attempt to improve the risk estimate for lung cancers due to cadmium exposure. Rather, this treatise should demonstrate how results of well-planned and -conducted animal inhalation studies and lung

Table 27-7 Upper Bound (95%) Human Lung Tumor Risk Estimates From Chronic Exposure to Environmental Airborne Cadmium[a]

Cadmium Air Concentration and Living Area	Extrapolation From: Animal Study	Extrapolation From: Epidemiological Study (EPA, 1985)
<1 ng/m^3—rural	$<1.1 \times 10^{-5}$	$<3.5 \times 10^{-6}$
3.5 ng/m^3—urban	3.9×10^{-5}	1.2×10^{-5}
200 ng/m^3—industrial	2.2×10^{-3}	7×10^{-4}

[a] Breathing conditions: at rest.

dosimetry can be combined to arrive at a human risk estimate. Obviously, this is most important for compounds for which no, or only limited, epidemiological data are available.

IV. SUMMARY AND CONCLUSIONS

Interpretation of results from long-term animal inhalation studies and extrapolation of those results to humans requires knowledge of regional deposition efficiencies in the respiratory tract, pulmonary retention half-times of the compound in question, the metabolism of the compound in the respiratory system, and the sites of action within the respiratory tract. Inhaled cadmium compounds have been shown to induce lung carcinomas in both rats and humans, however, it appears that the carcinogenic effect is greater in rats than in humans. The objectives of this chapter were to demonstrate the applicability of lung dosimetric principles to interpreting experimental results addressing two practical examples:

1. To determine whether variations in lung cancer localization between rat and human can partially be explained by differences in local deposition of inhaled cadmium aerosols.
2. To extrapolate results of the carcinogenic risk of inhaled cadmium from a rat inhalation study to humans.

Carcinomas induced in rats by inhalation of cadmium seemed to be mostly of bronchiolar–alveolar origin, whereas in humans inhaled cadmium induced mostly cancer of bronchogenic origin. Model deposition calculations were performed assuming both an 8-hr/day exposure (occupational exposure) and 24-hr continuous exposure (environmental exposure) to cadmium aerosols of different particle sizes, heterodispersity, and concentrations. For a submicrometer size cadmium aerosol, the particle size used in rat studies, the deposited dose per unit surface area in the transitional region of the lung when inhaled via the nose is about four times higher in the rat than in man. Inhalation of a larger sized cadmium aerosol, often encountered under workplace conditions (e.g., exposure to CdO dust with a median particle size of 2.5 μm) leads to a doubling of the surface area dose in the upper conducting airways of humans. Light activity, with switching to oro-nasal breathing, increases the deposited dose per unit surface area in generations 2–6 of the human lung by an additional factor of 8. Since pulmonary retention of cadmium differs significantly between rats ($t_{1/2} = 75$ days) and primates ($t_{1/2} = 780$ days), this also must be considered when calculating accumulation of cadmium in the lung during long-term exposure. Respective model calculations using first-order clearance kinetics show that the long-term accumulated dose of cadmium per g of lung is about the same in rats and humans under resting conditions when equilibrium values are reached, but is greater in humans by a factor of about 5 under conditions of light activity. However, the deposited dose rate per g of lung (ng/g/hr) during cadmium aerosol exposure is up to 15-fold greater in the rat. These results suggest that differences between rats and humans in the sites of tumor induction by inhaled cadmium may be partly due to differences in the deposited surface area dose and the deposited dose rate. The greater pulmonary turnover rate of pulmonary cadmium in rats involving cadmium-binding proteins may also partly explain a greater susceptibility of rats to the carcinogenic effects of inhaled cadmium.

Principles of lung dosimetry should also be considered when, based on animal studies, risk estimations are performed for environmental exposure of humans to inhaled carcinogenic compounds. A classical exposure–response relationship for lung tumor induction in rats due to inhaled cadmium had been established in the past. This relationship was converted into a dose (cadmium per g lung)–response relationship using dosimetric deposition and retention calculations. The lung dose of cadmium was assumed to be represented by the equilibrium value at the steady state of the accumulated cadmium in the lung which takes into account differing exposure times for rats (~18 months) and humans (~16 years) because of the approximately 10-fold longer pulmonary retention half-time for cadmium in primates. Extrapolation to low lung cadmium doses was then performed by applying a linear interpolation model. Using the human lung dosimetry data, human equivalent concentrations of cadmium in the air were derived which will lead to the same low lung cadmium concentrations at steady state. Resulting risk estimates for environmental airborne cadmium can be specified for different breathing conditions (at rest, light activity) and for different inhaled particle sizes. These estimates are slightly more conservative (i.e., less than one order of magnitude higher) than estimates extrapolated from results of an epidemiological study. In general, lung cancer risk from exposure to environmental cadmium is relatively low, e.g., $\sim 4 \times 10^{-5}$ for urban areas, taking into consideration the limitations outlined in Section C. However, the risk estimate resulting from this modeling approach with cadmium should not be viewed as an attempt to improve this estimate for lung cancer risk from inhaled cadmium. Instead, the result demonstrates that extrapolation modeling of findings from an experimental animal inhalation study together with lung dosimetry can be used to arrive at a human risk estimate for an environmental air pollutant. This is particularly important when epidemiological data are lacking or are only limited.

ACKNOWLEDGMENTS

I would like to thank J. Wojciak who wrote the particle deposition model in Basic for use with a microcomputer, and C. Cox for performing the logit analysis of lung carcinoma incidences in rats. This paper was partially based on work supported by grants ES01247 and ES01248 from the National Institutes of Health.

REFERENCES

Adalis, D., Gardner, D. E., Miller, F. J., and Coffin, D. L. (1977). Toxic effects of cadmium on ciliary activity using a tracheal ring model system. *Environ. Res.* **13,** 111–120.

Altman, N. H., and Goodman, D. G. (1979). Neoplastic diseases. *In* "The Laboratory Rat, Vol. I Biology and Diseases" (H. J. Baker, J. R. Lindsey, and S. H. Weisbroth, eds.), pp. 334–377. Academic Press, New York.

Armstrong, B. G., and Kazantzis, G. (1983). The mortality of cadmium workers. *Lancet* **1,** 1425–1427.

CEC: Commission of the European Communities (1978). "CEC Criteria (Dose/Effect Relationships) for Cadmium." Pergamon, Oxford.

Dorn, C. R., Pierce, J. O., Philips, P. E., and Chase, G. R. (1976). Airborne Pb, Cd, Zn, and Cu concentration by particle size near a Pb smelter. *Atmos. Environ.* **10,** 443–446.

(EPA) U.S. Environmental Protection Agency (1985). Updated mutagenicity and carcinogenicity assessment of cadmium. EPA/600/8-83/025F, Office of Health and Environmental Assessment, Washington, D.C.

Gaylor, D. W., and Kodell, R. L. (1980). Linear interpolating algorithm for low dose risk assessment of toxic substances. *J. Environ. Pathol. Toxicol.* **4,** 305–312.

Glaser, U., Kloeppel, H., and Hochrainer, D. (1986). Bioavailability indicators of inhaled cadmium compounds. *Ecotoxicol. Environ. Safety* **11,** 261–271.

Gunn, S. A., Gould, T. C., and Anderson, W. A. D. (1963). Cadmium induced interstitial cell tumors in rats and mice and their prevention by zinc. *J. Natl. Cancer Inst.* **31,** 745–759.

Gunn, S. A., Gould, T. C., and Anderson, W. A. D. (1964). Effect of zinc on carcinogenesis by cadmium. *Proc. Soc. Exp. Biol. Med.* **115,** 653–657.

Haddow, A., Dukes, C. E., and Mitchley, B. C. V. (1961). Carcinogenicity of iron preparations and metal-carbohydrate complexes. *Rep. Br. Emp. Cancer Campaign* **39,** 74.

Haddow, A., Roe, F. J. C., Dukes, C. E., and Mitchley, B. C. V. (1964). Cadmium neoplasia sarcomato at the site of injection of cadmium sulphate in rats and mice. *Br. J. Cancer* **18,** 667–673.

Hadley, J. G., Conklin, A. M., and Sanders, C. M. (1980). Rapid solubilization and translocation of 109-CdO following pulmonary deposition. *Toxicol. Appl. Pharmacol.* **54,** 156–160.

Harrison, P. T. C., and Heath, J. C. (1986). Apparent synergy in lung carcinogenesis: Interactions between N-nitrosoheptamethyleneimine, particulate cadmium and crocidolite asbestos fibres in rats. *Carcinogenesis* **7,** 1903–1908.

Hart, B. A. (1986). Cellular and biochemical response of the rat lung to repeated inhalation of cadmium. *Toxicol. Appl. Pharmacol.* **83,** 282–291.

Heering, H., Oberdörster, G., Hochrainer, D., and Baumert, H. P. (1979). Learning behavior and memory of chronically Cd exposed rats and organ distribution of Cd. *In* "Health Effects of Heavy Metals," pp. 61–62. Commission of the European Communities, Brussels (Doc X11/ENV/64/79).

Heyder, J. (1982). Particle transport onto human airway surfaces. *Eur. J. Respir. Dis.* **63** (Suppl 119), 29–50.

Hofmann, W., Koblinger, L., and Martonen, T. (1989). Monte Carlo modeling of aerosol deposition in human and rat lungs. *Health Phys.*, in press.

ICRP: International Commission on Radiological Protection (1975). No. 23: "Report of the Task Group on Reference Man," pp. 151–173. Pergamon, Oxford.

Kazantzis, G. (1963). Induction of sarcoma in the rat by cadmium sulphide pigments. *Nature (London)* **198,** 1213–1214.

Lemen, R. A., Lee, J. S., Wagoner, J. K., and Blejer, H. P. (1976). Cancer mortality among cadmium production workers. *Ann. N.Y. Acad. Sci.* **271,** 273–279.

Lippman, M. (1977). Regional deposition of particles in the human respiratory tract. *In* "Handbook of Physiology," Chap. 14, pp. 213–239. American Physiol. Society, Bethesda, Maryland.

Niinimaa, V., Cole, P., Mintz, S., and Shephard, R. J. (1980). The switching point from nasal to oro-nasal breathing. *Respir. Physiol.* **42,** 61–71.

Niinimaa, V., Cole, P., Mintz, S., and Shephard, R. J. (1981). Oro-nasal distribution of respiratory air flow. *Respir. Physiol.* **43,** 69–75.

NIOSH: National Institute of Occupational Safety and Health. (1984). Current Intelligence Bulletin 42: Cadmium. Cincinnati, Ohio (DHHS; NIOSH), Publication No. 84-116.

Nriagu, J. O. (1979). Cadmium in the atmosphere and in precipitation. *In* "Cadmium in the Environment" (J. O. Nriagu, ed.), pp. 71–114. Wiley, New York.

Oberdörster, G. (1986). Airborne cadmium and carcinogenesis of the respiratory tract. *Scand. J. Work Environ. Health* **12,** 523–537.

Oberdörster, G. (1988). Lung clearance of inhaled insoluble and soluble particles. *J. Aerosol Med.,* **1,** 289–330.

Oberdörster, G., and Hochrainer, D. (1980). Lung clearance of Fe_2O_3- and $CdCl_2$-aerosols during chronic CdO inhalation. *Ges. Aerosolforsch.* **8,** 198–203.

Oberdörster, G., Baumert, H. P., Hochrainer, D., and Stöber, W. (1979). The clearance of cadmium aerosols after inhalation exposure. *Am. Ind. Hyg. Assoc. J.* **40,** 443–450.

Oberdörster, G., Cox, C., and Baggs, R. (1987). Long-term lung clearance and cellular retention of cadmium in rats and monkeys. *J. Aerosol Sci.* **18,** 745–748.

Oldiges, H., Hochrainer, D., and Glaser, U. (1989). Long-term inhalation study with Wistar rats and four cadmium compounds. *J. Toxicol. Environ. Chem.,* in press.

Phalen, R. F. (1984). Particle size-selective sampling in the workplace. Introduction and recommendations. *Ann. Am. Conf. Ind. Hyg.* **11,** 23–26.

Raabe, O. G. (1980). Physical properties of aerosols affecting inhalation toxicology. *In* "Pulmonary Toxicology of Respirable Particles" (C. L. Sanders, F. T. Cross, G. E. Dagle, and J. A. Mahaffey, eds.), pp. 1–28. Technical Information Center, US DOE.

Raabe, O. G., Yeh, H. C., Newton, G. J., Phalen, R. F., and Velasquez, D. J. (1977). Deposition of inhaled monodisperse aerosols in rodents. *In* "Inhaled Particles IV" (W. H. Walton and B. McGovern, eds.), pp. 3–21. Pergamon, Oxford.

Schum, M., and Yeh, H. C. (1980). Theoretical evaluation of aerosol deposition in anatomical models of mammalian lung airways. *Bull. Math. Biol.* **42,** 1–15.

Smith, T. J., Anderson, R. J., and Reading, J. C. (1980). Chronic cadmium exposures associated with kidney function effects. *Am. J. Ind. Med.* **1,** 319–337.

Stahl, W. R. (1967). Scaling of respiratory variables in mammals. *J. Appl. Physiol.* **22,** 453–460.

Takenaka, S., Oldiges, H., Konig, H., Hochrainer, D., and Oberdörster, G. (1983). Carcinogenicity of cadmium chloride aerosols in Wistar rats. *J. Natl. Cancer Inst.* **70**, 367–373.

Task Group on Lung Dynamics (1966). Deposition and retention models for internal dosimetry of the human respiratory tract. *Health Phys.* **12**, 173–207.

Task Group on Metal Accumulation (1973). Accumulation of toxic metals with special reference to their absorption, excretion and biological half-times. *Environ. Physiol. Biochem.* **3**, 65–107.

Thun, M. T., Schnorr, T. M., Smith, A. B., Halperin, W. E., and Lemen, R. A. (1985). Mortality among a cohort of U.S. cadmium production workers—an update. *J. Natl. Cancer Inst.* **74**, 325–333.

U.S.-DHHS (1982). Department of Health and Human Services: The Health Consequences of Smoking: Cancer. A report of the Surgeon General.

Wojciak, J. F. (1988). Theoretical and experimental analyses of aerosol deposition in the lung: Implications for human health effects. Ph.D. thesis, Departments of Chemical Engineering, Pediatrics and Biophysics, University of Rochester, Rochester, New York.

Wojciak, J. F., Oberdörster, G., and Notter, R. H. (1987). Dosimetric aspects of particle deposition in the respiratory tract of rat and man: Implications for extrapolating from rodent studies. *In* "Deposition and Clearance of Aerosols in the Human Respiratory Tract" (W. Hofmann, ed.), pp. 47–52. Facultas, Salzburg.

Yeh, H. C., and Schum, M. (1980). Models of human lung airways and their application to inhaled particle deposition. *Bull. Math. Biol.* **42**, 461–480.

Index